HUMAN BIOLOGY

FOR GENERAL CERTIFICATE OF EDUCATION CANDIDATES, STUDENT NURSES AND INTENDING MEDICAL STUDENTS

HUMAN BIOLOGY

BY

C. J. WALLIS, M.A.(Cantab.)

Consultant, previously Principal, Eaton and Wallis, Tutors.
Formerly Master-in-Charge of Biology, University College School and Biology Master, Highgate School

A TEXT BOOK OF
HUMAN ANATOMY, PHYSIOLOGY AND HYGIENE

Covering the Syllabuses in Human Biology of the General Certificate of Education and other examinations of similar standard

SECOND EDITION

LONDON

WILLIAM HEINEMANN · MEDICAL BOOKS · LTD

1976

First Edition, 1969
Second Edition, Re-set, 1976

ISBN 433 34705 8

By the same author
Practical Biology (Fifth Edition)
Translated into Spanish, 1955
Practical Botany (Fifth Edition)
Practical Zoology (Sixth Edition)

Photoset, printed and bound in Great Britain by
R. J. Acford Ltd., Industrial Estate, Chichester, Sussex

FOREWORD TO THE SECOND EDITION

In the text dealing with anatomy and physiology little change has been found to be necessary though some alterations and additions have been made including that covering the recommended practical work. But in matters appertaining to human health a great deal of new and revised material has been brought in. Further facts and details of a number of diseases have been added and up-to-date information on organ transplants, antibiotics and anaesthetics has replaced the previous accounts.

The sections dealing with world food, drug addiction, smoking and lung cancer and the effects of radioactivity have been considerably expanded and an entirely new chapter on the important subject of environmental pollution has been added.

The introduction by the Oxford Board of a new AO syllabus to come into effect in 1977 in addition to the O Level syllabus "for more mature students who are pursuing a one year (or two year) course in the sixth form or in further education" has called for some additional matter such as an outline of the process of photosynthesis and its importance to man, methods of contraception, the use of Eldon Cards in blood grouping and the quantitative estimation of glucose and protein in abnormal urine. Finally, the opportunity has been taken to revise all the statistics in the book, using the latest available. Most of these are up to the end of 1974 but in some cases 1973 figures have had to suffice. As stated in the first edition, reference has been made, within the compass of the book, to bring in as many topics as possible affecting the biology of man.

It remains for me to thank many people for providing me with the latest information on a wide diversity of subjects—the World Health Organisation (World Health, disease and malnutrition), the U.N. Environment Programme (Pollution), the U.N. Narcotics Board (Drug addiction), the International Planned Parent Federation (Family planning in the Far East) and the International Food and Agriculture Organisation (World Food). At home—the Royal College of Physicians and the Royal College of Obstetrics and Gynaecology (Results of transference of human ova fertilised *in vitro* into human uteri and details of the Abortion Act), the Royal College of Surgeons (Organ Transplants), the Medical Research Council (Cancer Research), and the National Heart Hospital (Heart Transplants). I am again indebted to Dr. Ruth Irvine, L.R.C.P.,

L.R.C.S., F.F.A., D.A., Consultant anaesthetist, King Edward Memorial Hospital, Ealing and West Middlesex Hospital for an account of the modern methods of anaesthesia.

The Department of Health and Social Security went to an enormous amount of trouble to provide me with details of the reorganised National Health Service and the latest medical statistics, the Department of the Environment with details of pollution in the United Kingdom and the Office of Population Censuses and Surveys with the latest population statistics. To all the above I extend my most grateful thanks and appreciation for their help in enabling me to bring these matters up to date.

It gives me much pleasure again to express my gratitude to my publishers and to Mr. Richard Emery and Mr. Richard Warner in particular, for their customary courtesy and help in the production of the new edition and to my wife for many useful suggestions and for her help with the proof reading.

West Quantoxhead, C. J. WALLIS
Somerset.

PREFACE

There has been a very marked increase in the teaching of Human Biology over the last few years. This is an excellent trend in the right direction, for everyone ought to have some knowledge of the human body and of the ways in which it can be kept in working order, of the dangers to health to which it may be subjected and of the conditions which are necessary to provide a healthy environment.

This book covers the syllabuses in *Human Biology* of the various General Certificate of Education Examinations at Ordinary Level, A /O Level and the specialised parts of Advanced Level syllabuses in Biology. It is hoped that it will be of help to students preparing for these examinations and to student nurses. While the material in the book is based on these syllabuses, the contents have not been restricted to them and teachers of the subject will be able to decide what is relevant and appropriate to the needs of their students. It is hoped that the book may also have a wider appeal and prove to be of help and interest to potential medical students before entering a University or Medical College, in colleges of further education, to non-science students and to the general reader who wishes to learn something about his body and its health.

I have frequently been asked by students such questions as What is a thrombosis? What causes a duodenal ulcer? or What is a cancer? when dealing with the various body functions and I thought it would be of interest and not inappropriate if brief mention was made of some of the common disorders and diseases of the various systems at the end of the chapters in addition to the characteristics of certain specified diseases as required by some examination syllabuses. In this connection I have not hesitated to write at greater length on the all-important facts concerning venereal disease or S.T.D. (Sexually Transmitted Disease) as it is now universally known, to contraception and other matters appertaining to sex, and to lay emphasis on the serious dangers of drug-taking and to stress the relationship between cigarette smoking and lung cancer. In fact, an attempt has been made, within the compass of the book, to refer to as many topics as possible which concern the biology of man.

All measurements are given in the metric system and temperatures on the Centigrade scale as this country is now changing over to these more practical units. If the reader wishes to ascertain the values on the British system he will find a Conversion Table in Appendix I.

The importance of practical investigation must not be overlooked in any scientific study and suggestions for practical and experimental work are given at the end of the chapters, the human subject being used wherever it is practicable. The rabbit has been suggested for anatomical study (the rat is a suitable, if smaller, alternative) when human anatomical material or models are not available. The small human anatomical model called "The Visible Man" is well made and may prove quite helpful. No instructions have been given for the dissections as it is assumed that these will be demonstrations but if they are performed by the student, reference should be made to a suitable manual.

For the First Edition most of the illustrations were drawn specially for the book and I was indebted to Mr. André Laubier for his assistance in preparing my pencil drawings for the block-maker. Use was also made of a number of suitable illustrations from my *Practical Biology*. I am most grateful to a number of individuals and to various Bodies, duly acknowledged after this Preface and in each separate instance, for the use of various photographs, some of which were specially prepared for the book.

My very cordial thanks are due to Professor F. G. Young, D.Sc., Ph.D., F.R.S., Professor of Biochemistry in the University of Cambridge for reading through the Chapter on The Cell and the Concepts of Molecular Biology and to Dr. C. Howard Whittle, M.D., F.R.C.P., Associate Lecturer in Clinical Medicine in the University of Cambridge, formerly Hon. Consulting Physician to the United Cambridge Hospitals, for reading through all the disorders and diseases of the systems when I first wrote the book and I very much appreciated their extremely helpful comments and suggestions. My thanks were also due to Dr. Ruth R. Irvine, L.R.C.P., L.R.C.S., F.F.A., (R.C.S.), D.A., Consultant Anaesthetist at the King Edward Memorial Hospital, Ealing and the West Middlesex Hospital, for providing me with up-to-date information on anaesthetics.

As always, I am deeply indebted to my publishers, and to Mr. Owen R. Evans in particular, for their courtesy and co-operation in the production of the book and to my wife for her patience and help with the proof-reading and the lengthy task of preparing the index.

C. J. WALLIS

ACKNOWLEDGEMENTS

The author gratefully acknowledges the kindness and courtesy of the following for the provision of photographs for use in the book:—

Dr. A. J. Duggan, Director of the Burroughs Wellcome Medical Museum for photomicrographs of pathogenic bacteria (specially prepared).

Dr. D. Kay and Dr. J. C. F. Poole of the Sir William Dunn School of Pathology, Oxford for the electronmicrograph of an animal cell (specially prepared).

Dr. R. W. Horne of the John Innes Institute, Norwich for the electronmicrographs of the poliomyelitis virus and the Adenovirus.

Mr. C. James Webb of the London School of Hygiene and Tropical Medicine for the micrographs of the malarial parasite, Trypanosoma and Entamoeba histolytica from his private collection.

The Wright-Fleming Institute of Microbiology, St. Mary's Hospital, Paddington for the photographs of Sir Alexander Fleming's original culture plate and the culture plate of staphylococcal colonies.

The National Heart Hospital, London, for the electrocardiograms.

Mr. E. C. Butler, F.R.C.S., Curator of the London Hospital Medical Museum and Mr. P. C. Nunn for the photographs of endocrine diseases from the collection of Dr. Donald Hunter, former Curator of the museum.

Dr. R. G. Edwards of the Cambridge University Physiological Laboratory and Central Press Photos Ltd. for the micrograph of the test-tube fertilisation of a human ovum.

The World Health Organisation, Geneva for the photographs of the heart operation and of treatment with the artificial kidney.

Mr. A. J. Cadogan of the M.E.L. Equipment Company, Crawley, Sussex, for the photograph of an electron microscope and for providing information as to possible magnifications.

Mr. P. J. W. Hyde of Cambridge Scientific Instruments for the Stereoscan Electronmicrograph of an eyelash.

The Liverpool Water Board for the photograph of the Lake Vyrnwy Reservoir.

Messrs. W. Watson and Sons for the photograph of the binocular microscope.

And to one of my former students, G. Sprigge, for photographs of bacterial cultures in petri dishes and of the two views of the human brain from specimens in my laboratory.

Second Edition

To Dr. B. E. Heard, M.D., F.R.C.P., F.R.C.Path., Consultant Pathologist, Brompton Hospital and to the Royal Postgraduate Medical School, Hammersmith Hospital, University of London for the photograph of a lung showing a cancer and the photomicrograph of a lung cancer.

I am also indebted to the Drugs Branch of the Home Office for the 1974 statistics of recorded drug addicts.

CONTENTS

Chapter I

WHAT IS MAN?

What a piece of work is man!
How noble in reason! How infinite in faculties!

HAMLET

THE ORIGIN OF MAN

It was getting on for 2,000 million years ago, so it has been estimated, that the earth was formed but it was not until a great deal later that conditions under which life could exist had arisen. The earliest remains of any form of life date back about 500–550 million years but life began long before this, first as tiny microscopic forms and later as soft bodied organisms, none of which would leave any trace of their existence. From these simple forms, by a process of **evolution**, more complex ones arose as time went on and this was necessarily a very slow process. During this process of evolution many types of organisms which had arisen became extinct, probably because they were ill-adapted to their environment, but others survived, continued to reproduce and new ones arose.

About 60 million years ago, the first modern types of mammals came into existence and ultimately from some common stock arose the anthropoid apes and, along an entirely separate line, the ancestors of man. At first, some 500,000 to a million years ago, these were sub-human types, apes showing some man-like features. Gradually there evolved more human forms. What appear to be human remains recently found in Tanzania suggest that some predecessor of modern man existed between 3 and 4 million years ago. Some of these became extinct as time went on, until, around 10,000 years ago, the first true man appeared in Europe and it is from him, *Cro-Magnon man*, that modern man has evolved with little change. Thus man is but an infant in biological time. This, very briefly, is the story of man's origin. For a complete account of evolution and the evidence on which it is based, reference to books on evolution or general biology must be made.

Human biology is that branch of the science of life which is concerned with the life of man. This involves a study of the structure or **anatomy** of his body, the way in which it functions, its **physiology**, and all that concerns his welfare in health and disease, individually and collectively, in the environment in which he lives, a study known as **hygiene**.

MAN AN ORGANISM

Like every other organism, plant or animal, man's body is composed of living matter known as **protoplasm**, together with a certain amount of non-living material. This protoplasm is found in microscopical units called **cells** and man's body is composed of millions and millions of cells, though he begins his life as one single cell derived from his parents. Each cell, other than the red cells in the blood, contain a specialised portion of protoplasm known as the **nucleus** which controls its activities and is essential for its reproduction, the rest of the protoplasm being called **cytoplasm.** Unlike the cells of which plants are composed, animal cells have no bounding wall and are separated from one another by a **cell membrane** which is the modified outer layer of the cytoplasm. Cells differ very considerably in both size and shape and are adapted to the performance of particular functions. They are grouped together

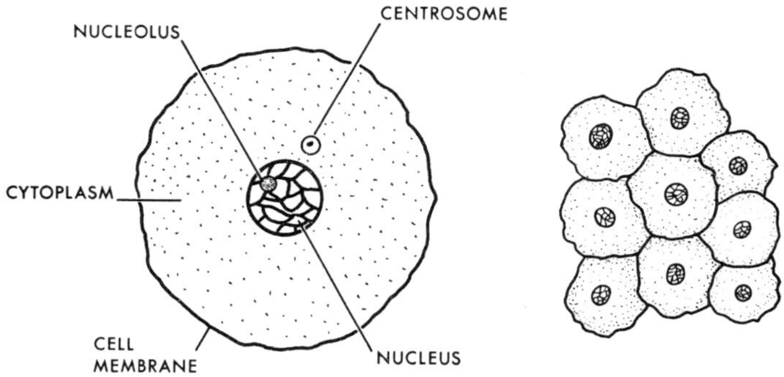

A GENERALISED ANIMAL CELL (DIAGRAMMATIC)

TISSUE CELLS LINING THE BODY CAVITY OR COELOM

Fig. 1.1. **Cells.**

to form **tissues**, each tissue consisting of cells of a particular kind, and tissues are aggregated together to form **organs**. These organs consist of more than one kind of tissue and usually form part of a **system**, each system being concerned with a particular function or functions.

Man is thus composed of a series of systems which in health work in complete harmony with one another. Sometimes this harmony is upset by the onset of disease or by man's way of living and if he is to maintain a healthy life, he must pay attention to his diet, the purification of his water supplies, the protection of his food and water, the proper ventilation of his buildings and the disposal of his refuse. Furthermore he must protect himself against infection by disease causing organisms. All these matters are the concern of the science of hygiene.

Living organisms possess certain definite characteristics which distinguish them from non-living objects and man shows all of these attributes. First and foremost is the presence of **protoplasm** and this is responsible for all the activities of life which may be summarised as follows:—

(1) **Movement**; (2) **Nutrition** which involves feeding and digestion and provides the materials for growth and repair and for the supplying of energy; (3) **Respiration** in which this energy is released to the body; (4) **Excretion** in which the waste products created in the body's physiological activities are removed from the body; (5) **Growth** by cell division and as the result of the assimilation of substances unlike those of which the body is composed; (6) **Reproduction** and (7) **Irritability** or the response of cells to stimuli.

MAN AN ANIMAL

Man is an animal, albeit the most highly developed and intelligent one, because (1) he feeds on organic food (plants or other animals or both); (2) he is capable of autonomous movement and locomotion; (3) he grows uniformly over his body and growth is limited to a certain period of his life; (4) the cells in his body have no cellulose walls and (5) there is a complete absence of chlorophyll in his body. He is a vertebrate animal because he has a backbone or vertebral column.

MAN A MAMMAL

Mammals are characterised by the fact that (1) their young develop in a uterus inside the body of the female; (2) when born the young are suckled by milk secreted by the mammary glands of the female; (3) they have hair on their bodies; (4) there are sweat glands in their skin; (5) the trunk is divided into a thorax and an abdomen separated internally by a muscular partition called the diaphragm; and (6) the temperature of their blood does not vary with that of their surroundings and they are therefore said to be *warm-blooded* or **homoiothermic**, a feature which they share with birds. There are a few more characteristics but these are sufficient to show that man is a mammal. Biologically he is known as *Homo sapiens* but in view of the behaviour of some members of the species and the present state of the world, the reader may wonder if his specific name (Lat. *sapiens*; wise) is always appropriate!

What is it that distinguishes man from other mammals? The simple answer to this question is his possession of a highly developed and specialised brain which gives him the power to reason.

THE BODY AS A WHOLE

The body of man is divided into head, neck, thorax, abdomen and limbs and is composed of the following systems:—

1. The **skeletal system** for support and protection.
2. The **muscular system** for movement of the body and of some of the organs.
3. The **alimentary system** for the digestion and assimilation of food.
4. The **cardio-vascular system** for the transport of digested food, oxygen, excretory products, hormones and heat and for the protection of the body against infection.
5. The **lymphatic system** which is accessory to the cardio-vascular system.
6. The **respiratory system** for the production of energy.
7. The **excretory system** for the elimination of the waste products of metabolism.
8. The **reproductive system** for the maintenance of the species.
9. The **nervous system** for the co-ordination and control of all the other systems and for effecting appropriate responses in the body to changes in its environment. This system includes the organs of special sense.
10. The **endocrine system** for the co-ordination and control of the body by chemical secretions known as **autacoids** or **hormones**.

To this list should be added the **reticulo-endothelial system** which is composed of isolated cells *e.g.* in the spleen and liver, which act as scavengers and remove unwanted substances which cannot be dealt with by the blood. This **morphological differentiation**, as the differentiation in structure of the various component organs and systems is called, enables them to perform specific functions and only those functions and this is known as **physiological division of labour**. We shall study the anatomy and physiology of these various systems in some detail.

The **head**, of course, contains the brain which is contained in that part of the skull known as the cranium. In the head are the jaws enclosing the mouth, the orbits in which lie the eyes and the nose and external ears leading to the nasal organs and the middle and internal ears respectively.

The **neck** is supported by seven vertebrae which, like those in the thorax and abdomen, enclose and protect the spinal cord which is continuous with the brain. The neck contains the windpipe or trachea, the gullet or oesophagus and the larynx or sound-box.

The **thorax** is somewhat cone-shaped and is bounded behind by twelve vertebrae and dorsal muscles together with twelve pairs of ribs which encircle the thorax and, with the muscles between them, protect the sides of the thorax. In front, the costal cartilages join the ribs to the breast-bone or sternum. The lower boundary of the thorax is the diaphragm and the upper boundary the floor of the neck.

The **abdomen** contains the largest cavity in the body and in it lie the stomach, intestines, liver, pancreas, kidneys and the supra-renal glands. The lumbar vertebrae with their accompanying muscles form the dorsal wall and the sides and front are protected by the abdominal muscles. The upper boundary is the diaphragm and beneath the abdominal cavity lies the pelvis. The cavity of the **pelvis** is continuous with that of the abdomen and it contains the lower end of the large intestine, the bladder and part of the reproductive system in both sexes. Dorsally are the sacral vertebrae fused together as the sacrum and the cavity is situated within the pelvic girdle of which the hip bones form the sides. There are muscles at the lower end.

All the body cavities contain blood vessels, lymphatics and nerves and all the cavities are lined by what are known as serous membranes. A **membrane** is a thin layer of tissue covering an organ or other structure. **Serous membranes** secrete a fluid which allows movement of the *viscera* (or organs) with a minimum of friction. Thus the thorax and the lungs are lined by the pleurae and the pericardium encloses the heart. The abdomen is lined by the **peritoneum** though some of the organs are partly covered by it and therefore not entirely within the **peritoneal cavity** which it forms. Support to the intestines and stomach is provided by the mesenteries and the omenta. All the internal passages of the body which communicate with the exterior are lined by **mucous membranes**. These secrete a fluid called **mucus** which lubricates their surfaces and gives them protection.

The **limbs** are supported by the long bones and by other smaller ones to which muscles are attached and joints of various kinds make movement possible. Like all other parts of the body, the limbs are supplied with blood vessels, lymphatics and nerves. **Synovial membranes** line the cavities of joints and provide them with a lubricating fluid.

As the result of the activities of the cells of the body, parts of the cellular structure are broken down and this process is known as **katabolism**. Repair is necessary and the building up processes which bring this about and which enable growth to take place constitute the process of **anabolism**. The sum total of all the chemical processes in these two stages is referred to as **metabolism**.

Before proceeding to the study of the various physiological processes and of the systems which are responsible for them, we must first examine the different kinds of tissues of which these systems are composed.

Chapter II

CELL STRUCTURE AND TISSUES

As already stated, the cells of which the body is composed are units of *protoplasm*, each containing a *nucleus* and bounded, in animal cells, by a *cell membrane*. This is probably of a fatty nature. The nucleus, too, is enclosed in a *nuclear membrane*. **Protoplasm** is a complex mixture of compounds, including proteins and carbohydrates, together with mineral salts and water. Some of these substances are dissolved in the water but the larger molecules are in very fine suspension and are therefore said to be *colloidal*. The water is present to the extent of 60 per cent. to 80 per cent. or more. Colloids may be in the form of an extremely fine suspension and this is known as a *sol* or they may be of a jelly-like nature known as a *gel*. Protoplasm is constantly changing its physical form from one to the other and, just as there are chemical differences in the protoplasm of different animals and plants, so the chemical nature of protoplasm itself is constantly changing.

The electron microscope has played an enormous part in revealing the true nature of protoplasm and the structure of cells to modern investigators and our present knowledge completely revolutionises the older theories. The **cytoplasm** is now known to be composed of a matrix known as *hyaloplasm* in which is a *reticulum* and

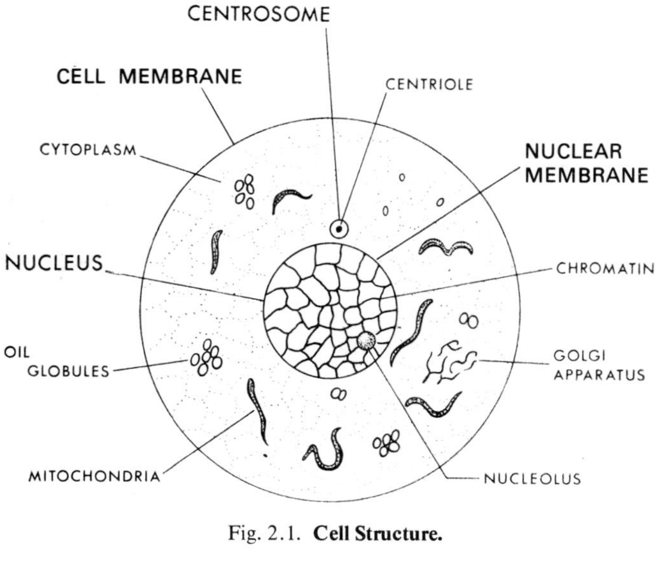

Fig. 2.1. **Cell Structure.**

cell inclusions such as a *centrosome* containing a *centriole*, *mitochondria* and the *Golgi Apparatus* (or *Body*) which, however, need not concern you at this stage of your studies. The **nucleus** contains sap, a small body called a *nucleolus* and *nucleo-protein* or *chromatin*, and two nucleic acids are present, *ribonucleic acid* (R.N.A.) and *deoxyribonucleic acid* (D.N.A.). A more detailed study of the cell and the functions of the cell inclusions will be found in Chapter XIX.

Cells grow as the result of the assimilation of substances derived from food and when they have reached a certain size they may divide into two cells. This process continues, of course, in the formation of tissues and differentiation may occur giving rise to cells of different forms and sizes and so to tissues of different kinds.

Cell division is preceded by nuclear division in which bodies derived from the chromatin and known as **chromosomes** develop, each composed of a chain of units called *genes*, and the cell membrane breaks down. The division is effected by a complex process called **mitosis** which keeps the number of chromosomes constant. This is called the **diploid** number and in man it is 46.

MITOSIS

The *centrosome* divides into two, one of these moves to the opposite pole and a *nuclear spindle* is formed between them. Fibrils radiate from the centrosome forming asters and these form the poles of the spindle. The apparent chromatin network of the resting nucleus (**interphase**) does, in fact, consist of chromosomes and these split longitudinally into two *chromatids* although the double nature is not visible at this stage. Meanwhile the nuclear membrane and the nucleolus disappear. This is the end of the first stage which is called the **prophase**.

In the **metaphase** which follows, the double structure of the chromosomes is evident and the chromatids arrange themselves on the equator of the spindle, the points of attachment being known as *centromeres*.

Then the chromatids move towards opposite poles, guided by the spindle and being pulled into V-shaped loops in the process. This is the **anaphase**.

Finally when the chromatids reach the opposite poles they become chromosomes, a nuclear membrane develops, the asters disappear and a single centrosome is left outside each nucleus. A constriction forms between the nuclei and a new cell membrane is formed. Thus two new cells are formed, each with its nucleus containing the original number of chromosomes. The final stage of mitosis is known as the **telophase**.

In the reproductive cells or gametes, however, the number is halved; this is called the *haploid* number. This method of nuclear division is known as **meiosis**. It will be examined in the chapter on Reproduction. The reason for this is that the number of chromosomes is constant for any one species of plant or animal but varies

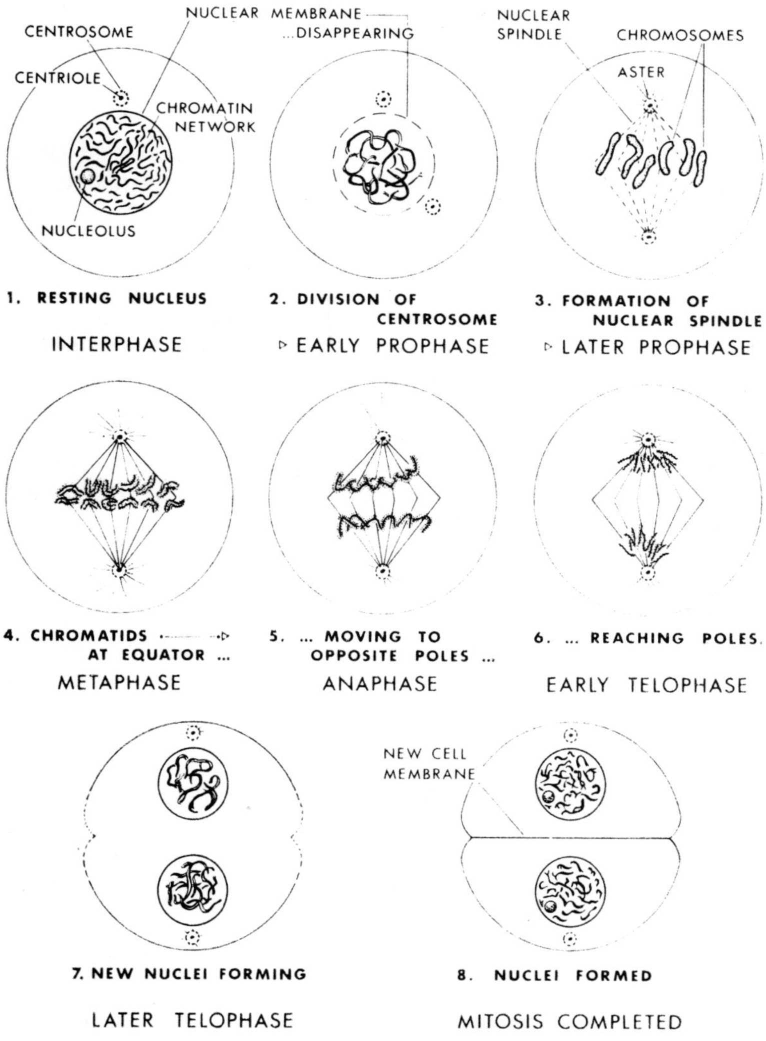

Fig. 2.2. **Mitosis.**

from species to species. In man the diploid number is twenty-three pairs (all chromosomes are in pairs). Thus, when two gametes fuse at fertilisation forming a cell (the zygote) which, by cell division, will give rise to the new organism, the gametes must contain the haploid number if the diploid number for the species is to be maintained.

Chromosomes are composed of proteins associated with D.N.A. and there is indisputable evidence that they are the bearers of inheritable characters and there is now considerable evidence to show that it is the D.N.A. which is responsible for this.

TISSUES

A **tissue** has been defined as an aggregation of cells similar in form and function and the study of tissues is known as **histology.** The chief tissues found in the body are as follows:—

Epithelial, Connective, Muscular and Nervous.

EPITHELIAL TISSUES

Organs and other structures are covered externally by **epithelium** and so, too, are the internal surfaces of the alimentary canal, blood vessels and other structures which have cavities within them. This internal epithelium is often called **endothelium.** There may be but a single layer of cells, when it is called **simple epithelium,** or there may be several when the term **compound epithelium** is used, and these two groups of tissue are of several kinds:—

Simple epithelia	*Compound epithelia*
Columnar	Transitional
Cubical	Stratified
Ciliated	Pseudo-stratified
Squamous (or Pavement)	
Glandular	

Simple Epithelia

Columnar epithelium—The mucous membranes of the stomach and intestine are lined by columnar epithelium which consists of column like cells, elongated at right angles to the surface, containing rather elongated nuclei and resting on a *basement membrane.*

Cubical epithelium—The cells in this tissue are cubicle in shape and have rounded nuclei. It is found, for example, in the linings of the kidney tubules and in the sweat glands.

Ciliated epithelium—In this tissue the free edges of the cells have hair-like protoplasmic processes, resembling eye-lashes, which are capable of movement. They are known as *cilia*. The tissue occurs freely in the air-passages and keeps them free of obstruction.

Squamous epithelium—Also known as **pavement epithelium**. The cells are flat and thin like paving-stones and fit into one another rather like "crazy paving". It is found in the peritoneum lining the body cavity or coelom and in the pleura. In some instances, as in the peritoneum, the edges of the cells are wavy and the tissue is said to be **tesselated**.

Glandular epithelium—As the term implies, this is found in the glands of the body. There are several kinds of gland varying from

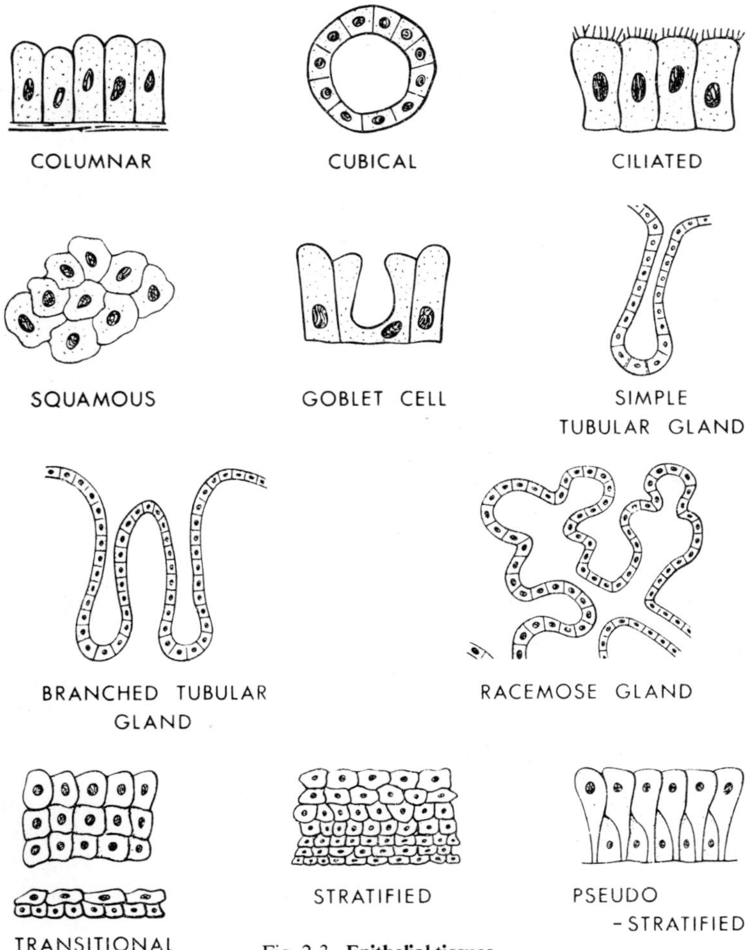

Fig. 2.3. **Epithelial tissues.**

the simple cup-shaped **goblet cells** in mucous membranes which secrete mucus, the **simple tubular glands**, shaped rather like test-tubes, in the intestine and the **branched tubular glands** of the stomach to the compound **racemose glands** such as the salivary glands and the pancreas. *Secretory cells* occur amongst columnar epithelial cells.

Compound Epithelia

Transitional epithelium—This consists of but three or four layers of cells and occurs in the bladder and ureters. It is capable of considerable stretching, when the number of layers may be reduced even to a single one.

Stratified epithelium—Several layers of cells are found in this tissue and it occurs in the skin, the buccal cavity and in the anus.

Pseudo-stratified epithelium—This is, perhaps, better considered as simple epithelium as, although it appears to be composed of several layers, all the cells lie on a basement membrane though some do not reach up to the free surface. It is found in the trachea.

CONNECTIVE TISSUES

Connective tissue is of various kinds and, as the name implies, it joins and supports various organs and structures but in some cases it is purely skeletal. Its characteristic feature is the presence of cells and fibres in a non-cellular *matrix*. There are six kinds of connective tissue, two of which are skeletal in function:—

White fibrous tissue	Adipose tissue
Yellow elastic tissue	Cartilage
Areolar tissue	Bone

Blood and lymph are sometimes considered as connective tissues in which the matrix is liquid but they are better considered separately and will be examined when studying the systems in which they are found.

White fibrous tissue—This is found where strength is needed as in tendons, which join muscles to bones. It is composed of white fibres consisting mainly of a protein called *collagen*. The fibres are unbranched and are in wavy bundles.

Yellow elastic tissue—As the name implies, these fibres are capable of stretching. They are straight and branched and anastomose, or join up, with other fibres. They do not run in bundles. It is found in ligaments, which join bone to bone, and in the lungs where considerable elasticity is required.

Areolar tissue—This is the most frequently occurring connective tissue in the body and is found in the dermis of the skin, in the mesentery, around the brain and spinal cord and elsewhere. It is composed of both white and yellow fibres and combines strength and elasticity.

Adipose tissue—This is areolar tissue in which the cells are filled with fat. There is little cytoplasm and this lies close to the cell membrane, the nucleus being embedded in it.

Cartilage—Popularly known as "gristle", this is one of the skeletal tissues and is of three kinds. In all three, cartilage cells known

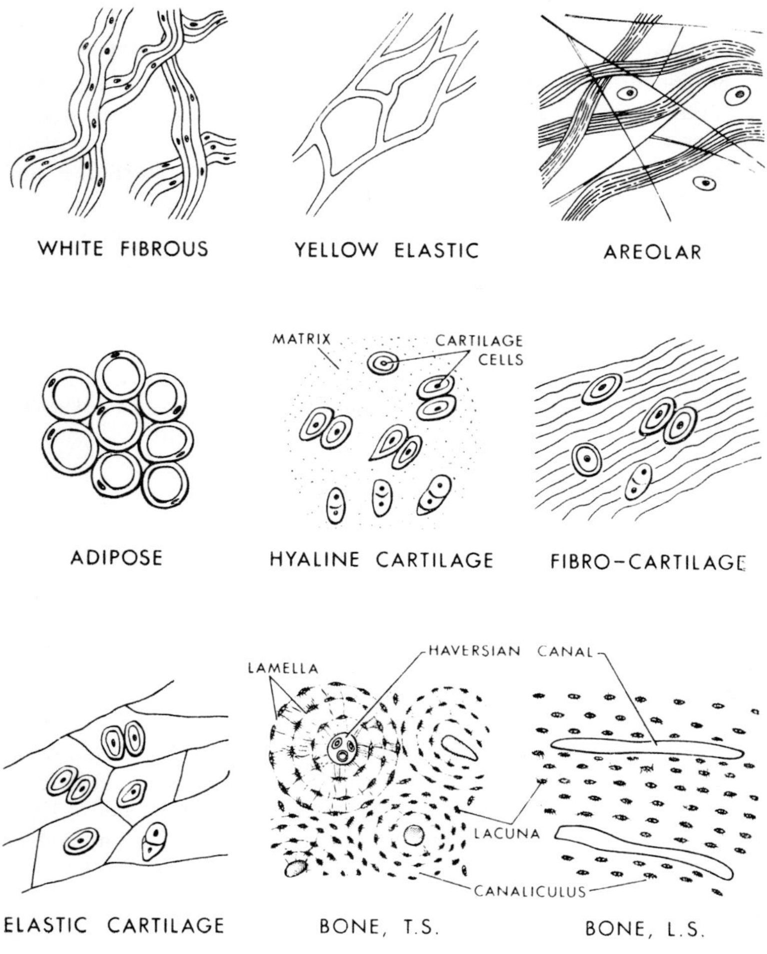

Fig. 2.4. **Connective tissues.**

as *chondroblasts* lie, singly or in groups, in spaces in the matrix called *lacunae.*

In **hyaline cartilage**, the matrix is translucent and is made of a substance known as *chondrin.* It is found in the larynx and in the walls of the air passages and as *articular cartilage* covering the ends of bones where they form joints.

Fibro-cartilage, found in the intervertebral discs, is similar but white fibres abound in the matrix.

Elastic cartilage occurs wherever elasticity combined with strength is required and is therefore found in the free end of the nasal septum, in the external ear and in the epiglottis at the entrance to the trachea in the pharynx. Yellow elastic fibres lie in the matrix.

Bone—Bone is one of the few body tissues in which inorganic material is found. It is composed of a series of what are known as *Haversian Systems* each consisting of concentric rings or plates called *lamellae* and in these lamellae calcium phosphate and carbonate are deposited, giving hardness to the bone. On the edges of each lamella is a series of *lacunae* in which bone cells or *osteoblasts* occur and in the centre of each system is an *Haversian canal* containing blood vessels and nerves. The whole system is put into communication by a series of canals radiating out from the Haversian canals to the edges of the *lamellae* connecting the lacunae; these are called *canaliculi.* The appearance of these constituents of bone will vary, of course, according as to whether it is being examined in transverse or longitudinal section.

MUSCULAR TISSUES

Muscles are composed of bundles of fibres and these are of three kinds:—

Striated (or striped)
Unstriated (or unstriped)
and Cardiac

Striated muscle—This occurs in voluntary muscles under the control of the will. The fibres are long and cylindrical and show alternate light and dark bands, each fibre being enclosed in a membranous sheath known as the *sarcolemma.* Oval nuclei occur irregularly at intervals and the fibres are not single cells. Each one is actually composed of a large number of much smaller ones called *myofibrillae* and the cytoplasmic *sarcoplasm* lies between them.

Unstriated muscle—The fibres in this case are spindle-shaped and they dovetail into one another. Each contains a single centrally placed nucleus. This kind of muscular tissue occurs in the involuntary muscles such as those found in the wall of the alimentary canal, the blood vessels, the bladder and so on.

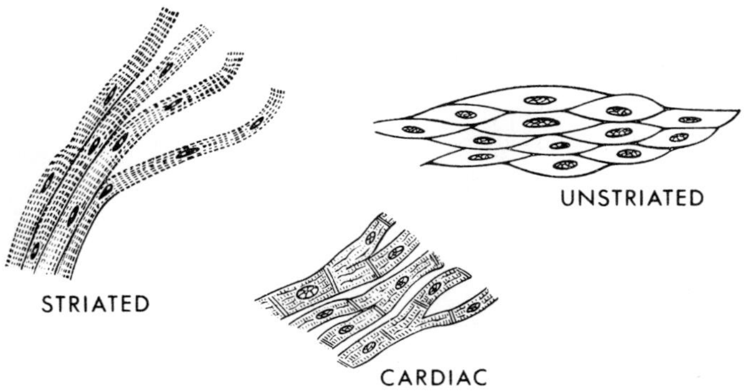

Fig. 2.5. **Muscular tissues.**

Cardiac muscle—This is peculiar to the heart and although it is striated it is, of course, involuntary. The fibres are much shorter than those in voluntary muscle and the striations are less clearly visible. Furthermore the fibres branch and anastomose with other fibres.

NERVOUS TISSUES

In the brain and spinal cord nerve cells or neurons occur while nerves are composed of nerve fibres.

Neurons—Each nerve cell is composed of a *cell body* containing a large nucleus and small elongated granules called *Nissl's granules* occur in the cytoplasm. From the cell body arise processes called *dendrons.* There may be only one *(unipolar),* or there may be two *(bipolar),* though this is rare in the human being, or there may be several *(multipolar).* The dendrons terminate in small twigs called *dendrites** and these form what are known as *synapses* with the dendrites of neighbouring neurons, interlocking with them without actual contact. Some of the neurons also bear a long stout process called an *axon* and this becomes the central axis cylinder of a nerve fibre. Neurons are embedded in a packing tissue composed of cells known as *neuroglia.* On the sensory nerves are groups of neurons constituting what are called *ganglia.*

* The entire process (dendron and dendrites) may be called a dendrite.

Nerve fibres are of two kinds, those with a fatty sheath and those without it. The former are called *myelinated* or *medullated fibres* and the latter *non-myelinated* or *non-medullated fibres.*

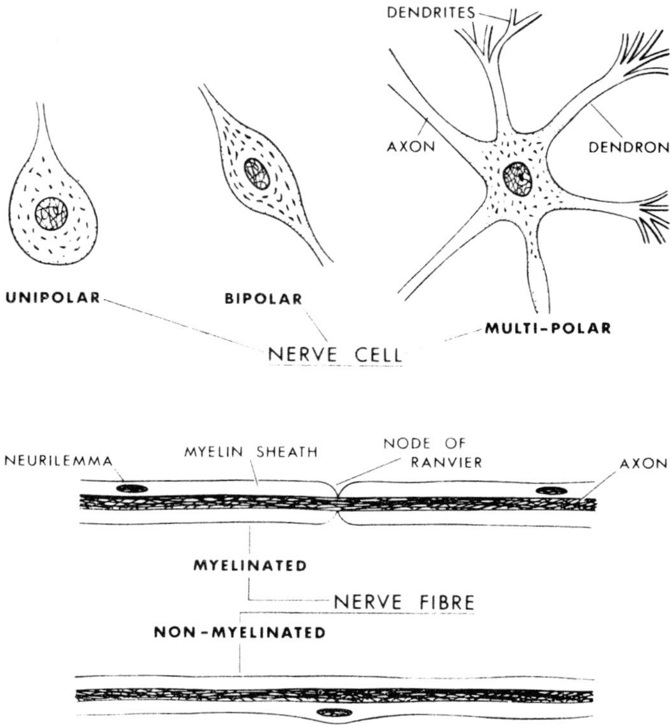

Fig. 2.6. **Nervous tissues.**

Myelinated fibres—The central core, the *axis cylinder* or *axon, is* enclosed in a fatty or *myelin sheath* outside which is a thin *neurilemma* which envelops the entire fibre. At intervals constrictions called *Nodes of Ranvier* occur in the myelin sheath and nuclei are found between the nodes. This kind of fibre is found in the cranial and spinal nerves.

Non-myelinated fibres—These fibres occur in the autonomic (sympathetic and parasympathetic) nervous system which is not under the direct control of the will. The fibres are similar to the myelinated fibres except that they lack the myelin sheath and they often branch and anastomose.

These various kinds of tissues are found throughout the body and we shall come across them when examining the various systems, a study which we shall now begin.

PRACTICAL WORK

Examine prepared microscopical slides of the tissues described in this chapter under the lower power of the microscope. Once the specimens are clearly focussed, it will be necessary, in most cases, to examine them under the high power in order to see sufficient detail. Many of the specimens will have been stained in order to show up their structure. The slides should be examined in conjunction with the descriptions given in the chapter. Suitable slides showing the different tissues are given below.

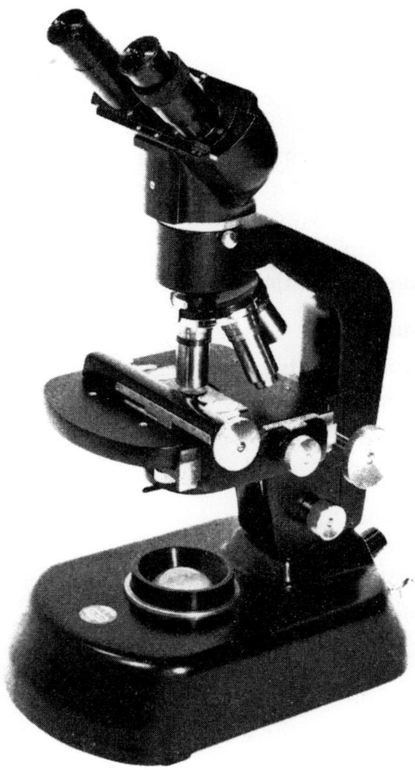

Fig. 2.7. **Optical microscope.**
(By courtesy of Messrs W. Watson & Sons.)

An instrument known as a **Bioviewer*** provides an excellent alternative to the use of a microscope and slides. This instrument is a convenient magnifying device which can be focussed as required and through which a strip of photomicrographs of stained specimens called a **Bioset** can be threaded. Accompanying each bioset is an explanatory folder of the eight specimens on the strip and magnifications are given. There are fifteen such biosets for Human Biology and these cover not only the tissues to be studied below but those of all the systems of the body.

Epithelial Tissues

Columnar. T.S. stomach. Columnar epithelium lines the mucous membrane.

Cubical. T.S. kidney. Cubical cells will be seen lining the cross-sections of the tubules.

Ciliated. Isolated cells.

Squamous *e.g.* from the inside of the cheek. Scrape it with the back of a scalpel and mount in a little physiological saline.

Glandular. T.S. small intestine for goblet cells and simple tubular glands.
T.S. stomach for branched tubular glands.
T.S. salivary gland for racemose gland.

Transitional. T.S. ureter or bladder.

Stratified. T.S. skin.

Pseudo-stratified. T.S. salivary gland. Lining the large ducts.

Connective Tissues

White fibrous. Teased fibres from tendon.

Yellow elastic. Teased fibres.

Areolar.

Adipose *e.g.* in lower region of dermis in T.S. skin.

Cartilage. Hyaline.
Fibro.
Elastic.

Bone. Examine both T.S. and L.S.

Muscular Tissues

Striated. Teased fibres.

Unstriated. T.S. bladder or stomach.

Cardiac.

* The name and address of the suppliers of Bioviewers and Biosets will be found in Appendix II.

Nervous Tissues

Neurons. Isolated nerve cells.

Myelinated nerve fibres. Teased.

Non-myelinated nerve fibres. Teased.

Mitosis

Examine a slide showing various stages of mitosis.

Chapter III

THE SKIN

The skin or **integument** which serves as the external covering of the body obviously protects the tissues which lie beneath it from damage and infection from outside, but it also performs other functions. An examination of its structure in a transverse section under the microscope shows that it consists of two layers, an external **epidermis (cuticle)** and a thicker **dermis (corium)** underneath. Beneath this is connective tissue which allows some free movement of the skin over the tissues covered by it.

The epidermis is a stratified epithelium composed of squamous cells and devoid of blood vessels and also for the main part of nerves, though some do appear in its innermost region. This is the **Malpighian layer (stratum Malpighi)** in which the cells are capable of division and regenerate the layer external to it. The layer also contains granules of a pigment known as *melanin* and

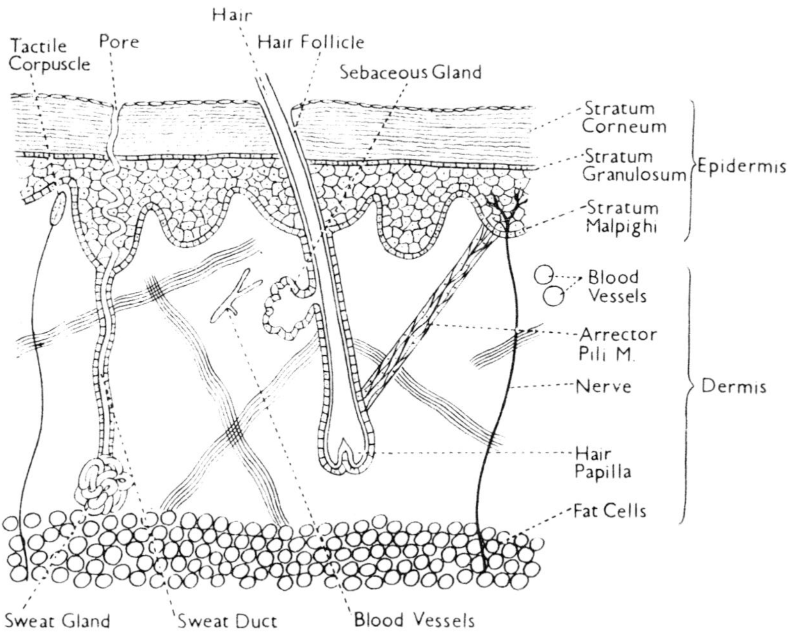

Fig. 3.1. **Transverse section of skin.**
(From Wallis—Practical Biology.)

which are abundant in dark-skinned races. This pigment protects against the harmful rays of the sun. Immediately outside the stratum Malpighi is the **granular layer (stratum granulosum),** the cells of which contain granules. The outermost layer is composed of squames, flattened and horny, and this is the **horny layer (stratum corneum)** and its outermost cells are continually being worn off.

The dermis is composed of connective tissue and its upper surface folds up irregularly into the Malpighian layer above to form **papillae.** The skin of the finger tips has a definite pattern of ridges and whorls as the result of this. They are different in every individual and this enables finger prints to be made as a means of identification. The papillae are well provided with blood vessels and many of them contain nerve-endings which terminate in **end-bulbs, tactile corpuscles (Meissner's corpuscles)** and **Pacinian bodies** which are sense organs. The dermis also contains sweat glands, hair-follicles and hairs with their associated sebaceous glands. In the deepest region of the dermis is a layer of **adipose tissue** in varying amounts. The structure of the sense organs in the skin will be examined in a later chapter dealing with the organs of special sense.

Sweat glands or **sudoriferous glands** consist of coiled up tubules from which a **sweat duct** runs upwards to leave the dermis between adjacent papillae and so through the epidermis to the surface of the skin. The **pores** of the skin are the openings of the sweat ducts. These glands are plentiful on the palms of the hands and the soles of the feet and are present over the entire skin except in certain places such as the lips. Incidentally mammary glands and the wax-producing glands in the external ear are modified sweat glands. **Sweating** is a process of evaporation which causes cooling and therefore regulates the temperature of the body; it also plays a part in water loss and is to some extent excretory. Body temperature is also controlled by the hair but the extent is very limited in man. **Sweat** is largely water containing various substances in solution of which common salt (sodium chloride) is the most abundant; hence its salty taste. Approximately 1·2 per cent. of solids are present and of these 0·8 per cent. are inorganic salts. The smell of sweat is due to volatile fatty acids. Invisible or **insensible sweating** goes on continuously and it is only when, due to various causes, sweat is produced at a greater rate than that at which it can evaporate that visible or **sensible sweating** occurs.

Hairs are epidermal structures enclosed in pits called **hair follicles** which dip down deeply into the dermis. The base of the follicle is infolded like the bottom of a wine bottle and this structure is the **hair papilla**. It is highly vascular and is supplied with nerves and is the growing point of the hair. Human hair grows about 2 mm. a day.

The bulbous root of the hair fits over the papilla. The hair itself is composed of a horny, fibrous material enclosed in a scaly cuticle and contains pigment, and the colour of the hair is dependent on the pigment. This diminishes in later years causing the hair to become grey or white but this may occur prematurely. The thickness of the shaft varies in different parts of the body and in different individuals. Attached to each hair follicle is a small muscle, the **arrector pili muscle**, consisting of unstriated fibres which is responsible for the erecting of the hair and a **sebaceous gland**, which secretes an oily substance called **sebum**, opens into the follicle. These glands are, in fact, found in all parts of the skin except in the palms of the hands and the soles of the feet and the sebum prevents the hairs from becoming brittle and the skin from becoming too dry.

Nails are localised thickenings of the stratum corneum and contain a high proportion of the protein **keratin**. They are securely attached to the dermis and grow from the base.

The functions of the skin are as follows:—

Protection—The tough stratum corneum clearly protects the underlying tissues. It also prevents loss of body fluids externally.

Regulation of body temperature—This is brought about by sweating and by the position of the hair. The hair encloses a belt of air which has an insulating effect. However, the latter method is not of much importance in the human being as it is restricted to certain parts of the body—the head, the axilla, the pubic region and in man also the face and chest. It is, of course, particularly abundant on the head where there may be 200 to 300 hairs per sq. cm. When the hair is erect it exposes a greater surface of the skin enabling heat to be lost. Furthermore the size of the blood capillaries in the dermis can be varied and when they are dilated, causing redness of the skin, more heat can escape. Conversely, when they are narrowed, heat loss is reduced. Again the subcutaneous adipose tissue acts as an insulator; it also serves as a reserve of food.

Excretion—The sweat glands are responsible for excretion though this is limited in extent.

Sensation—The tactile corpuscles, end-bulbs and Pacinian bodies are sense organs which respond to the stimuli of touch, temperature changes and pressure.

Another function of the skin is its ability to **synthesise vitamin**

D in sunlight. It also has a somewhat limited power of **absorption**; hence the use of lotions and ointments.

Disorders and Diseases of the Skin

The skin, being exposed to the outside environment, is subject to many disorders and infections. Some of these are purely functional while others are organic diseases. They may be indicative of internal disorders or diseases, for example excessive sweating occurs in tuberculosis though it may, of course, occur in normally healthy people. A few of the commoner affections are mentioned below:—

Baldness or *Alopecia*—caused by degenerative changes in the hair follicles. Though this is normally a thing which happens in old age, it may occur prematurely.

Corns—are thickenings of the epidermis.

Warts or *verucae*—are hard excrescences on the surface of the skin due to hypertrophy of the papillary region of the dermis. They are caused by a virus and may be removed by the application of caustics which should be applied with care.

Moles—are small, dark raised pigmented areas on the surface. A large reddish patch sometimes seen on the face is a **birth-mark** or *naevus* and is due to dilated blood capillaries.

Acne—caused by inflammation of the sebaceous glands and may be indicated by "blackheads" or by small red pimples which may become pustules which burst then heal.

Impetigo—a very contagious disease of the skin caused by bacteria, *Streptococcus pyogenes* and *Staphylococcus aureus*, which gives rise to pustules which become dry and form scabs. These eventually drop off.

Eczema—a skin inflammation causing great irritation. In some forms the skin becomes red and hot and minute vesicles appear. These may break and discharge. In other forms there is no discharge and the skin remains dry. The cause is often some form of irritation to the skin, or it may arise from within.

Seborrhoea—This is caused by disturbance of the secretion of the sebaceous glands in which the skin becomes greasy or dry and scaly. It often occurs on the scalp. When excess sebum collects in the ducts of the sebaceous glands it becomes admixed with dirt and "blackheads" develop.

Ringworm of the scalp—*Tinea*, as it is technically called, is caused by a fungus, *Microsporon*, and is very contagious. Circular patches develop on the scalp and these become covered with scales.

Scabies—commonly known as "itch", is the result of infection by a minute animal parasite, which burrows under the skin. Here the female lays eggs. Great irritation and hotness of the skin occurs, usually at night.

Shingles or *Herpes zoster*—an inflammation of the dermis which gives rise to vesicles in the vicinity of nerves. It may be extremely painful and is caused by a virus.

Leprosy—This terrible disease fortunately disappeared from this island in the late 18th century, but it still occurs in the Far East, Africa and elsewhere. It is caused by a bacterium. The skin becomes white and large nodules develop in advanced cases. The nerves may be affected and deformities may develop. Fortunately the disease can now be treated effectively and, if it cannot be cured, at least it can be kept under control.

Cancer or **epithelioma**—can be brought about by repeated excessive exposure to direct sunlight, to constant contact with mineral oils and other causes. If identified and treated early, the chances of a successful cure are high.

PRACTICAL WORK

Examine **Transverse sections of the skin** under the microscope. Identify the various structures mentioned in the text.

Chapter IV

THE SKELETAL SYSTEM

The study of this system, which provides support for the body, attachment for muscles and protection for some of the organs, is called **osteology**. The histological structure of the bone has already been examined but there are two forms of bone—**compact bone** on the outside of bones and **spongy** or **cancellous bone** inside.

The latter, which predominates in the ends of the long bones has, as the term suggests, spaces amongst the bone material. A substance called **marrow** is found in these spaces. It also occurs in the centre or *medullary cavity* of the long bones. This *yellow marrow* consists mostly of fat whereas the *red marrow* found in the spongy bone at the ends of the long bones and elsewhere

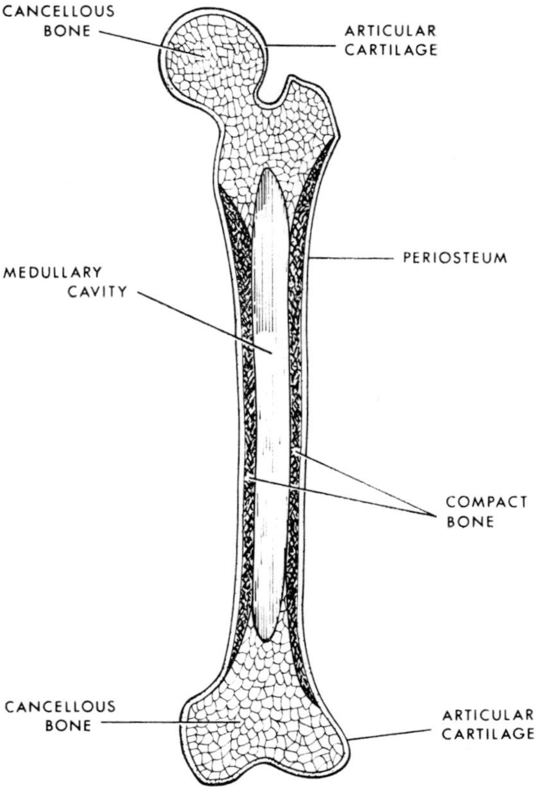

Fig. 4.1. **Longitudinal section through a long bone.**

has a much lower fat content and is rich in red blood corpuscles. It is here, in fact that the red and some of the white cells of the blood originate. A fibrous membrane covers the outside of the bones. It is called the **periosteum** and it supplies the bones with the necessary blood vessels. It is lacking at the ends of bones where they are covered with *articular cartilage* to form joints.

Bones develop in the embryo as the result of the action of bone-forming cells or **osteoblasts**. This formation may take place in connective tissue, when the bones so formed are known as *membrane bones*, or it may replace cartilage in which case we refer to them as *cartilage bones*. The difference is purely one of origin and there is no histological difference between membrane bone and cartilage bone. Most of the flat bones are membrane bones and the long bones cartilage bones. The process of bone formation is called **ossification**.

OSSIFICATION AND THE GROWTH OF BONE

A long bone, for example the upper arm bone or humerus, develops in the embryo first as a rod of cartilage. Then in the centre of the shaft ossification begins. This is at a point known as a **primary centre** from which ossification spreads along the shaft. Later a **secondary centre** develops at each end of the shaft and the bone formed from this is referred to as an **epiphysis**, that from

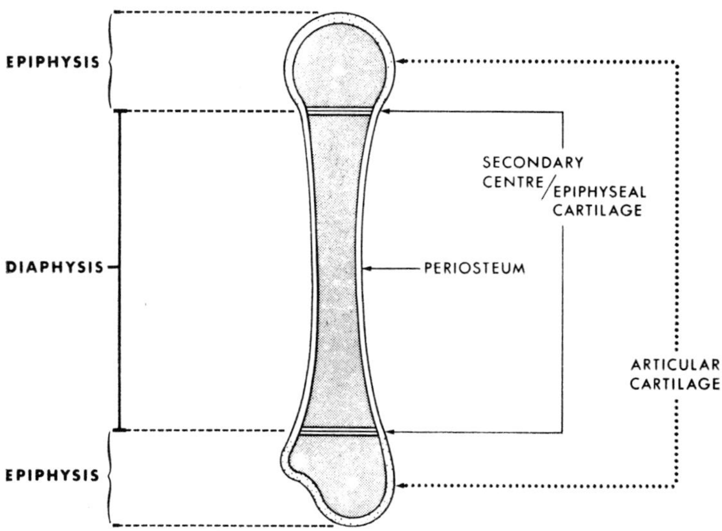

Fig. 4.2. **Ossification.**

the primary centre being the **diaphysis**. The two are separated from one another at each end by a thin layer of cartilage, the **epiphyseal cartilage**. It is here that growth in length of a bone takes place, the cartilage becoming ossified. Growth in girth of a bone is effected by the deposition of new layers of bone on the outside by the periosteum. As the bone continues to grow the epiphyseal cartilage eventually becomes replaced by bone. Complete ossification of all the bones in the body does not occur until about the twenty-fifth year and during childhood a great deal of the bone, particularly at the ends of the long bones, is still cartilage. Calcium phosphate is laid down in the matrix of the cartilage which is then called **calcified cartilage**, cells known as **osteoclasts** destroying the matrix in preparation for the work of the **osteoblasts** in laying down the bone lamellae.

The skeleton is composed of an **axial skeleton**—the skull, vertebral column, ribs and sternum—and an **appendicular skeleton**—the shoulder and hip girdles and the bones of the arms and legs.

THE SKULL

The skull consists of the *cranium*, which encloses and protects the brain, to which are articulated the *sense capsules*, the auditory and nasal capsules, and the bones of the *face* which includes the *upper jaw*. The *lower jaw* is the only movable bone in the skull.

The Cranium

The eight bones which form the cranium have immovable joints between them, the edges of the bones having irregular surfaces which fit into one another. These joints are known as **sutures**. On the upper side of the cranium are the two large **parietal bones** which reach down to form part of its sides and in front of them is the **frontal bone** protecting the forehead. These bones constitute the vault of the skull. At birth there is quite a large unossified region anteriorly where the two pariental bones and the frontal bone meet: it is known as the *anterior fontanelle*. There is also a much smaller *posterior fontanelle* where the two pariental bones meet the occipital bone behind. Ossification is normally complete in these regions by the age of two. The **occipital bone** lies posterior to the parietals and forms part of the base of the skull. It has a large aperture, the **foramen magnum**, through which the spinal cord passes from the brain into the vertebral column and on each side of this is a rounded process known as an **occipital condyle**. The two occipital condyles articulate with the first vertebra. The occipital bone articulates with the sphenoid and temporal bones.

The **sphenoid bone** forms the central part of the base of the skull and it has wing-like extensions which lie between the frontals and the temporals. It contains cavities called **sphenoid sinuses** which communicate with the nasal cavities. The **temporal bones** form sutures with the sphenoid in front and with the occipital behind

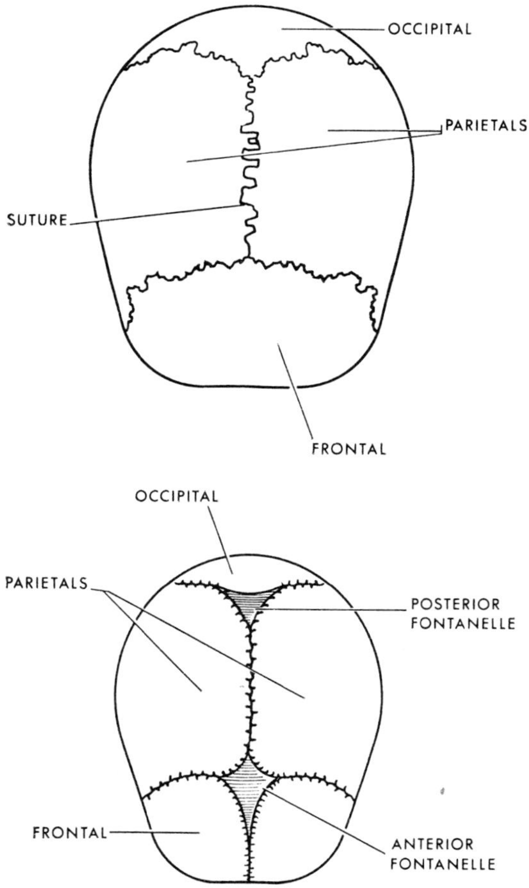

Fig. 4.3. **Skull from above.**
Top. **Adult.** Bottom. **New-born baby.**

as already stated. The flat parts of the temporals occupy those parts of the head referred to as the *temples*. A *zygomatic process* from this part of the bone joins up with the cheek bone on each side to form the **zygomatic arch**. Another part, the **petrous portion**,

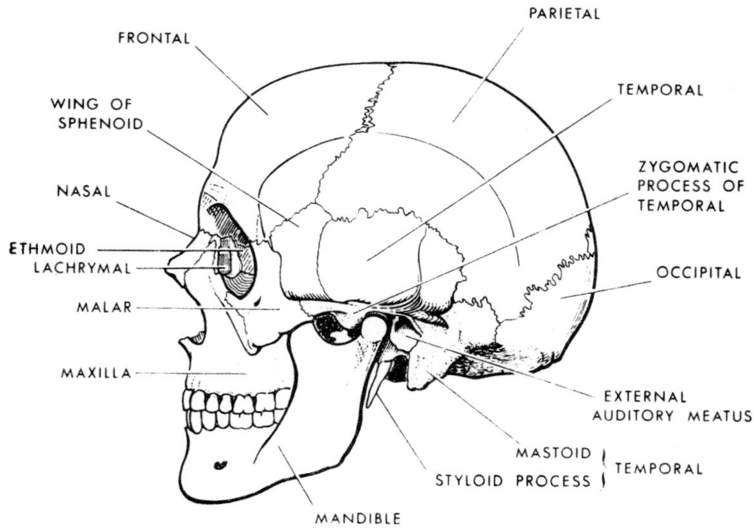

Fig. 4.4. **Skull. Side view.**

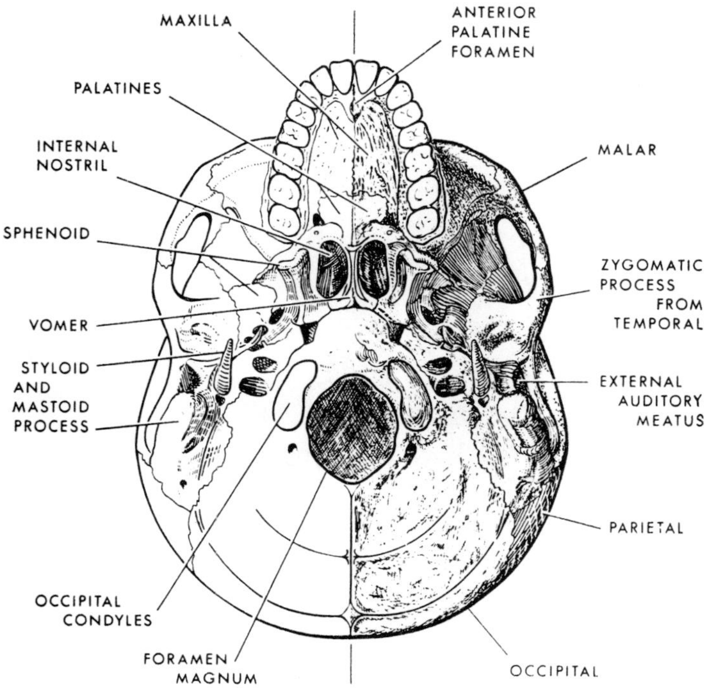

Fig. 4.5. **Skull. From below.**

encloses and protects the inner ear and a rounded part behind the ear called the **mastoid portion** encloses the *mastoid antrum* while a pointed **styloid process** runs forward for the attachment of ligaments which support the hyoid bone at the root of the tongue. Finally, the **ethmoid bone**, an irregularly shaped bone perforated by a large number of apertures or **foramina** for the entry of the many branches of the first cranial nerve from the nose, serves as an anterior wall to the cranium between the orbits.

The Face

The bones forming the facial skeleton are fourteen in number. The *upper jaw* is formed by the fusion of the two **maxillae**. These bear the teeth and form the greater part of the palate and each maxilla has a cavity within it known as the **maxillary sinus** or **antrum**. The posterior part of the palate is formed by the **palatine bones**. The **orbits** or eye-sockets are formed by several bones, most of which have already been seen. The bridge of the nose is formed

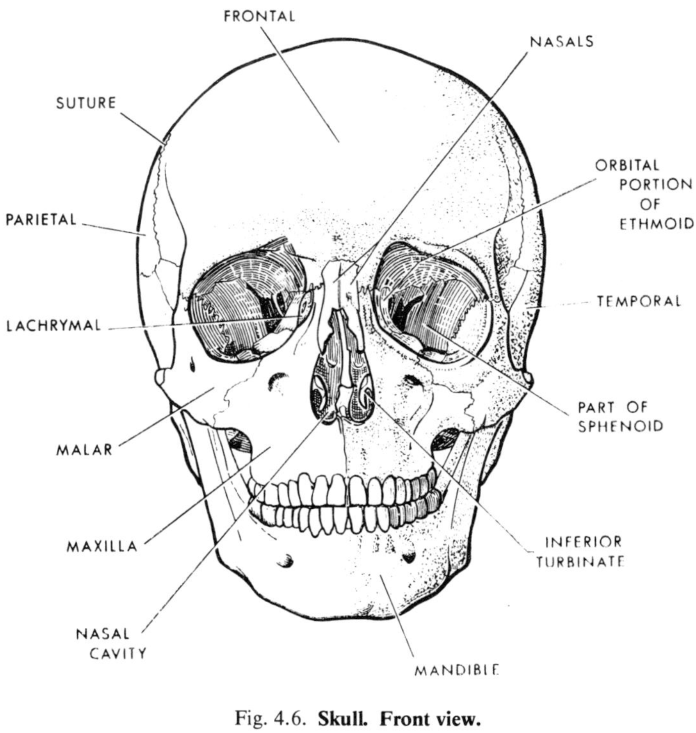

Fig. 4.6. **Skull. Front view.**

by the **nasal bones**, the lower part of the nasal septum being cartilaginous, and close to the nasal bones, in the orbits, are the **lachrymals*** which bear grooves for the lachrymal* (tear) ducts. A thin **vomer bone** forms the posterior part of the septum and on the outer walls of the two large nasal cavities are scroll-like bones known as the **inferior turbinated bones.**

The cheek bones are the **malar bones** which, as already seen, join with zygomatic processes from the temporal bones.

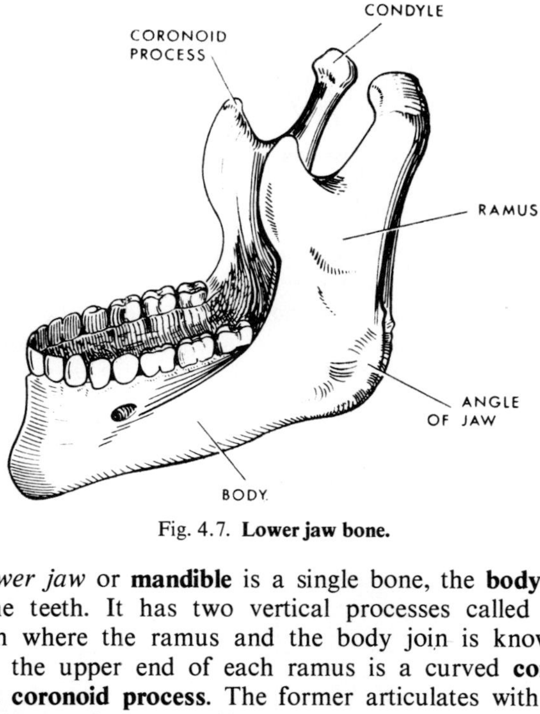

Fig. 4.7. **Lower jaw bone.**

The *lower jaw* or **mandible** is a single bone, the **body** of which carries the teeth. It has two vertical processes called **rami** and the region where the ramus and the body join is known as the **angle**. At the upper end of each ramus is a curved **condyle** and a pointed **coronoid process**. The former articulates with a corresponding fossa on the temporal bone.

There are several air cavities or **sinuses** in certain of the bones of the skull, as already seen. They lead into the nasal cavity and they make the skull light in weight. Sometimes they become infected causing a certain amount of discomfort or pain and it may be necessary to drain them.

Teeth

In common with other mammals, man's teeth are of different kinds and said to be *heterodont*, each consisting of a **crown** in

* or **lacrimal**

the mouth cavity, and a **root**, composed of one or more **fangs**, embedded in sockets in the jaw. Between the two at gum level is a slight constriction called the **neck**. There are four different shapes of teeth. First, in front, are the **incisors** with chisel-like edges for biting and cutting. These are followed by the **canines**, which correspond with the pointed teeth in this position in the dog (hence their name) and related animals, used for tearing flesh. In man the pointed shape of the crown has been lost and these teeth are popularly known as "eye teeth". Next come the **premolars,** also called *bicuspids* because they have two somewhat pointed ridges on the crown, and these are followed by the **molars** which have large grinding surfaces. The roots of the incisors and canines consist of one fang but the pre-molars and the lower molars have two and the upper molars three. The first molar is the largest and the last one the smallest. The latter are the so-called "wisdom teeth" since they do not appear until sometime between the seventeenth and twenty-fifth years.

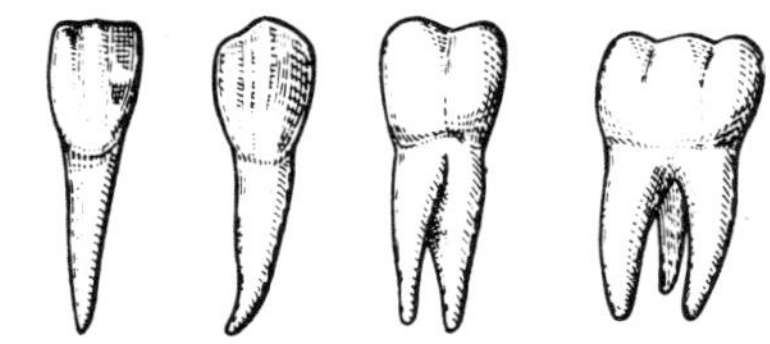

Fig. 4.8. **Kinds of teeth.**

Man has two sets of teeth, the first set beginning to erupt at about six months after birth and reaching completion by the end of the second year. In children suffering from rickets this development may be delayed. These are known as **deciduous teeth** and are popularly known as "milk teeth". There are twenty all told, four incisors, two canines and four molars in each jaw; there are no premolars. The arrangement is as follows:—

$$\frac{M2.C1.I2. \quad I2.C1.M2.}{M2.C1.I2. \quad I2.C1.M2.} \quad \begin{matrix}\text{Upper jaw}\\ \text{Lower jaw}\end{matrix}$$

These deciduous teeth begin to fall out at about the age of six years and are replaced by **permanent teeth**, the deciduous molars being replaced by permanent premolars, the molars being developed behind them. There are thirty-two teeth in all. The arrangement may be represented by what is known as the *dental formula* in which only one half of each jaw is shown

$$\frac{I2.C1.P2.M3.}{I2.C1.P2.M3.} \text{ or just } \frac{2.1.2.3.}{2.1.2.3.}$$

Structure of teeth—The bulk of a tooth is composed of a hard, bone-like substance called **dentine**. It is, in fact, harder than bone as it contains a higher percentage of calcium phosphate and it is not composed of Haversian systems, though a very large number of extremely minute tubules pass through it and lead to what is known as the **pulp cavity** in the centre. The **pulp** in this cavity consists of connective tissue, blood vessels and nerves, entry for which is provided by a minute aperture, the **apical foramen**, at the end of each fang. The crown is covered by a very hard, protective **enamel** (the hardest substance in the body) and the root is covered by **cement**. Covering the root and lining the socket in the jaw is a **periodontal membrane** which is fibrous and helps to hold the tooth in position.

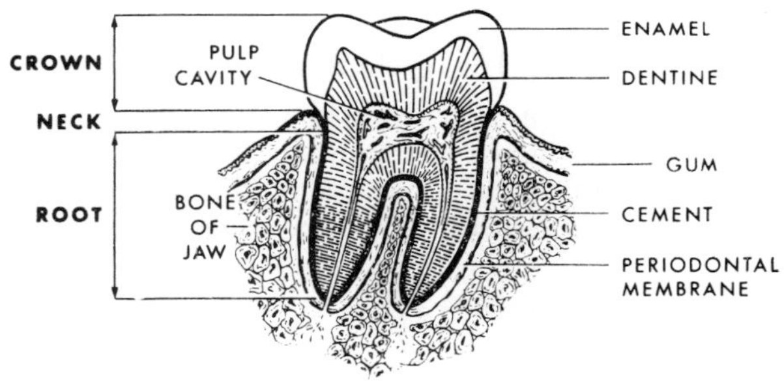

Fig. 4.9. **Structure of a tooth.**

Dental decay or **caries** is caused by bacteria, *Lactobacillus acidophilus*, in particles of food left in crevices in the teeth which decompose the food producing acids which attack the enamel, dissolving away its calcium salts. The dentine may, in this way become exposed and attacked by bacteria and if the nerves in the pulp become affected, the pain of toothache in varying degrees will be experienced.

Infection may spread to the gum, an *abscess* may develop on the root and inflammation of the gum (*gingivitis*) may be set up, resulting in what is called a *gumboil.*

There is evidence which shows that a very small (1 in 1,000,000) fluorine content of drinking water in the form of fluoride is necessary to prevent caries in children, at least up to the age of twelve and possibly older. It has been shown to reduce caries in children by as much as 60 per cent. Although fluorides occur in many natural waters this is not universal and in districts where it is deficient, fluoridation of the water is the obvious remedy. This is recommended by some health authorities but it is not enforceable by law and some local authorities object to it for various reasons. Tablets containing the appropriate amount of fluoride can be purchased. Children suffering from rickets are more prone to dental decay. The World Health Organisation has called on all member States to introduce fluoridation of community water supplies where the fluorine intake is below the optimum level of 1 in 1,000,000.

Pyorrhoea is an infection of the gum which results in the development of packets of purulent matter around the roots of the teeth and the consequent loosening of the teeth. The purulent matter is the cause of the bad breath (*halitosis*) of people suffering from this infection.

Tartar is a hard deposit which forms on the teeth near gum level. It is chiefly calcium phosphate and is derived from the saliva. Its presence may lead to loosening of the teeth and it should not be allowed to accumulate.

The Hyoid Bone

The **hyoid bone** is a roughly U-shaped bone situated at the root of the tongue. Though not part of the skull it is connected by ligaments with the styloid process of the temporal bones. To it are attached the muscles which are responsible for swallowing.

THE VERTEBRAL COLUMN

Thirty-three vertebrae take part in the formation of the vertebral column and they are classified into five distinct regions:—

	Vertebrae
Cervical in the neck	7
Dorsal in the thorax	12
Lumbar in the Loin	5
Sacral, fused together as the sacrum and articulated with the pelvic girdle	5
and	
Coccygeal at the lower end of the column, corresponding with the tail of lower mammals ..	4

Each vertebra except the first is composed of a solid anterior and ventral portion called the **body** and passing backwards are two processes each called a **lamina** which unite and thus form a **neural arch**. This encloses the neural canal or vertebral foramen. Thus the whole vertebral column is hollow, except at its lower end, and encloses the **spinal** or **vertebral canal** for the protection of the spinal cord. Projecting dorsally from the neural arch is a **spinous process** and projecting laterally are the **lateral processes**. These serve as anchorages for the muscles and for the ligaments which bind the vertebrae together. Between each pair of vertebral bodies is an **intervertebral disc** made of fibro-cartilage which give the backbone a limited amount of flexibility. These discs have been known to get out of position, a condition known as "slipped disc", causing considerable discomfort and pain owing to pressure on the spinal cord and the nerves.

This generalised structure of a vertebra is modified in the various regions of the vertebral column, the vertebrae in the different regions having different specialised functions.

Cervical region—The first vertebra or **atlas** supports the skull and therefore has large **occipital facets** on its anterior face in which the occipital condyles of the skull rest. It is on this vertebra that the head nods. It will be seen that it has no body. The second vertebra known as **axis** bears a forwardly projecting tooth-like peg known as the **odontoid process** or **dens** which is, in fact, the body of atlas which has become fused with that of axis. A powerful ligament holds it in position inside atlas and thus the head is enabled to turn sideways, the skull rotating on the dens. The five remaining cervical vertebrae are similar to each other. All are relatively small and in each the spinous process is forked and the vertebral foramen relatively large. In all the vertebrae in this region is a **vertebrarterial**

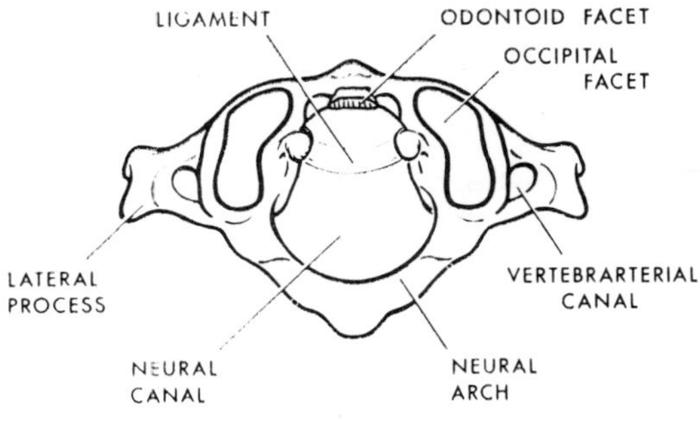

Fig. 4.10. **Atlas. From above.**

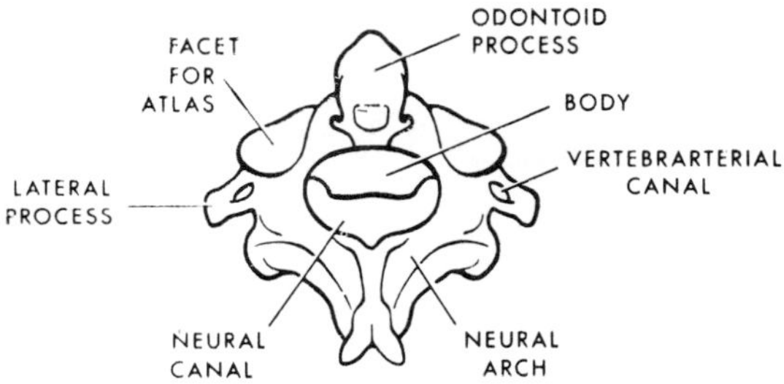

Fig. 4.11. **Axis. From above.**

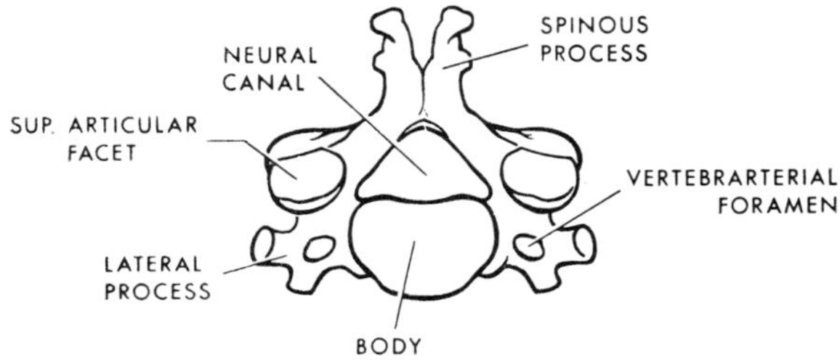

Fig. 4.12. **Typical cervical vertebra.**

foramen in each transverse process for the passage of the vertebral artery. These are peculiar to the cervical vertebrae.

Thoracic region—As these vertebrae serve for the articulation of the ribs, they have special **facets** for this purpose. There are two pairs on each vertebra, one pair on each body and one on each transverse process. The spinous processes are long and the vertebrae larger than those in the cervical region.

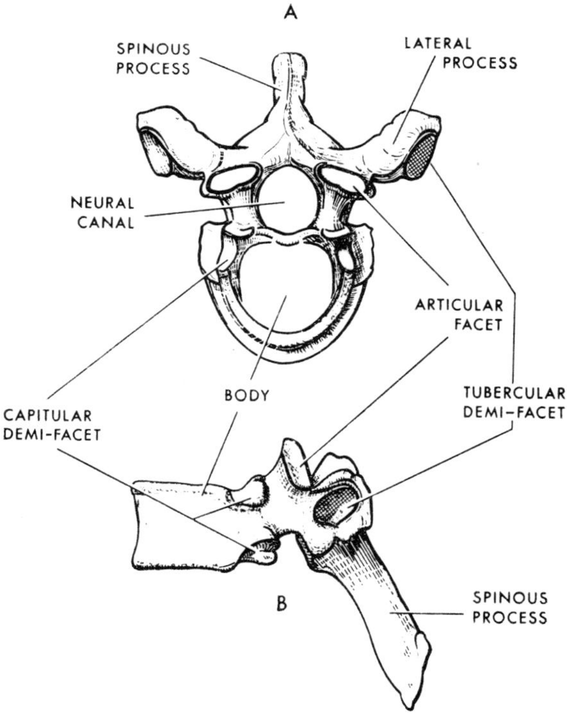

Fig. 4.13. **Thoracic vertebra.**
A. **From above.** B. **Side view.**

Lumbar region—These vertebrae are much larger than those in any other region. They have square spinous processes and rather slender transverse processes.

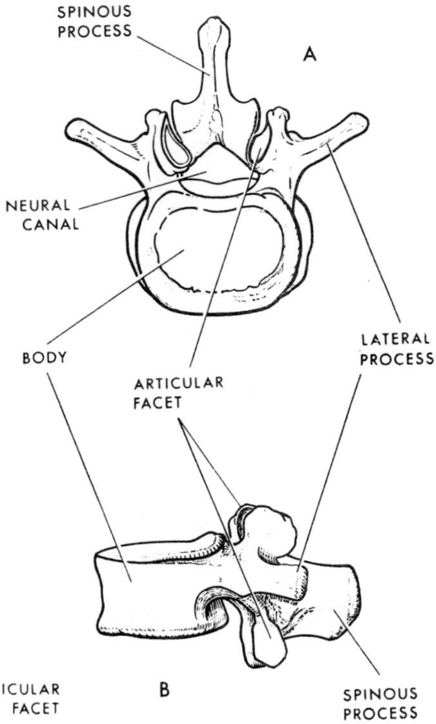

Fig. 4.14. **Lumbar vertebra.**
A. **From above.**
B. **Side view.**

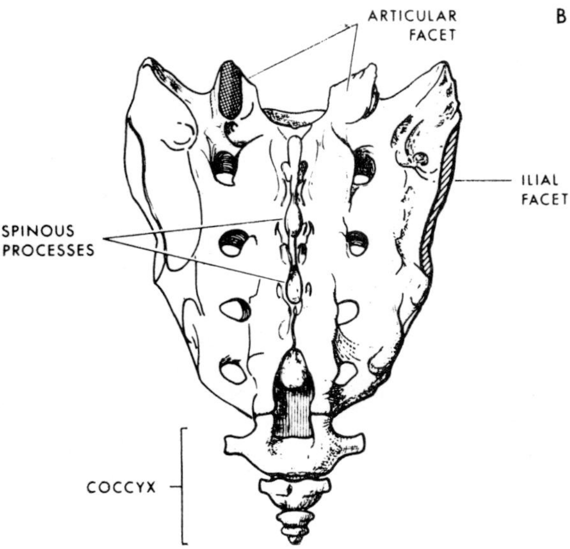

Fig. 4.15.
Sacrum and coccyx.

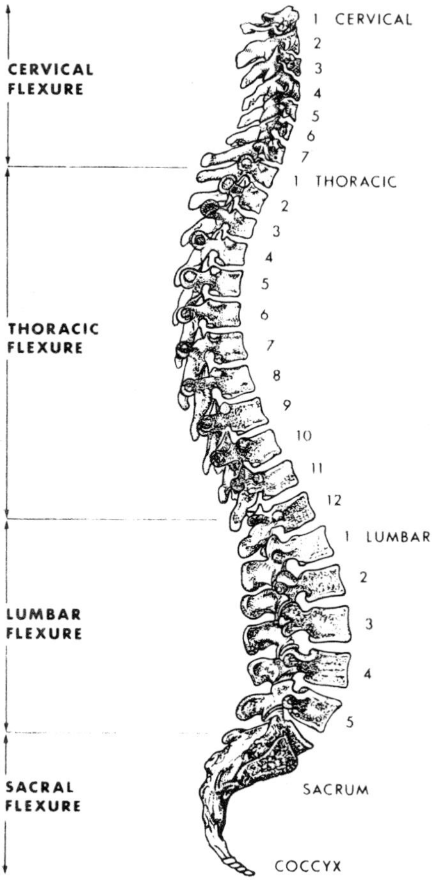

Fig. 4.16. **Vertebral column. Side view.**

The Sacrum—This is a wedge-shaped bone which articulates with the ilium or hip bone of the pelvic girdle on each side. It is composed of five fused vertebrae and with the hip bones forms a basin-shaped cavity called the pelvis.

The Coccyx—The four small bones which constitute the coccyx are fused together and are just vertebral bodies, no other parts of a vertebra being present.

The vertebral column as a whole is curved in four places, convex ventrally or frontwards in the cervical region, concave ventrally in the thoracic region, convex again in the lumbar region and concave in the sacral and coccygeal regions.

THE RIBS

There are twelve pairs of ribs all of which articulate with the thoracic vertebrae dorsally. They are curved flat bones, the **head*** of each articulating with demi-facets on the bodies of adjacent vertebrae and small processes alongside the head known as **tubercles** with demi-facets on the transverse processes. The opposite end of the rib is attached by a **costal cartilage** to the sternum in the case of the first seven ribs which are referred to as *true ribs.* In the next three pairs the costal cartilages join up with those in front

* Or **capitulum.**

while the eleventh and twelfth, known as *floating ribs,* terminate in the muscles of the back. These last five pairs of ribs are referred to as *false ribs.* All the ribs run downwards and forwards from the vertebral column and form a bony cage in the thorax which protects

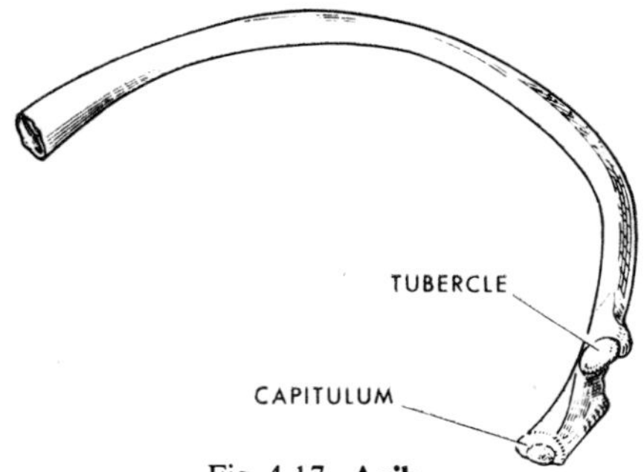

Fig. 4.17. **A rib.**

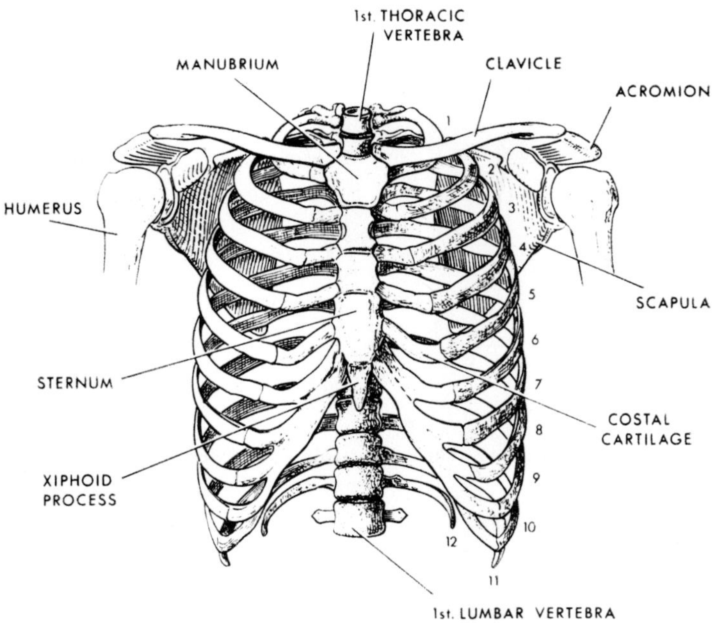

Fig. 4.18. **The rib cage and clavicle.**

the heart and lungs and the arches thus formed increase in size from top to bottom. The rib cage is movable and this movement, effected by muscles, is concerned in breathing since it alternately increases and decreases the capacity of the thorax.

THE STERNUM

This is the breast bone to which, as already stated, the ribs are joined. The uppermost part of the sternum to which the collar bone or clavicle is attached is called the **manubrium**. At the lower, free end is a tapering **xiphoid process**, made of cartilage.

THE SCAPULA

Each shoulder-blade or **scapula** is a large, flat triangular bone situated on the posterior side of the thorax at its upper end and with its base uppermost. The flat part of the bone is called the **blade** and on the posterior side of this is a ridge called the **spine** which ends in a widened portion known as the **acromion**. The anterior surface of the blade is rather concave and projecting from

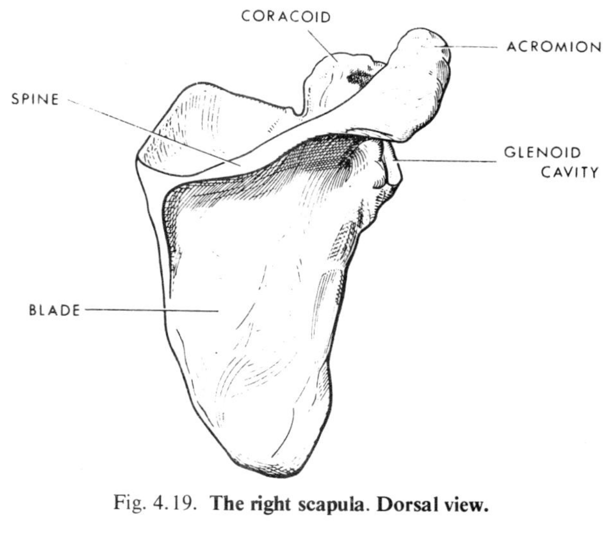

Fig. 4.19. **The right scapula. Dorsal view.**

the upper edge of the blade is a process called the **coracoid**. These two processes serve for the attachment of the collar-bone or clavicle which articulates with the acromion and is joined to the coracoid by ligaments. The other end of the clavicle articulates with the manubrium which is the upper end of the sternum. A shallow cup known as the **glenoid cavity** in the upper and outer region of the scapula serves as the socket of the ball and socket joint of the shoulder. The head of the upper arm bone, the humerus, fits into it. It should be noted that the scapula does not articulate with the vertebral column and therefore has freedom of movement.

THE CLAVICLE

As already seen, one end of the collar bone or **clavicle** articulates with the acromion of the scapula and the coracoid to which it is also joined by ligaments, while the other end articulates with the manubrium which is the upper end of the sternum. The clavicle is a long and slightly curved bone and gives support to the arm.

THE PELVIC GIRDLE

Unlike the scapulae, the pelvic girdle articulates with the vertebral column. It is composed of the two hip bones or **ilia**, larger in the female than in the male, which, with the **pubis** and **ischium** on each side form a deep cup, the **acetabulum**, which is the socket of the ball and socket hip joint, the head of the thigh bone or

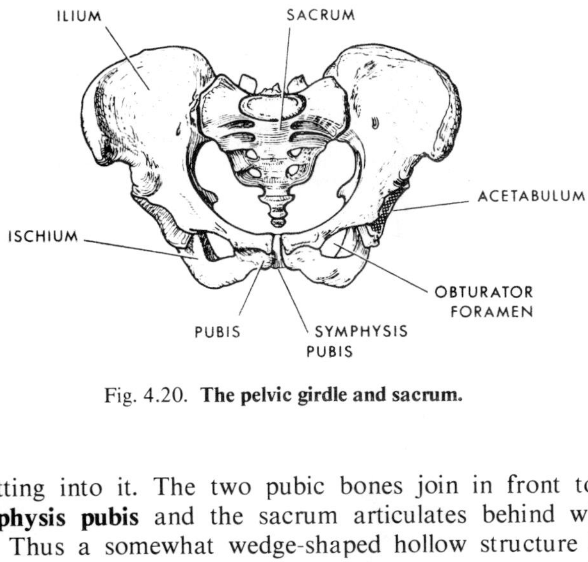

Fig. 4.20. **The pelvic girdle and sacrum.**

femur fitting into it. The two pubic bones join in front to form the **symphysis pubis** and the sacrum articulates behind with the two ilia. Thus a somewhat wedge-shaped hollow structure known as the **pelvis** is formed and inside this lie some of the abdominal organs. The cavity of the pelvis is larger and somewhat differently shaped, being wider and shallower, in woman than it is in man to allow for the birth of a baby.

BONES OF THE ARM

The *upper arm* bone is the **humerus** which is composed of a long cylindrical **shaft** bearing a rounded **head** and three processes or **tuberosities** at its upper end and two **epicondyles**, a **trochlea** and a rounded **capitulum** at the lower end which help in the formation of the elbow joint.

The *forearm* is composed of two long bones, **radius** on the outside and **ulna** on the inside, the latter having a prolongation called the **olecranon process** which fits into the end of humerus, limiting the backward movement of the forearm. The radius lies alongside

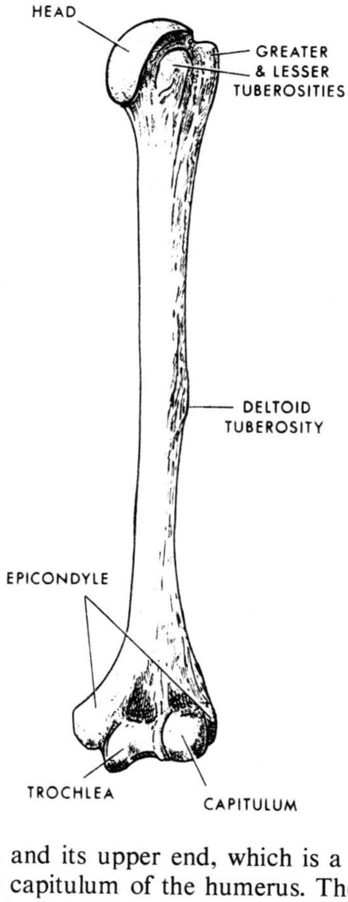

Fig. 4.21. **Humerus.**

the ulna and its upper end, which is a shallow cup, also articulates with the capitulum of the humerus. The lower end of radius articulates with the wrist. In the turning movement of the hand from a position with the palm of the hand uppermost in the horizontal position and known as *supination* to a position with the palm facing downwards, called *pronation*, the lower end of radius turns and crosses over ulna.

The *wrist* or **carpus** consists of eight **carpals** in two rows, those

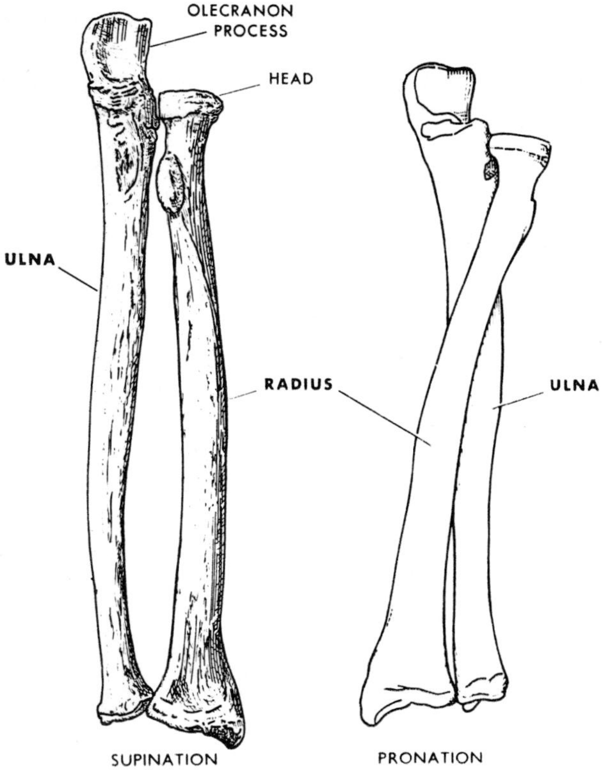

Fig. 4.22. **Bones of left forearm.**

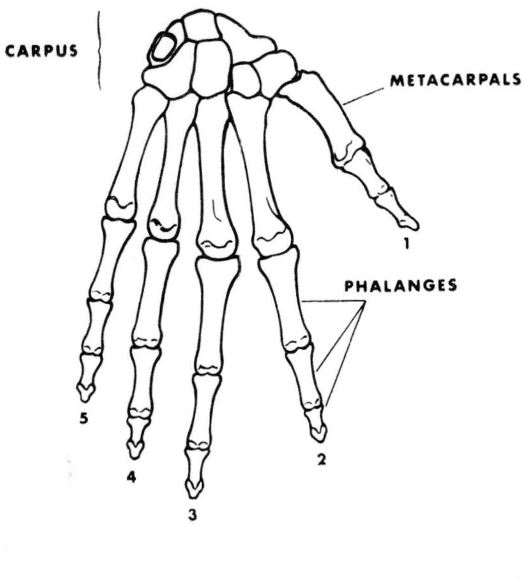

Fig. 4.23. **Bones of wrist and hand.**

in the upper row articulating with the radius and those in the lower row with the bones forming the palm of the hand.

The *palm of the hand* is supported by long bones called **metacarpals** which, as already seen, articulate with the carpals in the lower row.

Each of the four fingers is composed of three **phalanges** and the thumb of two and there is free movement at the joints.

BONES OF THE LEG

The *thigh bone* is known as the **femur** and, like the humerus, consists of a long **shaft** (larger and stronger than humerus) with a **head** (more rounded and prominent than that on humerus) beneath which is a distinct neck. There are three processes beneath the

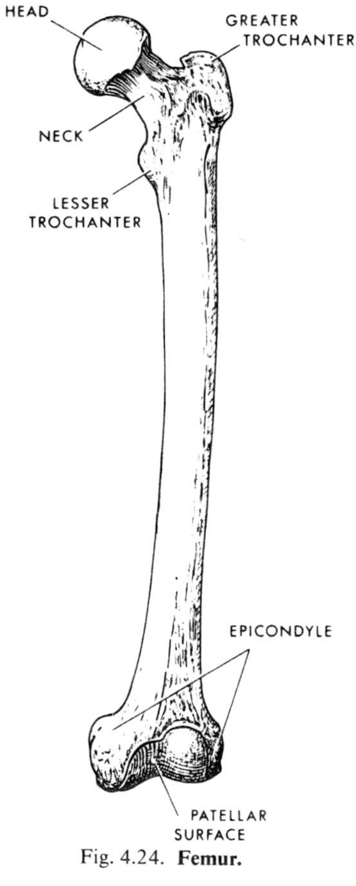

Fig. 4.24. **Femur.**

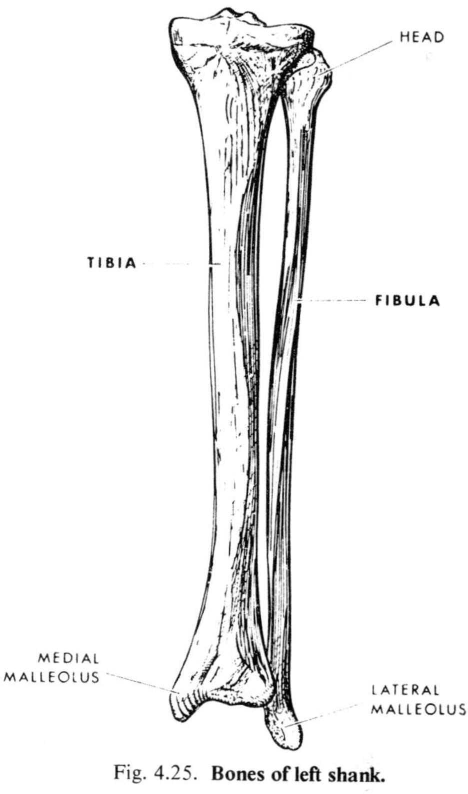

Fig. 4.25. **Bones of left shank.**

head known as **trochanters** and at the lower end of the bone are **epicondyles** for articulation with one of the lower leg bones. Between the epicondyles is a groove known as the **patellar surface**. This allows the *knee-cap* or **patella**, which is a small circular bone embedded in tendon, to be guided over the knee joint. Bones such as that described for the patella are called *sesamoid bones*.

The *lower leg*, like the forearm, has two bones, **tibia** and **fibula**. The former is the *shin bone* and it has a flat surface. It articulates with the femur at its upper end. Fibula is a more slender bone

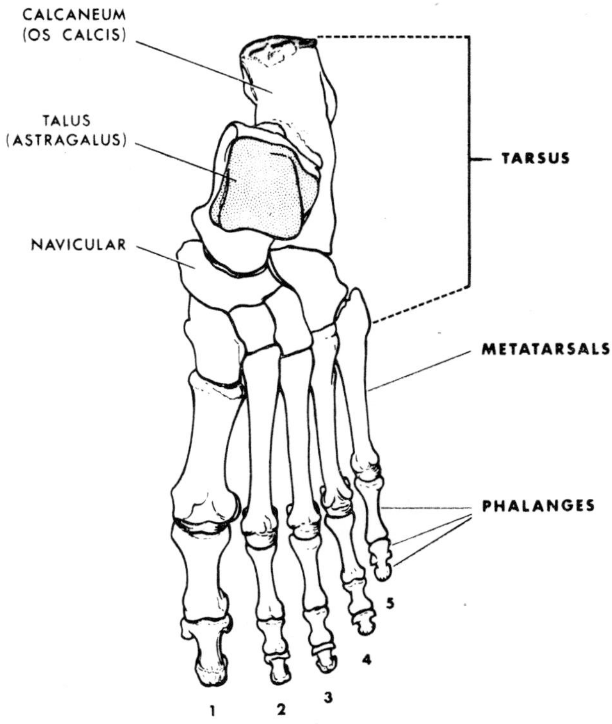

Fig. 4.26. **Bones of ankle and foot.**

but unlike its counterpart in the forearm it is unable to cross over the tibia to which it is joined at both ends but free from it otherwise along its length.

The *ankle* or **tarsus** consists of seven **tarsals** in two rows. In the upper row is the **calcaneum** or **os calcis** forming the *heel* and having a **talus** or **astragalus** on its upper surface and the **navicular**

bone beneath it to one side so that it actually lies in between talus and the four bones of the lower row.

The *sole of the foot*, like the palm of the hand, is formed by five long **metatarsals** and the toes consist of three **phalangs** apiece with the exception of the big toe which, like the thumb, has only

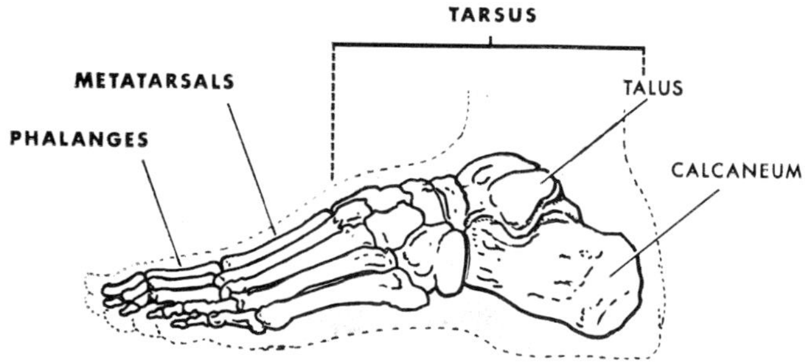

Fig. 4.27. **Side view of skeleton of ankle and foot to show arch.**

two. Since the foot takes the weight of the body transmitted through the leg bones, the sole of the foot is arched to give strength. Movement of the toes is very limited as compared with the fingers, except for the big toe.

Broken Bones

The breaking of a bone is known as a **fracture**. If the skin is broken and the bone, or broken parts of it, pierce the skin, it is called a **compound fracture**. This is a serious condition as infection may enter from outside, there may be considerable bleeding and healing is a slow process. When the broken bone does not penetrate the skin, even though there may be severe laceration of the surrounding tissues it is known as a **simple fracture**. It should be remembered that there are bone-forming cells, *osteoblasts*, in the lacunae of bone so that repair of a broken bone is possible. This of course requires the proper alignment and setting of the parts of the bone concerned and their maintenance in the correct position by means of splints or plaster until repair is complete. Damaged and dead bone is broken up by *osteoclasts* and the fresh tissue which develops on the broken surfaces resulting in the repair of the fracture is known as *callus*.

Diseases of Bone

Sometimes bones become diseased and this may be an inflammation of the surface of the bone, of the bone itself or of the medullary cavity. It may be *acute*, progressing rapidly and in some cases recovering rapidly, in others causing death or *necrosis* of the bone, or it may be *chronic*, progressing slowly and lasting for a long time. The term *osteitis* is used for bone inflammation.

Acute inflammation of the medullary cavity—may be caused by a bacterium, *Staphylococcus aureus*, and this can be treated successfully with penicillin or, if need be, by surgery.

Chronic inflammation—may be caused by injury to the bone, or by tuberculosis or syphilis.

Rickets—usually occurs in early childhood and this disease causes a softening of the bone due to a deficiency of vitamin D in the diet, which in turn prevents absorption of calcium from the food.

Acromegaly—is a disease in which the bones become considerably enlarged and it is caused by excessive secretion of a hormone from the pituitary gland at the base of the brain. If it sets in in early life, *gigantism* with excessive stature, results. The opposite condition of *dwarfism* is due to a deficiency of this hormone.

Spina bifida—is a congenital malformation of the bony neural arch in which this is incomplete in some vertebrae. Part of the spinal cord is therefore exposed and unprotected. The effects of this are explained under diseases of the nervous system on p. 282.

PRACTICAL WORK

Examine the **skull** and **other bones of the human skeleton** if available. If these are not obtainable, excellent plastic copies are made, some on a much reduced scale. The miniature skeleton in the anatomical model called "The Visible Man" is quite well made. If none of these is available, the bones of a *rabbit's* skeleton should be used. Differences in shape will be observed but there is sufficient similarity to enable a reasonably satisfactory study and identification to be made.

Chapter V

JOINTS

Joints or articulations occur in the body wherever bones are in contact and these joints are of three kinds:—

Fibrous
Cartilaginous
and Synovial

Ligaments, some of which are elastic and others non-elastic, bind the articular surfaces together, giving the joints protection and strength.

Fibrous joints—These are found where the teeth fit into the jaw bones and in the sutures of the skull where the edges of the bones are irregular and jagged and dovetail into one another. There is no cartilage between the bones and the joints are *immovable.*

Cartilaginous joints—In this type of joint, hyaline cartilage separates the bones and though actual movement of the bones is not possible, there is a certain amount of limited flexibility. The intervertebral discs and the junction of the two pubic bones at the symphysis pubis in the pelvic girdle are examples of cartilaginous joints.

Synovial joints—The majority of joints in the body allow movement between the bones concerned. The articulating surfaces are covered with **articular cartilage** and the whole joint is enclosed in a **capsular ligament** and lined by a **synovial membrane**. This

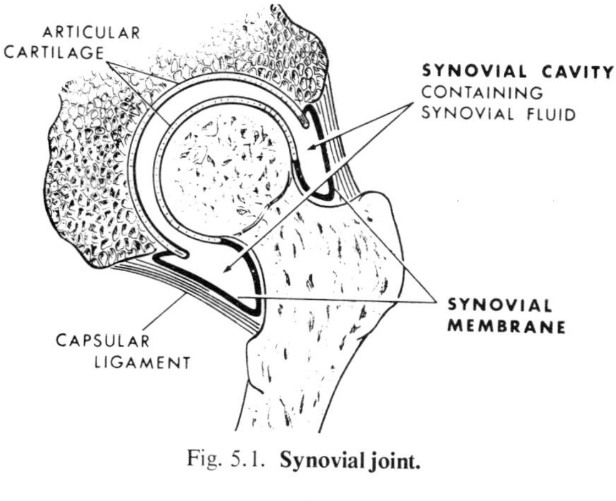

Fig. 5.1. **Synovial joint.**

secretes a fluid known as **synovial fluid** which lubricates the joint in the **synovial cavity**. There are four kinds of synovial jcint as follows:—

Ball and socket joints—occur at the shoulder and hip, the rounded head of humerus and femur fitting into the glenoid cavity of the scapula and the acetabulum of the pelvic girdle respectively. There is great freedom of movement, more than in any other kind of joint.

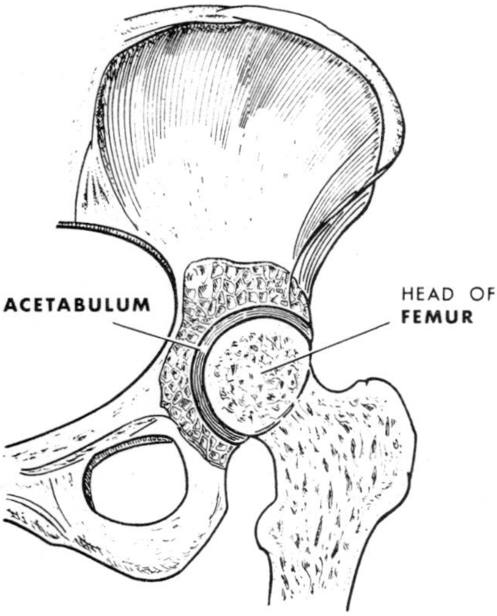

Fig. 5.2. **Ball and socket joint at hip.**

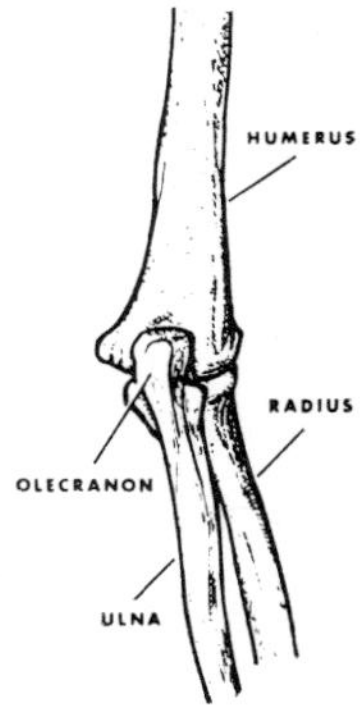

Fig. 5.3. **Hinge joint at elbow.**

Hinge joints—occur at the elbow, knee and ankle and between the phalanges of the fingers and toes. In this type of joint, movement is possible in one plane only and even so its extent is limited.

Gliding joints—are found in the ankle and wrist where the small bones glide over one another.

Pivot joints—are clearly illustrated by the articulation between atlas and axis where a projection on the latter, the odontoid process or dens, acts as a pivot on which atlas rotates. In the pronation of the hand, too, radius turns on ulna, it will be remembered. This is another example of a pivot joint.

Dislocations and Diseases of Joints

Dislocations—in which the ends of one of the bones forming a joint are displaced thereby interfering with the proper working movement of the joint, may be the result of accidents. The ligaments at the joint may be torn as the result of this displacement and this is the cause of a **sprain**. Repositioning of the dislocation without further damage being done calls for considerable skill and once the bone is back in position it must be held in place for some time to prevent a reoccurrence. At the same time, disuse of the joint must not be too prolonged. Stiffness can result from a dislocation or from fracture of one of the bones forming the joint.

Synovitis—The synovial membrane can become inflamed as the result of a sprain or a blow. Considerable pain may be experienced and the joint may become swollen. Furthermore there may be excessive secretion of synovial fluid resulting in such a condition as "water on the knee".

Ganglion—The swelling known as a **ganglion** which sometimes develops at joints in the back of the hand and wrist and which obviously contains a fluid may be caused by a strain. It is due to a small pouch in the synovial membrane in which synovial fluid collects. Weakness of the joint may be experienced but in some cases there appear to be no ill effects. Pressure by suitable bandaging may cause the ganglion to burst with consequent dispersal of the fluid: otherwise minor surgery can be applied. However, the ganglion sometimes clears up of its own accord.

Arthritis—is a chronic inflammation of a joint and may be caused by wear of the articular cartilage and the development of bony spurs. This is known as **osteo-arthritis** (or **arthrosis**). It causes great pain and discomfort and leads to reduced movement and even to crippling. It frequently occurs in the hip joint, the knee and the fingers. A surgical operation called an *osteotomy* in which the bone is cut through and realigned is sometimes successful in arresting the disease, if taken in its early stages. But the most successful method, pioneered in this country and which has been in use for several years in cases of arthritis of the hip, is the *total joint replacement.* In this operation a new metal head is fixed on the femur and a tough plastic socket firmly fixed in the acetabulum by a very strong cement. This operation removes all pain and restores normal movement to at least a very considerable degree. Replacement of knee and ankle joints has been successfully performed and it is hoped that before long surgeons will be able to fit artificial elbow and shoulder joints. **Rheumatoid arthritis** is a more serious condition due to a different cause and this can be treated, though not cured, by injection of *cortisone.* Unfortunately cortisone sometimes produces undesirable side effects. Other methods of treatment are possible. Recent research suggests that the wearing of the cartilage in rheumatoid arthritis is due to enzyme action but the source of this enzyme has not yet been discovered. 13,000,000 working days a year are lost by arthritic sufferers.

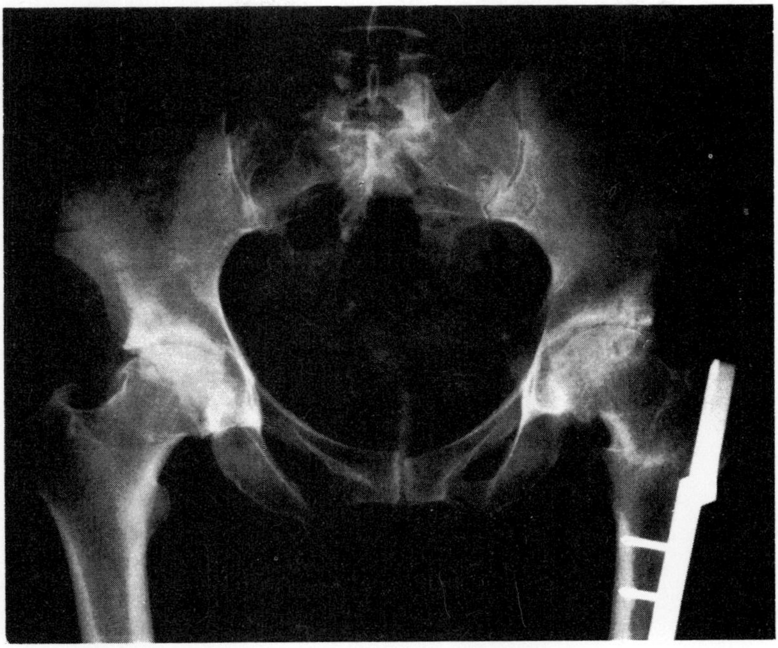

Fig. 5.4. **X-ray of female pelvis. An osteotomy has been performed on the left hip joint and a metal plate screwed into the femur.**

PRACTICAL WORK

Ball and Socket Joints

Fit the head of *humerus* into the *glenoid cavity* of the scapula and the head of femur into the acetabulum of the pelvic girdle. Note the very considerable freedom of movement in all directions.

Hinge Joints

Fit the *radius and ulna* into their correct position on the lower end of humerus. Flex and extend the forearm and observe that movement is restricted to one plane and that the olecranon process on ulna restricts the limit of this movement.

Pivot Joints

Examine an *atlas* and fit an *axis* into it, taking care that you place it on the lower end of atlas, with the odontoid process of axis in its facet. Now rotate atlas on the odontoid process. With the skull resting on atlas this is how the head is moved from side to side.

Chapter VI

MOVEMENT AND THE MUSCULAR SYSTEM

We saw in our examination of the tissues that the voluntary muscles are composed of striated fibres. The voluntary or skeletal muscles constitute the muscular system and we shall confine our study to these muscles in this chapter.

Each muscle consists of bundles or *fasiculi* of long striated fibres bound together by a thin fibrous tissue called **endomysium**, each bundle enclosed in a thicker **perimysium** and the whole enclosed in an outer fibrous sheath known as the **epimysium**. This constitutes

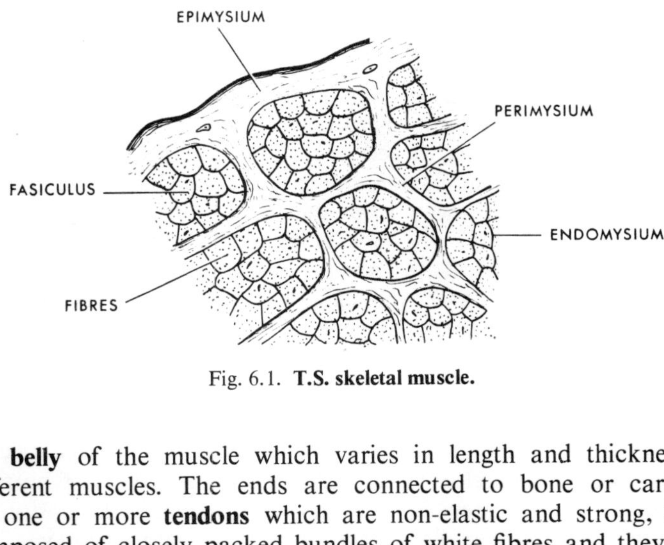

Fig. 6.1. **T.S. skeletal muscle.**

the **belly** of the muscle which varies in length and thickness in different muscles. The ends are connected to bone or cartilage by one or more **tendons** which are non-elastic and strong, being composed of closely packed bundles of white fibres and they may be cylindrical or flat. One end of each muscle is joined to an immovable bone and this is known as the **origin** or the muscle. The other end, called the **insertion**, is attached to a movable bone. Most skeletal muscles operate at the joints where movement is effected by several muscles acting together and in the movement of bones the principle of the *lever** comes into use. To each group of muscles there is always a group of opposing muscles, thus making movement in opposite directions possible. One group, called the

* It will be remembered that a lever is a rigid rod or bar capable of being turned about a point known as the fulcrum. Fig. 6.2, p. 55.

prime movers effect movement in one direction, the opposing **antagonists** bringing about in the opposite direction and so relaxing when the prime movers are contracting. It will be understood that whereas a muscle is an antagonist in one movement it will become a prime mover in the opposite one. Now when any particular movement is being performed, it is essential that other muscles should remain unaffected so that the body as a whole remains stable. For example, when carrying a heavy weight in one hand, the body tends to lean over to that side. This is controlled by muscles known as **synergists**. This co-ordination between specific groups of muscles enables fine and delicate movements to be performed. It does not follow that because a muscle contracts that all the fibres in that muscle contract. The number contracting at any one time depends on the work which has to be done and when a fibre contracts it does so to the extent of about 50 per cent. of its length and it cannot contract less than this. It either contracts to its fullest extent or not at all. This is referred to as the **all-or-nothing response**.

Movements are of different kinds. We have already come across **pronation** and **supination** in the turning of the radius on ulna when the palm of the hand is turned over. The bending of a joint such as at the elbow is called **flexion** and the straightening of the forearm **extension**. Movement away from the body such as the lifting of the arm in a sideways direction is known as **abduction** and the return of the limb towards the body is **adduction**. Finally the turning or rotation of a joint, as at the shoulder and hip, is called **rotation**. It should be understood, however, that many movements are combinations of these different kinds.

Muscles are never completely relaxed and are always in a condition called **muscular tone** or **tonus** in which the fibres are always under a slight tension so that they are ready to contract quickly on demand. The equal contraction of opposing groups of muscles resulting in no movement is necessary in maintaining the posture of the body but movement of any part can be made without delay.

Throughout the body are layers of fibrous connective tissue containing elastic fibres known as **fasciae**. The muscles are covered by what is known as **deep fascia**; this also lies between the groups of muscles and round the muscles. **Superficial fascia** lies over the body beneath the skin and contains a considerable number of fat cells as well as blood vessels and nerves. This has an insulating value as well as providing a reserve of food.

Let us now examine a *few of the common movements of the body* and see how the muscles bring them about. It has already been stated that the *principle of the lever* comes into play in movement.

There are *three orders of levers* and we shall see to which of these orders the different instances we examine belong.

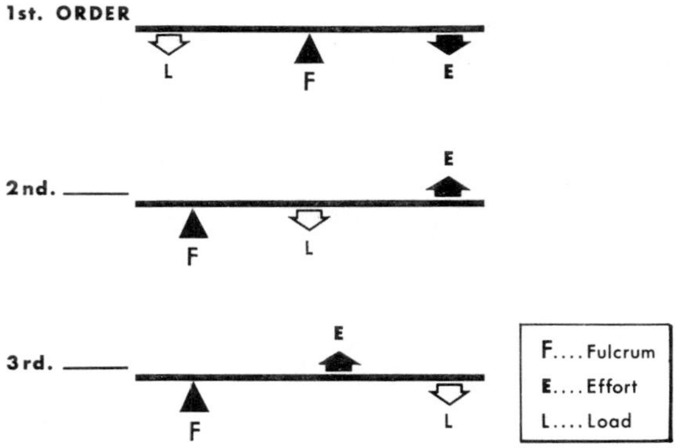

Fig. 6.2. **Orders of levers.**

Flexion and extension of the forearm—flexion is brought about by the *biceps muscle* on the front of the humerus and by the *brachialis.* For simplicity's sake we shall consider the single action of the biceps. This muscle has its two origins on the scapula and its insertion on the radius. When the muscle contracts it moves the forearm towards the upper arm by movement at the elbow joint. Here the fulcrum is at the elbow, the load the weight of the forearm and the effort the biceps between them. Thus it belongs to the *3rd order of levers.* When the forearm is straightened again, a case of **extension**, the *triceps muscle* on the back of humerus contracts. This muscle has its three origins on the tuberosities near the head of humerus and on the scapula and its insertion on the olecranon process of ulna. The fulcrum being between the load and the effort, this belongs to the *1st order of levers.* Fig. 6.3, p. 56.

The raising of the heel from the ground—In this movement the toes act as a fulcrum which is thus at one end, the *gastrocnemius (calf) muscle* at the other end being the effort and the weight of the body through the leg being the load between them. The gastrocnemius muscle has its origin on the condyles on the lower end of femur and the upper end of tibia and its insertion on the calcaneum to which it is joined by the long *tendo Achillis.* Here the load is between the fulcrum and the effort and it belongs to the *2nd order of levers.* Fig. 6.4, p. 56.

The nodding and raising of the head—The fulcrum is at the point where the occipital condyles of the skull rest on atlas and

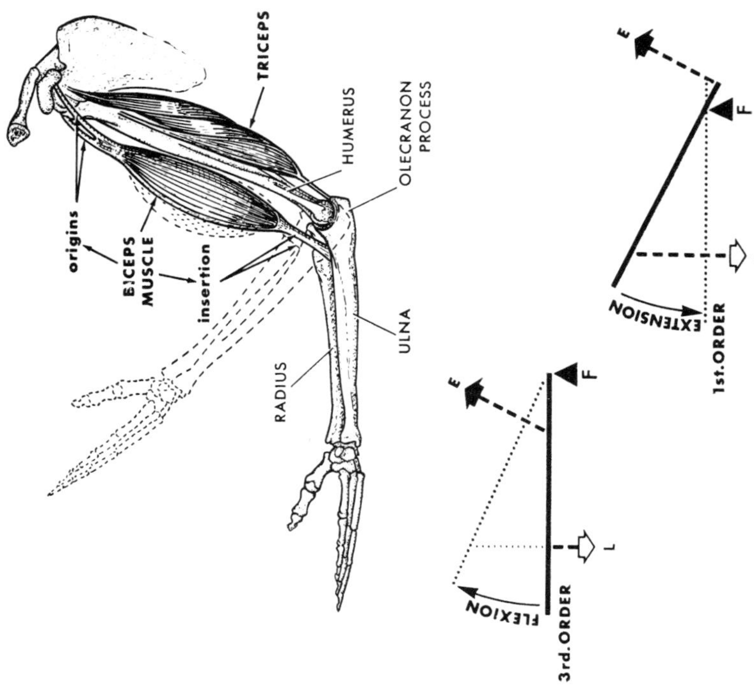

Fig. 6.3. **Flexion and extension of forearm.**

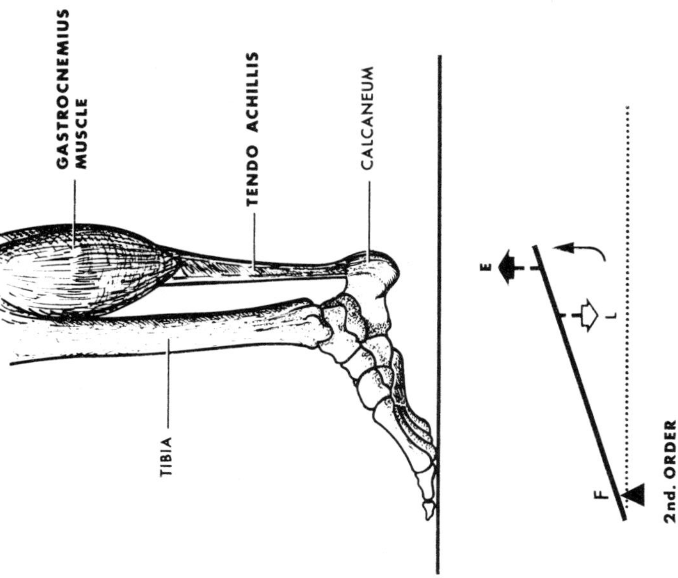

Fig. 6.4. **Raising of heel from ground.**

the load is the weight of the head. In **nodding** the effort is provided by the large, long *sterno-mastoid muscles* in the neck *acting together.* These have their origins on the upper end of the sternum and on the clavicle and their insertions on the temporal bones of the skull. **Raising of the head** is due to the *trapezius muscles* in the

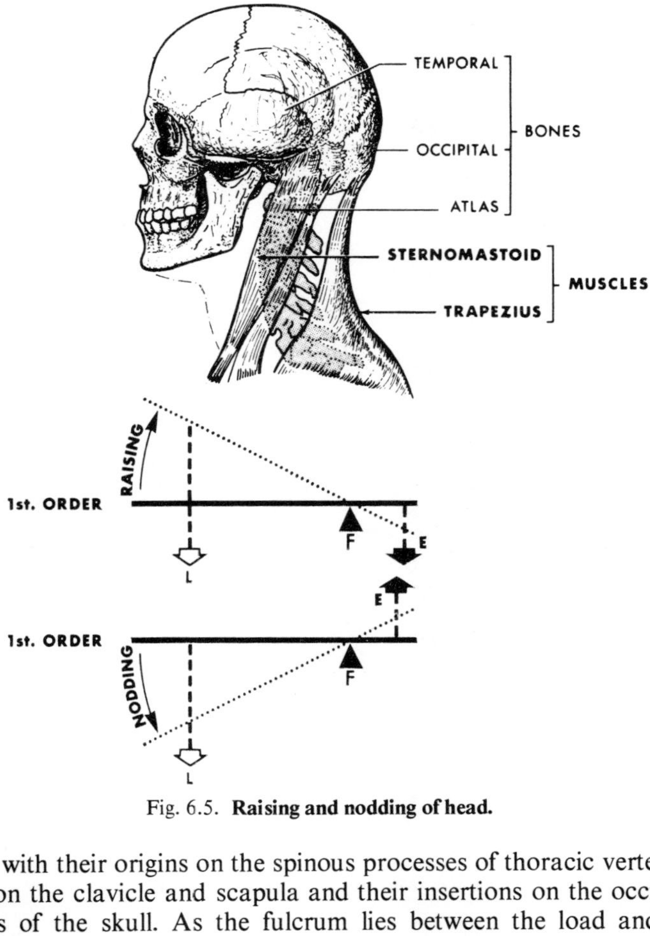

Fig. 6.5. **Raising and nodding of head.**

neck with their origins on the spinous processes of thoracic vertebrae and on the clavicle and scapula and their insertions on the occipital bones of the skull. As the fulcrum lies between the load and the effort, we have an instance here of the *first order of levers.*

Turning of the head—This is brought about by the *sterno-mastoid muscles acting separately.*

THE MUSCLES OF THE BODY

We must now make a superficial examination of some of the

muscles in the various parts of the body and you will, no doubt, be able to work out for yourselves, at any rate in some cases, to which order of levers the movements belong. There are over six hundred voluntary muscles in the body forming more than two-fifths of its weight and we shall consider only a few in order to obtain an overall picture of the system.

Muscles of the Head and Neck

The chief muscles of the head and neck will be seen in Fig. 6.6. Their functions are given below.

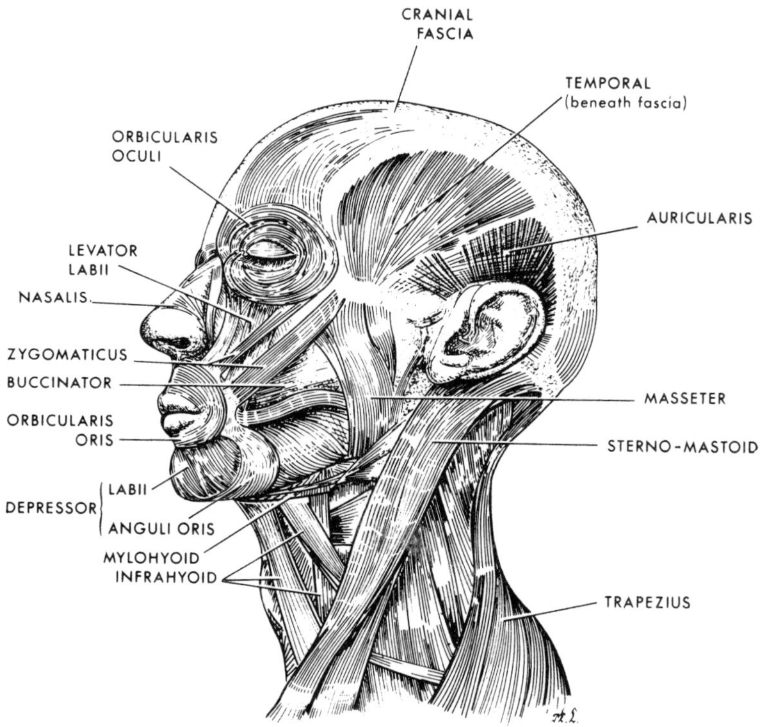

Fig. 6.6. **Muscles of the head and neck.**

Muscles of Mastication

Buccinator—force food between the teeth and help in mastication.

Masseter—bring the teeth of the lower jaw into contact with those of the upper jaw.

Pterygoid—(not visible in the figure) help to bring the jaws into contact and effect sideways movement of the lower jaw.

Temporal—raise the lower jaw.

Mylohyoid—forms the floor of the mouth.

Muscles of the Mouth

Levator labii—raise the lips in such expressions as sneering.

Zygomaticus—raise the corners of the mouth in laughing.

Orbicularis oris—pout the lips.

Depressor anguli oris— } Lower the corners of the mouth and
Depressor labii— } lips in such expressions as discontent.

Some of the muscles in the facial region, known as the *facial muscles,* are clearly responsible for the various expressions we assume.

Muscles around the Eye

Orbicularis oculi—close the eyes and lower the eyebrows when frowning.

Muscles around the Ear

Auricularis—for moving the external ear but are usually incapable of functioning in man. (Some people are able to "wag" their ears!).

Muscles for Moving the Head

Sternomastoid—When acting singly move the head from side to side and are thus responsible for turning the head. When acting together they lower the head as in nodding.

Trapezius—raise the head.

Muscles of the Tongue

These are partly responsible for swallowing and speech and other movements of the tongue.

Muscles of the Pharynx

These muscles complete the act of swallowing, forcing the food down into the oesophagus.

Muscles of the Thorax

It was stated earlier that the *trapezius muscles* inserted into the skull have their origins on the clavicle and scapula and on the spinous processes of the thoracic vertebrae. There is a large fan-shaped muscle known as the *pectoralis major* reaching from the front of the thorax to the arm. It moves the rib cage upwards in inspiration and it is used in moving the arms across the chest.

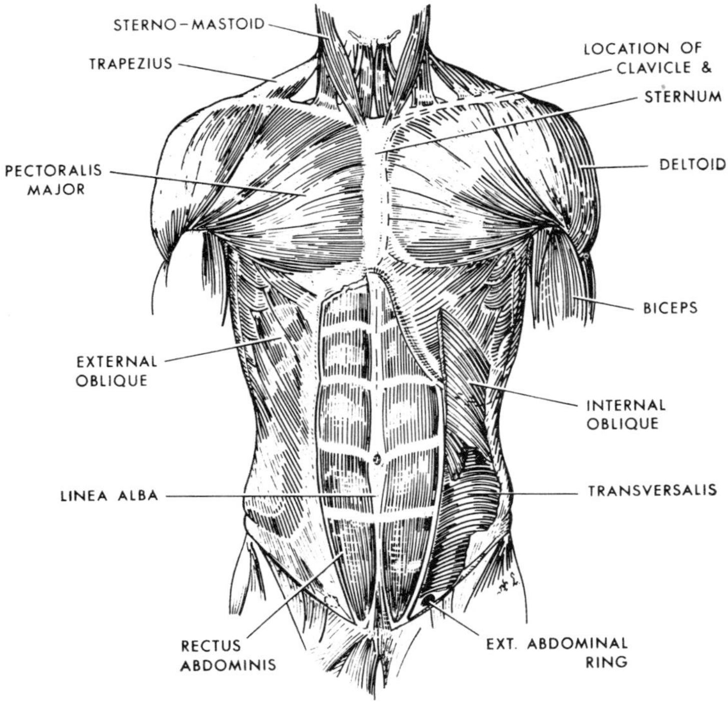

Fig. 6.7. **Muscles of the thorax and abdomen.**

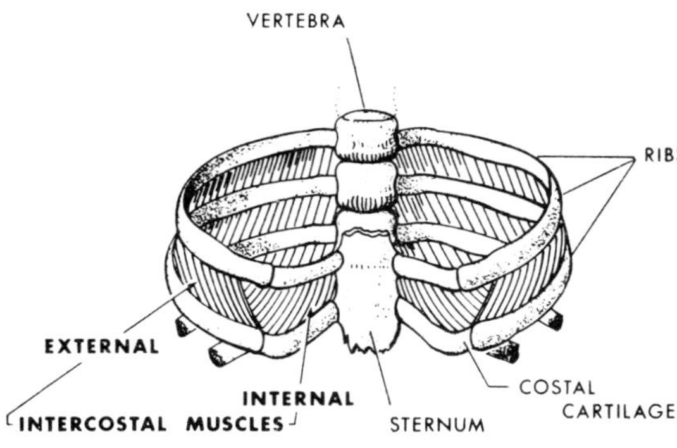

Fig. 6.8. **Internal and external intercostal muscles.**

A more deeply seated *pectoralis minor* pulls the scapula forward. Over the shoulder is the triangular *deltoid muscle* by which the arm is moved upwards and outwards. At the back of the shoulder, and reaching down as far as the lumbar region is a large *latissimus*

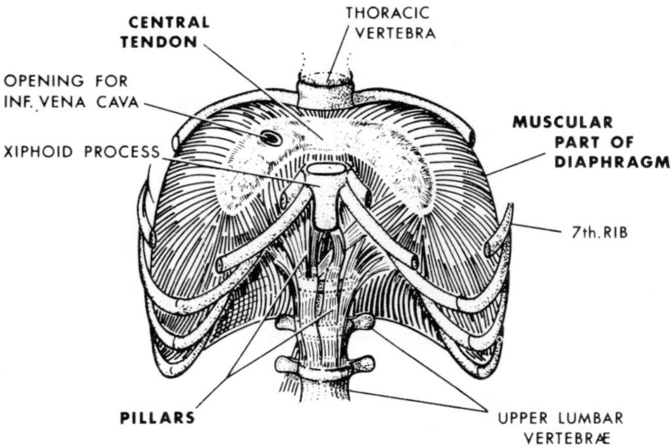

Fig. 6.9. **The diaphragm.**

dorsi muscle which moves the arm downwards and outwards. The muscles in the intercostal spaces between the ribs are responsible for the movement of the rib-cage in breathing; they are known as the *external* and *internal intercostals.* The most important respiratory muscle, however, is the *diaphragm.* It is attached to the inner sides of the lower ribs and costal cartilages, the xiphoid process and the upper lumbar vertebrae. Its fibres are joined to a *central tendon* in the diaphragm itself.

The Abdominal Muscles

The large flat muscles reaching from the lower costal cartilages and the xiphoid process above to the pelvis below and which form the ventral wall of the abdomen are known as the *rectus abdominis muscles* and where they join in the mid-line is the *linea alba*. These muscles are responsible for the flexion of the trunk and protect the abdominal organs. The *external* and *internal oblique muscles* running obliquely upwards from the pelvic girdle to the lower ribs and which cover the rectus muscles in front also take part in the formation of the linea alba and in the flexion of the trunk. The lower part of the tendon of the external oblique is very strong and is called *Poupart's ligament.* The tendon of the internal oblique lies in the outer part of this ligament. Just above the ligament is an opening, the *external abdominal ring*, through which passes the spermatic cord in the male and the round ligament of the uterus in the female. It is situated just above the pubis and is the end of what is known as the *inguinal canal.* Lying beneath the oblique muscles and running across the abdominal wall is the *transversalis muscle.* The abdominal muscles also play their part in respiratory movements and help in defaecation. The floor of the pelvic cavity is composed of muscle and this gives support to the organs contained in the cavity and helps in defaecation.

Muscles of the Back

We have already noted the large muscles of the thorax and abdomen. Other muscles, the *erector spinae*, on both sides of the spinal column enable it to be straightened.

Muscles of the Upper Limb

The *biceps,* assisted by the *brachialis*, has already been stated as responsible for the flexion of the forearm at the elbow. The

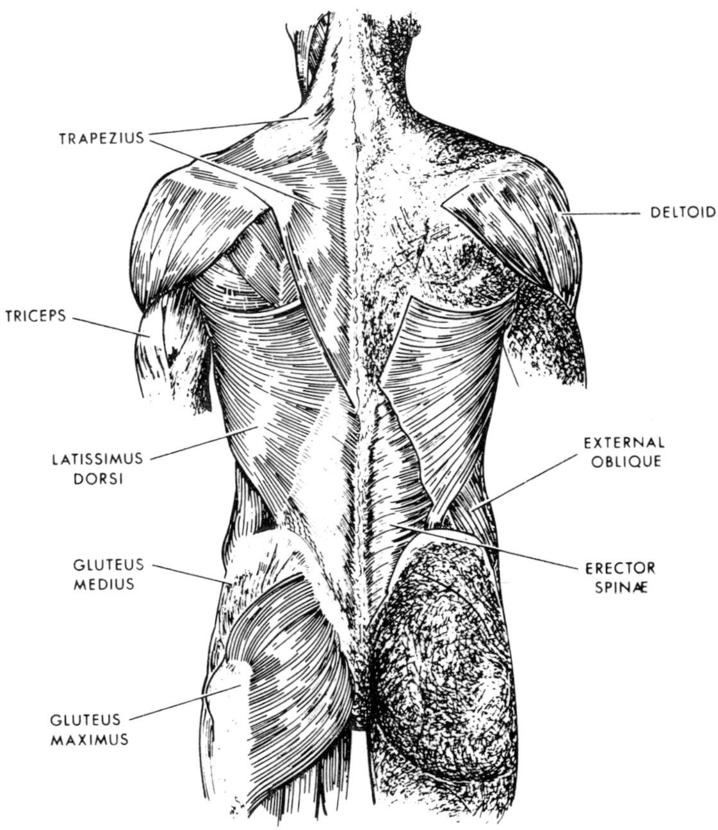

Fig. 6.10. **Muscles of the back.**

triceps, too, has been seen as the muscles extending the forearm (Fig. 6.3) and the *deltoid* in abducting the arm. The *coraco-brachialis* adducts it. Further movements of this limb are brought about by several groups of muscles and it will be obvious that different groups are concerned in such movements as flexion and extension of the wrist, pronation and supination of the hand and the flexion and extension of the fingers and thumb.

Muscles of the Lower Limb

As in the upper limb, various groups of muscles take part in the various movements of the legs, ankle and feet. The large calf-muscle or *gastrocnemius* is one which, by flexing the foot, throws the body forward when walking. It is joined by the long and powerful *Achilles tendon* to the calcaneum. The other muscle concerned is

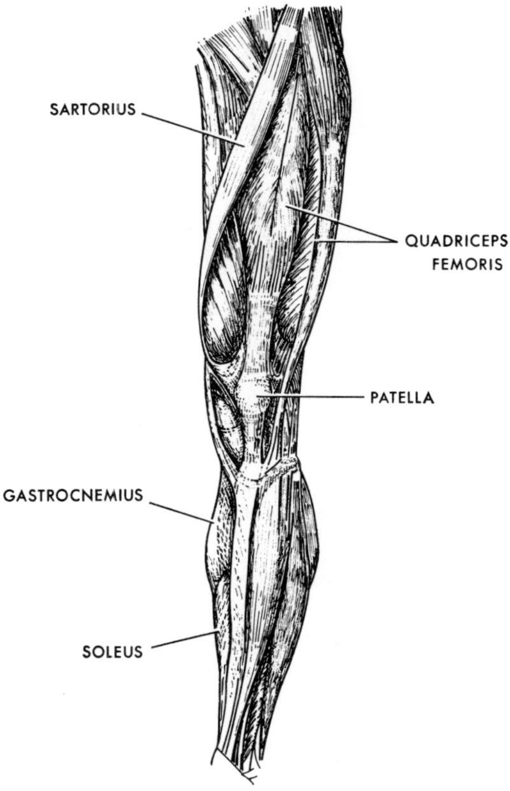

Fig. 6.11. **Muscles of the front of the leg.**

the *soleus* which raises the heel and its tendon also joins the Achilles tendon. The *quadriceps femoris* in front of the thigh with its origin on the pelvic girdle and its insertion on tibia flexes the thigh at the hip joint. The large muscles of the buttocks abduct

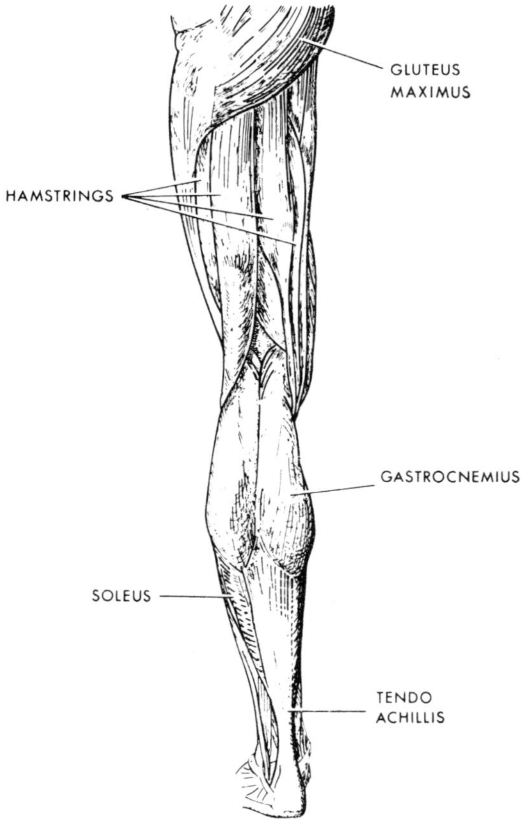

Fig. 6.12. **Muscles of the back of the leg.**

the thigh and have their origin on the ilium and their insertions on the trochanters of femur. They are known as the *gluteal muscles*, the largest being the *gluteus maximus*. The *sartorius*, the longest muscle in the body, flexes and lifts the thigh and at the back of the thigh are the *hamstring muscles* which are its extensors.

Walking, Running and Jumping

When walking, one foot is always on the ground and the heels of the right and left feet are brought to the ground alternatively whereas in running the heels never reach the ground and the feet are off the ground for a very short period. In jumping both legs are off the ground together and act simultaneously.

Let us now examine the *sequence of movements in walking*, stepping off with the left foot (see Fig. 6.13).

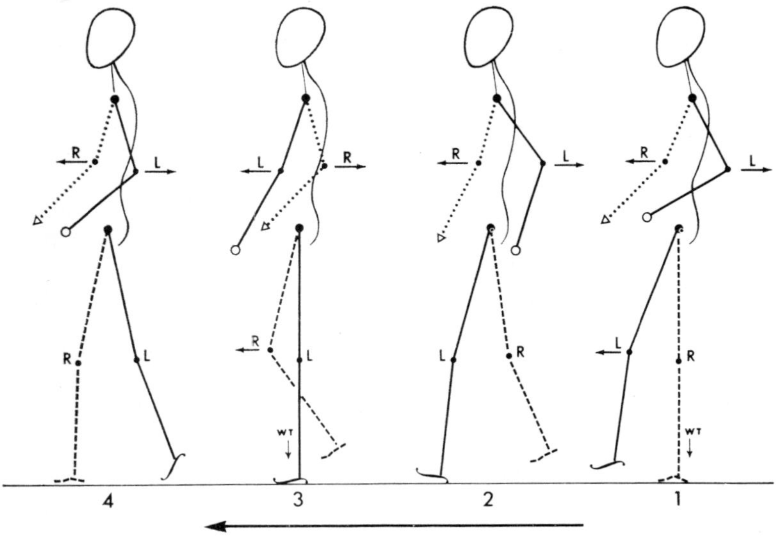

Fig. 6.13. **Movements in walking.**

1. The weight of the body, equally shared between both legs when standing still, is transferred to the right foot, the left heel being raised by the *gastrocnemius* which also pushes the body forwards. The left leg is then flexed at the hip by the *quadriceps femoris* and is raised from the ground, the knee being slightly bent while the right leg is, of course, straight. At the same time the right arm is moved forwards and the left arm backwards.

2. The left leg is then moved forwards and gradually straightened by extension at the knee and the heel is placed on the ground. Thus the weight of the body is transferred to the left foot and the right knee is bent and the heel raised. Movement of the arms is continued to the limit.

3. Now the right leg, with the left foot taking the body's weight, is moved past the left with the right knee slightly bent. At the same time the left arm moves forwards and the right arm backwards.

4. The right heel reaches the ground, the foot is flattened and

the left heel raised while the right arm swings forwards again and so the movement described in (2) is repeated.

In walking upstairs, there is, of course, some modification of the above movements. Let us again start off with the left foot. First the leg flexes at the knee and is raised by flexion at the hip while the right foot takes the weight of the body. The body is then raised by the muscles on both sides, the left foot taking the weight as the right foot is raised, the right knee and hip being flexed and the left leg straightened. Similar movements follow in succession. In walking downstairs less energy is expended as gravity is chiefly responsible for the downward movement of the body, flexion at left and right knees following in quick succession.

It will be obvious that in all these movements a whole series of leg, ankle, foot and arm muscles come into play and, in addition, muscles of the back and neck which move the trunk into appropriate positions for each movement.

Posture

The maintenance of posture is effected by a whole series of muscles in the neck, trunk and legs. The muscles which tend to pull the body forwards are equally opposed by others which tend to pull it backwards. Similarly those which would make the body incline to one side are balanced by opposing muscles pulling equally in the opposite direction. All these muscles have just that right degree of tension to maintain the body in an upright position without movement. Equilibrium is lost when the nervous control of these muscles is affected so that muscular tone is lost. When a person loses consciousness his body collapses.

The importance of correct posture when standing and sitting and of exercise in keeping the muscles in proper working order cannot be overemphasised. Muscles which are not used become weak and those which are exercised become strong. Hence the necessity for exercise and for training for sports events, when the muscles particularly required are strengthened and so produce the required movements with a minimum expenditure of energy. However, once a certain degree of efficiency has been reached there is always a risk of "staleness" setting in. This is partly due to muscle fatigue and partly psychological. The coach of a boat-race crew knows this only too well and has to keep a careful watch for it, for in this form of sport a highly concentrated effort is required for a prolonged period and, furthermore, it is essential for success that the crew as a whole work perfectly together. Proper attention to diet and adequate rest during training are as necessary as practice in the art and the trainer or coach endeavours to bring his team

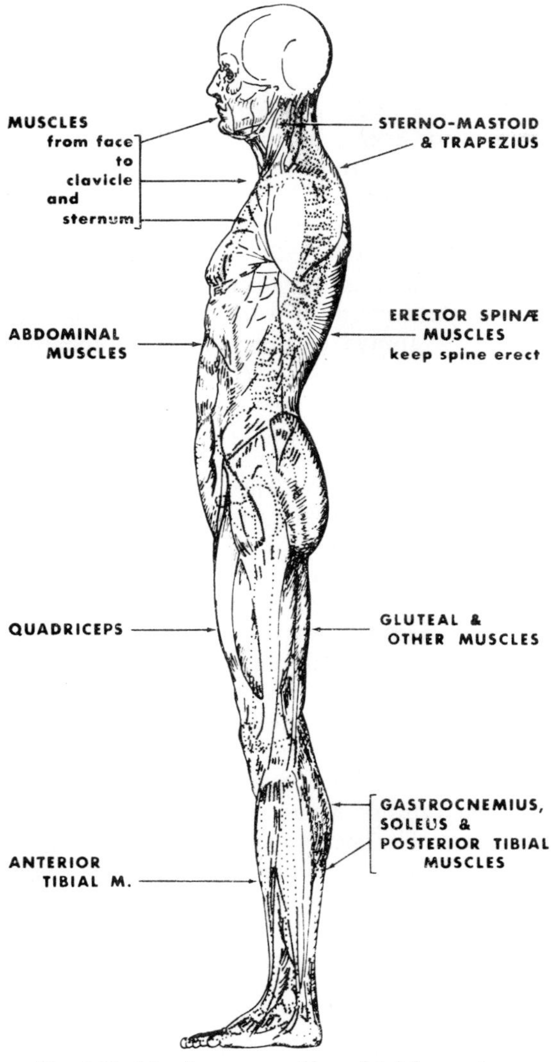

Fig. 6.14. **Muscles concerned in maintaining posture.**

or crew to the height of perfection at the time the event is to take place.

THE MECHANICS AND PHYSIOLOGY OF MUSCULAR MOVEMENT

When movement takes place, whether in an inanimate machine

or in the body, *work* is done and this involves the *expenditure of energy*. This work is measured in units known as *ergs* on the metric system and *foot-pounds* on the British system. These units are the product of the force applied and the distance through which it acts. The unit of energy is the *joule* which is equivalent to 10^7 ergs and on the International System (S.I.)* of Units this is also a unit of work and quantity of heat. The rate at which work is done is known as *power* and on the British system the unit is one *horse power* which is 550 ft. lbs. per minute. Clearly the horse power produced is inverse to the time in which a movement is performed but the amount of work done remains constant. The intake of energy is always greater than the output which is the work performed and the ratio $\frac{Output}{Input}$ is known as *efficiency*. It is usually expressed as a percentage. The reason for this apparent wastage of energy is that friction must be overcome and some of the energy is released as heat. The *efficiency* of a machine varies according to the rate of working and there is an optimum rate at which efficiency is greatest. It will decrease at a higher rate which means, as far as the body is concerned, that the particular movement cannot be continued for as long a time as when it is working at its maximum efficiency and fatigue will set in. Compare walking a mile, running a mile and running in a hundred yards race. Consider what happens during these events, how you feel at the end of them and work this out for yourself.

Whence comes the energy and how is it set free and utilised by the muscles? Muscle is itself composed of a protein called **myosin** and about 75 per cent. of water and within the muscle are other chemical substances. The energy for muscular contraction is derived from a series of chemical reactions in which biochemical catalysts or enzymes take part. Glucose reaches the muscles by the blood stream and is broken down. As this is used up more must be provided. One of the substances stored, though not in large quantities in muscle, is animal starch or *glycogen* and this is converted into *glucose* and ultimately into *lactic acid*. Some of this lactic acid passes into the blood and is reconverted into glycogen in the muscles during the recovery period and oxygen is necessary for this. But some of the lactic acid is carried to the liver where a similar conversion takes place. The glycogen in the liver, which is a larger store than that in the muscles, provides a further supply of glucose for the muscles. The remainder of the lactic acid is oxidised to *carbon dioxide* and *water* by the oxygen reaching the muscle in the blood and the blood takes away the carbon dioxide

* S.I. = Système International d'Unités.

so produced. Some of the energy produced, however, is liberated as heat. *Shivering* is a series of involuntary muscular contractions which generate heat when the body is cold and sweating, a process of evaporation and therefore of cooling, is a method of losing excess heat.

Oxygen is needed for recovery in order to remove the excess of lactic acid and this is slower in the early stages of exercise than it is later. Hence the increase in the rate and depth of breathing and in the rate of heart beat during and after exercise. When the oxygen supply is adequate to deal with the oxidation of the lactic acid in the blood, the exercise can be continued for a long time. If, however, the exercise is extremely vigorous, shortness of breath follows owing to the accumulation of lactic acid in the blood due to insufficiency of oxygen and this is referred to as "oxygen debt". So the energy for muscular contraction is ultimately derived from carbohydrate but in severe exercise fats, and even proteins, may be broken down to provide sufficient energy and muscular contraction can take place without the breaking down of carbohydrates. In fact, it appears that the energy for contraction is provided in the first instance by a compound known as **adenosine triphosphate** (ATP) which is broken down to *adenosine diphosphate* (ADP) and the enzyme concerned in this reaction is the myosin of the muscles. However, the full story of the chemical changes which take place in contraction are too involved to enter into here. Electrical changes also occur. Muscular recovery takes place during the resting period when the blood takes away the lactic acid but *fatigue* results if this accumulates.

After death when there is no oxygen supply, some of the protoplasm in the muscle fibres coagulates. Consequently the muscles no longer relax. The contraction causes the whole body to become rigid and the condition is known as **rigor mortis.** It usually begins between four and ten hours after death and may last up to forty-eight hours or longer but the time of onset and the period of duration are affected by the cause of death and other conditions. It eventually passes off as degenerative changes take place and the muscle becomes softened.

Several interesting and useful experiments can be performed to demonstrate muscular contraction. Cold-blooded animals are used because their muscles survive for longer periods when removed from the body and because they function at ordinary temperatures and therefore do not need to be kept warm. Nor do they need oxygenation as the gas is adequately absorbed over the surface. The gastrocnemius muscle from the leg of the frog is frequently used. One such experiment can be carried out as follows:—

A frog is pithed (*i.e.* its brain is destroyed so that it loses consciousness though its tissues continue to live). This is done by the insertion of a seeker into the cranial cavity. The gastrocnemius muscle with the sciatic nerve which innervates it are then carefully dissected out from the leg, cutting through the femur and tibia in order to leave the knee joint, and so the origin of the muscle, intact. The **muscle-nerve preparation**, as it is called, is then laid on a board and kept moist with physiological saline (0·65 per cent. sodium chloride) or Ringer's solution which approximates to the plasma of the blood.

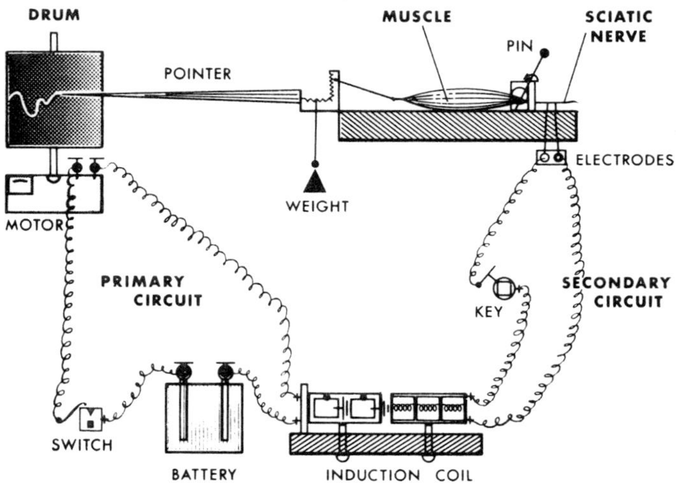

Fig. 6.15. **Muscle-nerve preparation experiment.**

The bone is fixed to the board by a pin and the free end of the muscle is joined to a light lever by means of thread, the lever being kept taut by the suspension of a light weight from it. The other end of this lever is pointed and is in contact with a drum covered with smoked paper and rotated by clockwork or an electric motor. Stimulation of the muscle is effected by means of two electrodes on which the nerve is laid. These electrodes are connected to the secondary circuit of an induction coil and this provides the excitation when the key is closed. The amplified contraction of the muscle is recorded on the slowly rotating drum by movement of the lever and this marking on the smoked paper is known as a **muscle twitch** or **myogram.**

The figures below show the results of (1) a single stimulus (2) a second stimulus applied before the relaxation from the first is completed so that the second contraction is added to the first (3) a series of stimuli, when very slight contraction occurs, between the applications and (4) the prolonged application of a series of rapid stimuli. In the last instance the contraction is sustained and this is known as **tetanus**. Note that there is a short period after application of the stimulus before contraction begins in the first two cases. This is called the **latent period**.

(1) **Single stimulus**

(2) **Second stimulus before relaxation from first completed**

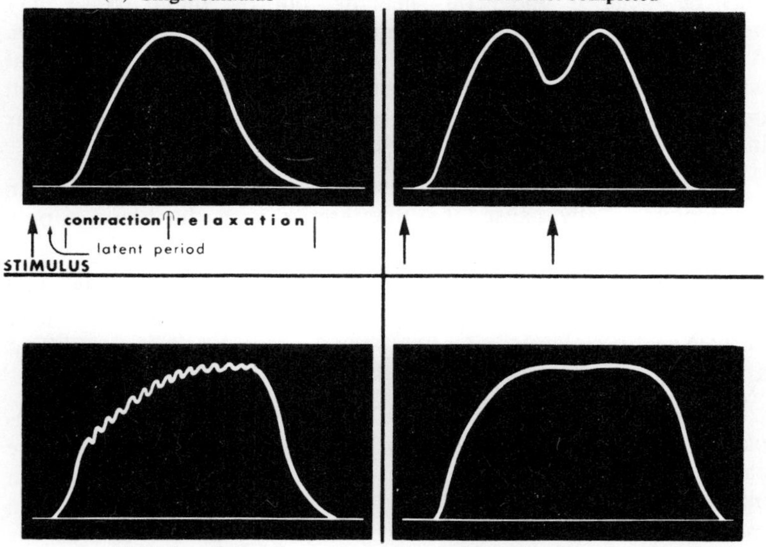

(3) **Series of stimuli**

(4) **Prolonged series of rapid stimuli (tetanus)**

Fig. 6.16. **Myograms.**

The same apparatus can be used to find the effect of load on muscle contraction by hanging a series of increasing weights from the lever, when it will be seen that the greater the load the less is the contraction.

Disorders and Diseases of the Muscular System

Hernia—A hernia is the protrusion of part of an internal organ covered by peritoneum through the wall of the cavity in which it lies and the commonest form is one through the abdominal ring.

In this case, it is part of the intestine which protrudes and it is known as an **inguinal hernia** and popularly called a *rupture.* It occurs more frequently in men than in women. Relief can be obtained by the application of a pad fitted to a support known as a *truss* but resort to surgery is necessary to effect a cure. The danger with a hernia lies in interference with the blood supply of the protruded part of the intestine owing to the constriction. This condition is called a **strangulated hernia**: it is extremely dangerous and requires immediate surgery to prevent the onset of gangrene which would prove fatal. The precipitating cause of a hernia is the application of a strained effort and severe constipation may be included under this heading, but the initial cause may be a congenital weakness in or injury to the wall of the abdomen. A hernia may develop in some other position in the abdominal wall, for example where an incision has been made in a surgical operation; this is called a **ventral hernia**. Again a hernia may occur in the umbilicus or navel or in the diaphragm.

Cramp—This is a painful muscular contraction, particularly in the limbs, usually caused by interference with the blood supply to the muscles concerned.

Convulsions—These manifest themselves as alternate and irregular contractions and relaxations of voluntary muscles. It may be caused by epilepsy but often occurs in young children without any serious cause or effect.

Rheumatism—This is a general term which covers many complaints and diseases including arthritis, already mentioned in the chapter on joints. As far as the muscles are concerned, it is an inflammation which causes both pain and stiffness. It has been estimated that 30,000,000 working days a year are lost as the result of sufferers of rheumatic diseases of various kinds.

Spasm—When an involuntary muscular contraction occurs, it is called a spasm and it may be quite painful as in the condition known as **cramp**. Its cause usually lies in some disturbance of the nervous system though cramp may be the result of constant usage of muscles for a particular movement and which does not result from the use of the same muscles for a different kind of movement.

Paralysis—Paralysis is loss of muscular power and is the result of interference with a muscle's nervous control. It may be caused by damage to the brain. Frequently one side of the body becomes completely affected (*hemiplegia*) and this is due to brain damage on the opposite side. In **total paralysis** both sides are affected.

On the other hand paralysis may occur in which only one part, for example one limb or one side of the face, is affected: this is called a **local paralysis. Creeping paralysis** or *locomotor ataxis,* caused by a disease of the spinal cord, is a form of paralysis in which there is lack of co-ordinated movement, principally in the arms and legs when walking. In **shaking paralysis** or *paralysis agitans* involuntary tremors and stiffening of the muscles of the limbs takes place. Its causitive origin lies in the degeneration of certain parts of the brain. In **tetanus**, the muscles of the jaw become fixed and rigid making the opening of the mouth difficult or impossible, a condition known as *lock-jaw,* and this is followed by rigidity and muscular spasms of other parts of the body. It is caused by a bacterium, *bacillus tetanus,* which normally lives in the soil and in dirt and gains entry into the body through a wound. Immunisation can be effected by injection of the tetanus toxoid.

Trichinosis—In this case a parasitic worm, *Trichinella spiralis,* in the intestine, obtained by eating infected meat, finally lodges in the voluntary muscles where it becomes encysted in a coiled-up form.

Muscular dystrophy—Degeneration of muscles occurs with accompanying weakness. Its cause is not known for certain but it may be due to deficient nutrition. It does not respond to treatment and is always fatal, death usually occurring between the ages of twenty and thirty.

PRACTICAL WORK

The practical work on muscles can be deferred until a rabbit is dissected for the study of the digestive system in order to economise in material.

Completely skin a rabbit, cutting through the connective tissue which attaches it to the muscular body wall.

1. Examine the **muscles of the fore-limb** and identify the **biceps**. Find its *tendons* and the *origins* on the scapula and *insertion* on the radius.

Now find the **triceps** at the back of the limb and examine in a similar way, finding the *origins* on the tuberosities of humerus and the *insertion* on the olecranon process of ulna.

2. Examine the **muscles of the hind-limb** and find the large **gastrocnemius** or calf muscle. Find the *origins* on the condyles at the lower end of the femur and the long **Achilles tendon** with its *insertion* on the calcaneum.

3. An examination should now be made of **muscles of the thorax**,

abdomen and **back**. Try and identify at least some of those mentioned in the text.

4. Finally make various movements of your body and limbs and feel with your hands which muscles come into play. This examination can be made with another student as your "guinea pig".

Chapter VII

THE BIOCHEMISTRY OF FOODSTUFFS

It was stated in Chapter I that the process of nutrition makes available to the body the materials for growth and repair of living cells and for the production of the energy which is necessary for all life processes. We must now examine the biochemical nature of the essential foodstuffs which are provided in the foods we eat but we shall defer the investigation into food and diet until a later chapter.

The most important elements required by the body are carbon, hydrogen, oxygen, nitrogen, sulphur, phosphorus, potassium, sodium, calcium, magnesium and chlorine while fluorine, iodine, iron, copper, manganese and cobalt are also required in very minute quantities and are therefore called *trace elements.* Of these elements only one, oxygen, is absorbed as such, the remainder being obtained in compounds. The compounds which provide the necessary elements are as follows:—

proteins
carbohydrates
fats and oils
mineral salts
water
vitamins

It is the biochemistry of these compounds which we are now about to investigate, after which we shall see how they are made available to the cells of the body.

PROTEINS

Proteins are essentially compounds of carbon, oxygen, hydrogen and nitrogen. Many of them also contain sulphur and some contain phosphorus. All have a high nitrogen content and the molecules are exceptionally large and they are therefore of very high molecular weight. One of the simplist, *insulin,* has a formula $C_{254}H_{377}O_{75}N_{65}S_6$, giving a molecular weight of 5,727 and *edestin,* found in hemp seeds, a comparatively simple one, has a probable formula $C_{622}H_{1020}N_{193}O_{20}S_4$ with a molecular weight of 14,530.

Protein molecules are made up of many units of what are known as **amino-acids** of which the simplest known is amino-acetic acid or glycine—$CH_2NH_2.COOH$. Some twenty-five enter into the composition of proteins though only ten are essential to life. Since all plant and animal cells contain proteins and, in fact, they form an essential part of protoplasm, plant and animal food provides natural sources of them, though both the quality and quantity varies considerably in different foods. Many proteins are insoluble in water and form colloid sols and most of them are coagulated by heat and by the addition of various chemical substances. An important factor in their synthesis is that amino-acids are amphoteric *i.e.* they contain both an acid (—COOH) group and a basic (—NH_2) group and by repeated reaction between the acid group of one amino-acid with the basic group of another, the protein molecule is built up by *condensation* (removal of the elements of water), —CO·NH—being the *peptide linkage* between the amino-acids.

As plant proteins do not contain all the essential amino-acids required by the human body, they are known as **second class proteins,** whereas animal proteins, which do provide the essential amino-acids, are called **first class proteins**. However, gelatin, an animal protein obtained from bone, lacks all the essential amino-acids and is therefore of little use as a foodstuff.

The usual method of classification of proteins is into the following groups:—

Simple Proteins

Albumins—Soluble in water and coagulated by heat. Found in blood, milk, eggs and seeds.

Globulins—Insoluble in water but soluble in solutions of neutral salts. Coagulated by heat. Found in muscle *(myosin)* and in seeds (such as *legumin* in peas and beans).

Gliadins—Insoluble in water. Soluble in aqueous alcohol. Not coagulated by heat. Found in maize (*zein*), wheat (*gliadin*) and other cereals.

Glutelins—Insoluble in water. Not coagulated by heat. *Glutenin* occurs in wheat flour.

Scleroproteins—Soluble only in concentrated solutions of acids and alkalies. An example is *keratin* which is found in nails and hair.

Elastin occurs in the elastic fibres of ligaments and similar tissues.

Compound Proteins

Phosphoproteins—compounds of proteins with the phosphoric acid. *Casein* in milk and cheese and *vitellin* in egg yolk are typical examples.

Nucleoproteins—compounds of protein and nucleic acid. These are, of course, found in nuclei.

Mucoproteins—Soluble in water and not coagulated by heat, they occur in *mucin* and are characteristic of mucous secretions. They are compounds of protein and mucoitic acid.

Chromoproteins—these are coloured and are compounds of protein with a pigment. A common example is *haemoglobin* which gives the blood its red colour. It is a compound of the protein *globin* with an iron compound called *haem.*

Derived Proteins

In the process of protein digestion, a process of hydrolysis, substances known as derived proteins are formed. These are **proteoses**, **peptones** and **peptides**. Amino-acids, which are the end-products of protein digestion, are included under this heading.

All proteins give very definite reactions with certain reagents and these are due to the amino-acid content. These tests will be found in the practical work at the end of the chapter.

CARBOHYDRATES

Carbohydrates are compounds of carbon, hydrogen and oxygen with the hydrogen:oxygen ratio the same as in water (*i.e.* 2:1). They therefore have a general formula of $C_m(H_2O)_n$ and include the sugars, starches and cellulose. They are broken down into carbon dioxide and water in the process of respiration and thus energy is liberated. They are divided into three main groups as follows:—

Monosaccharides

These are *simple sugars* with the general formula $C_n(H_2O)_n$ and obviously the simplest of these would be $C.H_2O$ but, in fact, this is not a sugar; it is *formaldehyde* (H.CHO). *Ribose* which is contained in R.N.A. is a *pentose* ($C_5H_{10}O_5$) but the most important monosaccharides as far as nutrition is concerned are the *hexoses* ($C_6H_{12}O_6$), glucose and fructose.

Glucose—This is also known as *grape-sugar* and *dextrose* and it occurs freely in plant juices. It is the end-product of carbohydrate digestion and is therefore found in the blood. It is very soluble in water and is less sweet than "ordinary" sugar (which is a disaccharide). Yeast readily ferments it to ethyl alcohol and carbon dioxide. When boiled with alkaline solutions of copper salts such as Fehling's Solution and Benedict's Reagent, it reduces them to cuprous oxide and therefore gives a red precipitate. Any sugar which brings this about is called a *reducing sugar.*

Fructose—Also called *fruit-sugar* and *laevulose,* fructose occurs with glucose in fruits and honey. It is sweeter than glucose and is a reducing sugar.

Disaccharides

These are compound sugars and are formed by the combination of two hexose (monosaccharide) groups by condensation.

$$2C_6H_{12}O_6 - H_2O = C_{12}H_{22}O_{11}.$$

Maltose—As this occurs in germinating barley which is known as *malt* it is called maltose. It is formed during the process of the digestion of starch, due to hydrolysis and on further hydrolysis it yields glucose

$$C_{12}H_{22}O_{11} + H_2O = 2C_6H_{12}O_6.$$

It is very soluble in water and is a reducing sugar.

Lactose—This is *milk sugar* and it occurs in different quantities in the milk of different mammals. It is less soluble and less sweet than are other sugars. On hydrolysis it yields glucose and another hexose called *galactose*. The souring of milk is due to the presence of lactic acid into which the lactose is fermented by a specific bacterium.

Sucrose—This is "ordinary" sugar and is obtained from the stem of the sugar-cane (*cane sugar*) and from the roots of the sugar-beet (*beet-sugar*). It is very soluble and very sweet. It is hydrolysed to a mixture of glucose and fructose called *invert sugar*

$$C_{12}H_{22}O_{11} + H_2O = C_6H_{12}O_6 + C_6H_{12}O_6$$

and it is a non-reducing sugar.

Polysaccharides

These *complex saccharides* are all non-sugars and the most important are starch, glycogen and cellulose. The general formula is $(C_6H_{10}O_5)_n$. They are all of high molecular weight and "*n*" is therefore a large number. All are insoluble in water and starch and glycogen form colloid sols. They do not reduce alkaline solutions of copper salts and are formed by the condensation of several monosaccharides.

Starch—An old name for starch is *amylum*. It occurs in vegetable foods, being a food-reserve in plants, but is never found in animals. The starch is stored in grains which vary in shape and size in different plants and which are characteristic of those plants. On the addition of iodine solution a dark blue colouration is obtained. Hydrolysis results in the formation of glucose

$$(C_6H_{10}O_5)_n + n\,H_2O = n\,C_6H_{12}O_6.$$

Glycogen—Animal starch is the form of carbohydrate stored in the liver and muscles. On hydrolysis it is converted to glucose and the addition of iodine produces a port-wine colour.

Cellulose—This is peculiar to plants in which it forms the structural material of young cell walls, though in older cells other substances become incorporated into the walls. Consequently we can eat a great deal of cellulose. Herbivorous animals digest it in the large intestine by bacterial action but we are unable to do so and it is useless to us as food. Nevertheless it is a necessary constituent of our diet as it serves as "roughage" which stimulates the action of the muscular walls of the alimentary canal enabling the food to pass along. Cellulose is insoluble in water but it dissolves in certain specific reagents.

FATS AND OILS

Fats and oils belong to a group of compounds known as *lipids* and contain carbon, hydrogen and oxygen but with a lower oxygen content than is found in carbohydrates. They are compounds called *esters* of certain higher fatty acids. An ester is formed when an acid reacts with an alcohol and the acids concerned are *stearic* $C_{17}H_{35}\cdot COOH$, *palmitic* $C_{15}H_{31}\cdot COOH$ and *oleic* $C_{17}H_{33}\cdot COOH$ while the alcohol is glycerol, commonly called glycerine $C_3H_5(OH)_3$. The esters are known as stearin, palmitin and olein respectively. The difference between a fat and an oil is that a fat is solid and an oil liquid at 20°C and the hardness of a fat depends on the proportion of these glycerides (as the esters are called), stearin and palmitin being solid and olein liquid at ordinary temperatures. Beef and mutton fat contain all three glycerides, human fat is largely olein, butter is also a mixture of all three but contains in addition esters of butyric and other acids. Fats and oils also occur in fish, milk, eggs, cheese, margarine and nuts.

Though insoluble in water, they are soluble in hot alcohol, ether, chloroform and other organic solvents which are consequently known as "fat solvents". With other liquids they form emulsions *i.e.* colloid sols. On hydrolysis they yield fatty acids and glycerol and when the hydrolysis takes place in the presence of alkali, the fatty acids react with it to form salts which are known as *soaps.* (Ordinary soap is sodium stearate.) Fats and oils make paper translucent and are stained a salmon pink colour by a dye called Sudan III and black by osmium tetroxide ("osmic acid"). Like carbohydrates they provide energy. (In fact, one gram of fat produces more than twice as much energy as one gram of carbohydrate.) But they cannot be completely broken down in the absence of carbohydrate. They also serve to insulate the body against heat loss and act as a reserve of food to be called upon in case of necessity *e.g.* in starvation.

In the digestion of animal fats, a closely related waxy substance called *cholesterol* is formed. A high cholesterol content of the blood causes it to be deposited on the endothelial lining of the arteries, narrowing them down and thus slowing down the circulation of the blood and increasing blood pressure. Furthermore some cholesterol may be set free from the arterial lining forming a clot and thereby blocking the artery. Vegetable fats and oils do not produce cholesterol. It is now an accepted fact, at least by most authorities, that a high cholesterol level in the blood is the cause of arteriosclerosis and they recommend a limited use of animal fats in the diet.

MINERAL SALTS

In addition to proteins, carbohydrates and fats, mineral salts are necessary in our diet. Calcium phosphate is a constituent of bone, iron forms part of the haemoglobin of the blood, iodine is contained in the secretion of the thyroid gland, calcium is concerned in the clotting of the blood and in the making of teeth and so on. These elements, together with others known as *trace elements* because they are required in very minute quantities (such as zinc, copper, cobalt, manganese and fluorine), are obtained in the form of mineral salts.

WATER

About 60 to 70 per cent. of the body's weight consists of water. It is contained in the muscles and other tissues and in the blood and other body fluids and it is an essential constituent of protoplasm. Furthermore it is the medium in which the various metabolic reactions take place and in which the various secretions of the body are transported. Again, its evaporation in sweating helps to regulate body temperature. It is therefore essential to life. The body loses a great deal of water every day by sweating, in urine and faeces and in expiration and this water must obviously be replaced. A man can live for weeks without food but if he is deprived of water he will die in a few days. Some 50 per cent. of the body's water requirements is provided in food, the remainder being from drinks.

Thus a man or woman weighing 10 stone (63·5 kg.) contains at least 84 lb. (38·1 kg.) or 8·4 gallons (38 litres) of water.

VITAMINS

These are complex organic substances of high molecular weight which are essential for the maintenance of health and which are, therefore, necessary in our diet. They may be called *accessory food factors.* They are, however, required only in very minute quantities and not all known vitamins are necessary to the human body. They occur in foods, the original source, in most cases, being plants. Some are water soluble, others dissolving only in fats. They are usually designated by letters but as their chemical constitution becomes known they are given appropriate chemical names. They will be studied in detail in Chapter XXIII.

ENZYMES

The proteins, carbohydrates and fats in food have to be broken down into simpler soluble and absorbable substances before they can be utilised by the cells of the body and the chemical agents responsible for this are known as **enzymes** and are contained in the various digestive juices. Enzymes may be briefly defined as *biological catalysts* since they are substances produced by living cells and they accelerate chemical reactions without being used up in the process. They are of colloidal dimensions and have certain definite characteristics:—

1. They are highly specific in their action *e.g.* an enzyme acting on proteins has no effect on carbohydrates or fats.
2. They act only within a certain temperature range and are inactivated by high temperatures.
3. They are specific in their requirements as to the H-ion concentration (*p*H value)* of the medium in which they act, some requiring an acid medium, others an alkaline one.
4. A small quantity of enzyme acts upon a large quantity of substrate.
5. Some are secreted in an inactive form and require combination with *co-enzymes* in order to activate them.
6. They are sensitive to certain chemical substances which inhibit their action.

Digestive enzymes act on foodstuffs by *hydrolysis* but not all enzymes act this way.

* **Hydrogen-ion concentration** is a measure of acidity and alkalinity and is expressed as a ***p*H value**. Pure water dissociates into equal amounts of hydrogen ions† $\mathbf{H^+}$ and hydroxyl ions OH^-, though the dissociation is a feeble one. $H_2O \longrightarrow [H^+] + [OH^-]$, the concentration of each being IO^{-7} gm-ions per litre. This is expressed as *p*H7. A neutral solution therefore has a H-ion concentration of *p*H7. Any solution with a *p*H greater than *p*H7 will be alkaline and any with one less than *p*H7 will be acid, the *p*H value indicating the degree of acidity or alkalinity.

† An **ion** is derived from an atom or a group of atoms and is charged with either positive or negative electricity.

Classification of enzymes is according to the substrate on which they act thus:—

Proteolytic enzymes are concerned with the various stages in the hydrolysis of proteins and are collectively known as **proteinases.**

Amylolytic enzymes hydrolyse carbohydrates and may be called **carbohydrases.** Starches are hydrolysed by **polysaccharidases** (or **amylases**) to maltose and **saccharidases** hydrolyse disaccharides to monosaccharides.

Lypolytic enzymes hydrolyse fats and oils to fatty acids and glycerol and are known as **lipases.**

It should be noted that not all enzymes are concerned with digestion. Others play their part in respiration and other processes but at the moment we are restricting our study to those related to nutrition.

PRACTICAL WORK

Proteins

Either break an egg into a basin, separate the yolk and throw it away and use the white for the following tests or make a sol with dried albumin.

1. **Millon's test**—Add a few drops of Millon's Reagent. A white precipitate forms. Heat and the ppt turns a dull red colour.

2. **Biuret test**—Add a few drops of sodium hydroxide solution. Then carefully add a few drops of very dilute (1 per cent.) copper sulphate solution. A violet colour is produced.

3. **Xanthroproteic test**—Add concentrated nitric acid. A white precipitate is formed. Heat. It turns yellow. Cool under the tap and add excess of 0·88 ammonium hydroxide. It turns orange.

Carbohydrates

1. **General test for all carbohydrates**—Prepare an aqueous solution of any carbohydrate. Add a few drops of a-naphthol solution. Slope the test tube and slowly pour a few millilitres of concentrated sulphuric acid down the side. A violet ring will appear at the interface.

2. GLUCOSE. (i) **Fehling's test**—Mix equal volumes of Fehling's Solution A and B and shake. Add this to a solution of glucose and heat. A red precipitate is obtained.

(ii) **Benedict's test**—Add a few millilitres of Benedict's Qualitative Reagent to a solution of glucose and heat. A red precipitate forms.

3. MALTOSE, LACTOSE and SUCROSE. (i) Perform **Fehling's test** with solutions of the three sugars and note the positive results with maltose and lactose. No precipitate is obtained with sucrose.

(ii) **Hydrolysis of sucrose**—Boil a solution of sucrose with a few ml. of dilute hydrochloric acid for a few minutes. Then neutralise the acid with sodium hydroxide, using red litmus paper to ensure neutrality. When this is obtained, perform Fehling's test. A positive result will be obtained.

4. STARCH. (i) **Iodine test**—Prepare some starch sol by making a paste with a little cold water, then adding hot water and shaking. To a portion of this sol add iodine solution. A blue colour will result.

(ii) **Hydrolysis of starch**—Hydrolyse some starch by the method given for the hydrolysis of sucrose above and then apply Fehling's test when a red ppt will form.

5. GLYCOGEN. (i) **Iodine test**—Perform the iodine test with a solution of glycogen. Note that a port-wine colour results.

(ii) **Hydrolysis of glycogen**—Hydrolyse some glycogen by the method used for starch and then perform Fehling's test. A positive result will be obtained.

Fats and Oils

1. **Emulsion formation**—Add a little olive oil to some water in a test tube and shake. An emulsion is formed but the two liquids quickly separate, the oil forming a layer on top of the water. Repeat the experiment using sodium hydroxide solution instead of water. The emulsion formed lasts much longer.

2. **Fat solution**—Add some ether to a little olive oil and shake. The oil dissolves.

3. **Effect on paper**—Pour a few drops of oil or some of the ethereal solution from the last experiment on to some paper. The ether quickly evaporates and the oil makes the paper translucent.

4. **Osmic acid test**—Add a few drops of "osmic acid" (osmium tetroxide) to a little olive oil in a watch glass. Leave for a few minutes when a black colouration will be seen.

Enzymes

Practical work on the effect of enzymes should be done during the study of Digestion and details of this will be found at the end of Chapter VIII.

Chapter VIII

DIGESTION AND THE ALIMENTARY SYSTEM

I. THE MOUTH, PHARYNX AND OESOPHAGUS

The food we eat is solid material and before it can be utilised by the body it must be changed into a liquid, absorbable form. This involves physical and chemical processes which are covered by the term digestion and for this purpose we have a digestive or **alimentary system.** In this system we have an alimentary canal in which food is *ingested, digested* and *absorbed* and from which the indigestible part is expelled. Associated with the canal are various glands which produce the necessary enzymes and other substances concerned in digestion. The alimentary canal consists of the *buccal* or *oral cavity,* the *pharynx, oesophagus, stomach, small intestine, large· intestine* and *rectum.* The digestive secretions and other activities of the whole system are under the control of the nervous and endocrine systems and this aspect of the subject will be dealt with later.

DIGESTION IN THE MOUTH

Mastication—In the first instance food has to be prepared and this usually involves cooking in order to change it into a more easily digested and more palatable form though some foods are, of course, eaten raw. Even so, when food has been ingested into the mouth it is necessary for it to be chewed into smaller particles to give a greater surface area over which the digestive juices can act. This process of mastication is effected by the incisor and canine teeth gripping the food and cutting it into small pieces after which the pre-molars and molars grind it up into much smaller particles, the movements of the lower jaw being brought about by the masseter, temporalis, internal and external pterygoid muscles.

The **mouth** or **buccal cavity** is divided into the **vestibule** between the lips and teeth and the **oral cavity** behind and is lined by a stratified mucous membrane. The roof is composed of the **hard palate** in front and the **soft palate** behind and hanging down from the latter is the **uvula**. The floor of the mouth is supported by the mylohyoid muscles and on it lies the **tongue** which is a muscular organ covered by mucous membrane and with its root in the pharynx or throat. The dorsal side of the tongue bears a number of papillae and contains **taste-buds**, sense organs which serve to distinguish between sweet, bitter, acid and salty tastes. They contain nerve endings. The tongue assists in moving the food about during mastication. The side walls of the oral cavity are composed of the buccinator muscles which form the cheeks.

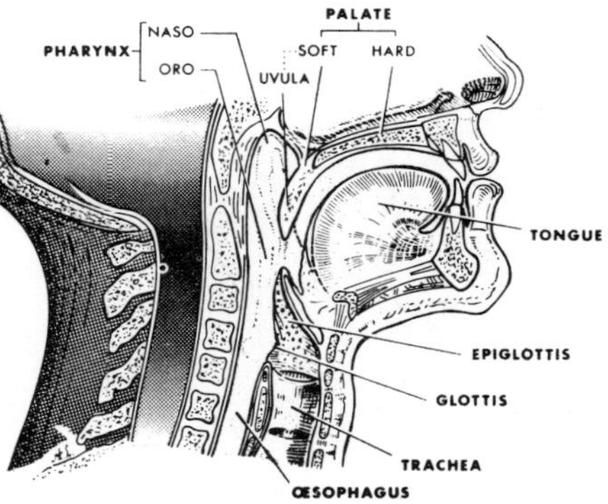

Fig. 8.1. **V.S. through mouth and neck.**

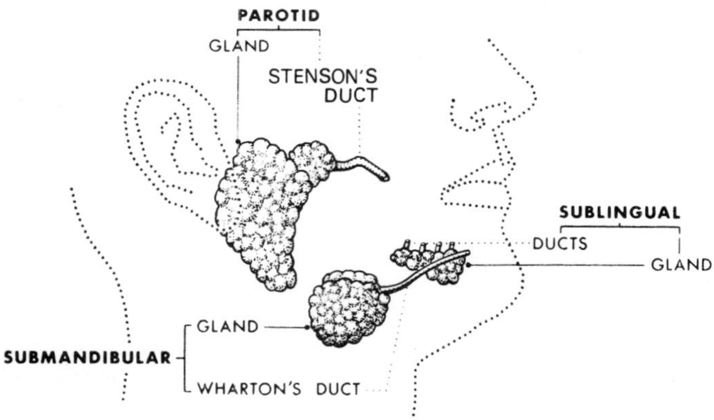

Fig. 8.2. **Salivary glands.**

During mastication the food is thoroughly mixed with a watery fluid, **saliva**, secreted by three pairs of racemose glands, the **salivary glands**. One pair, the **parotid glands**, lie in the cheeks just below the ears and each parotid duct (Stenson's duct) leads into the oral cavity opposite the second upper molar tooth. The **submandibular** (or **submaxillary**) **glands** are placed beneath the mandible and their ducts (Wharton's ducts) open into the floor of the oral cavity under the front of the tongue. The third pair, the **sublingual glands,** also in the floor of the oral cavity, have several ducts which also open beneath the tongue. The lobules of which the salivary glands are composed consist of small sac-like *alveoli* lined by *secretory cells.* Small ducts from these alveoli join up with those from others and eventually the salivary ducts are formed.

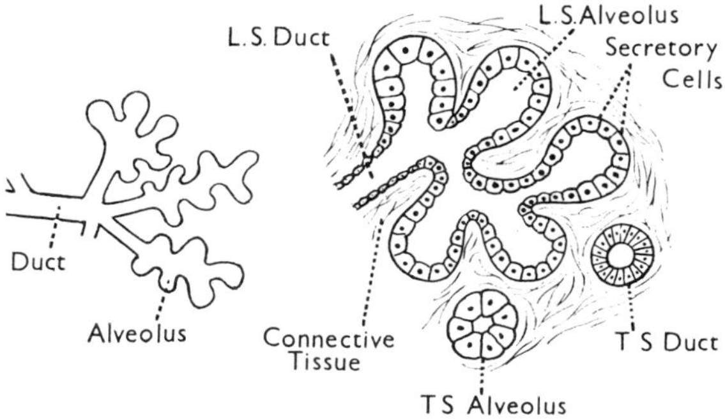

Fig. 8.3. **T.S. salivary gland.**
(From Wallis—Practical Biology.)

Saliva is a viscid, slimy, watery fluid containing **mucin**, an enzyme **ptyalin** (or **salivary amylase**), and various mineral salts, chiefly sodium chloride and bicarbonate, the solid content amounting to about 0·5 per cent. The secretion from the three pairs of glands varies in composition. The mucin lubricates the food and the mucous membrane of the mouth, which itself contains mucous glands, and the movement of the tongue in mastication and speech is helped by this secretion. Saliva is secreted continuously but the rate of flow is increased when food is in the mouth and, in fact, the mere smell of appetising food has a similar effect. This is a nervous response known as a *reflex action.* When the body loses too much water, for example by excessive sweating, salivation stops and the sensation of thirst then develops. Ptyalin is an amyolytic enzyme which hydrolyses starch to maltose in a slightly alkaline medium but the food does not remain in the mouth for a sufficiently long time to enable this action to be completed. However, it continues in the stomach until the acid gastric juice has penetrated the food sufficiently to stop its action.

Deglutition or *swallowing* is an act which is voluntary in the first instance but it afterwards becomes reflex, and the food is pushed into the pharynx or throat. This consists of two parts, the **naso-pharynx** which lies above the soft palate and into which open the posterior nostrils and the Eustachian tube from the middle ear, and the **oro-pharynx** beneath. The opening of the mouth into the latter is known as the **isthmus of the fauces** on each side of which are small lymphoid structures the **tonsils,** which are guardians against infection. There is another pair of tonsils in the naso-pharynx. The oro-pharynx continues downwards to the level of the sixth cervical vertebra where it enters the **oesophagus**. This is a muscular tube with striated fibres at its upper end and unstriated fibres lower down and is about 25 cms. in length. The muscles are in two layers, longitudinal externally and circular within. The oesophagus lies behind the windpipe or trachea, passes down the neck and through the thorax and then enters the stomach in the abdomen. The food is passed down the pharynx into the oesophagus by a series of muscular contractions known as **peristalsis**. This passage of food takes place in three stages. First it is forced to the back of the tongue by its organ becoming arched. Then the larynx is raised and the epiglottis closes the glottis, as the entrance to the trachea is called, to prevent food from passing into it. Thus the food is forced into the pharynx. Lastly a **bolus** of food (as it is called) is passed into the oesophagus. Like the buccal cavity, both the pharynx and the oesophagus are lined by stratified epithelium.

II. THE STOMACH

The remaining digestive organs are situated in the abdomen and for simplicity in describing their positions, this part of the trunk is divided into nine regions as shown in Fig. 8.4. The abdominal cavity is lined by a serous membrane, the **peritoneum**, which also envelopes the abdominal viscera to various extents and gives them support and attachment to one another. Its secretion lubricates all the surfaces, allowing free and smooth movement. The arrangement of the peritoneum will be seen in Fig. 8.5.

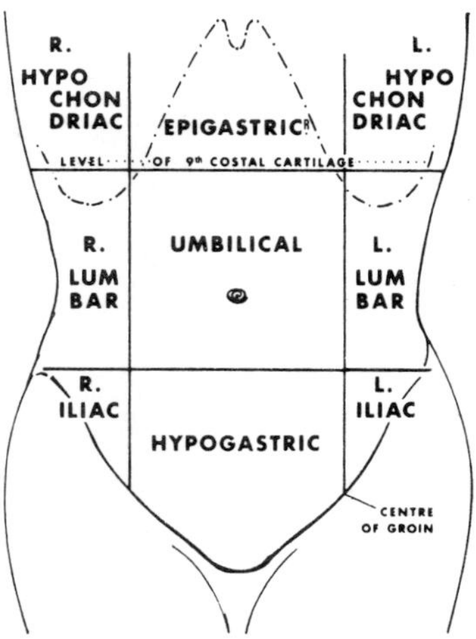

Fig. 8.4. **Regions of Abdomen.**

The **stomach** is a muscular organ situated in the epigastric and left hypochondriac regions of the abdomen, mainly in the latter. It is connected to the liver above by the **lesser omentum** at the upper **lesser curvature** and the **greater omentum** attaches its lower **greater curvature** to the spleen. Its shape varies according to

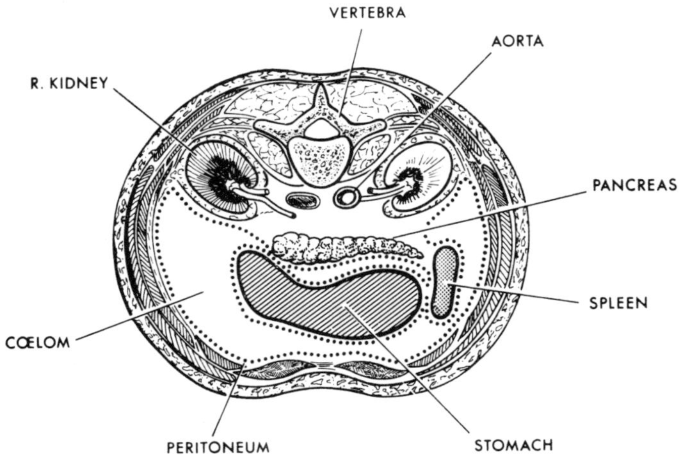

Fig. 8.5. **T.S. abdomen in region of lower part of stomach. (Diagrammatic.)**

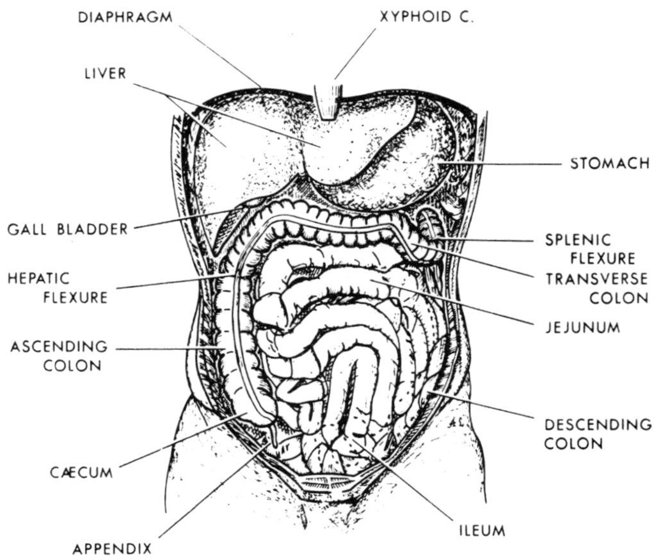

Fig. 8.6. **Abdominal viscera in situ.**

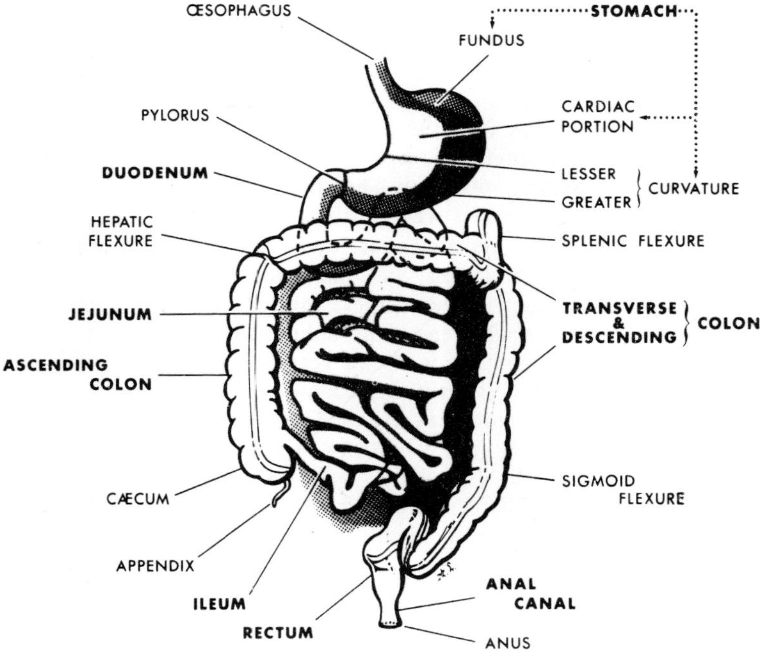

Fig. 8.7. **The alimentary canal.**

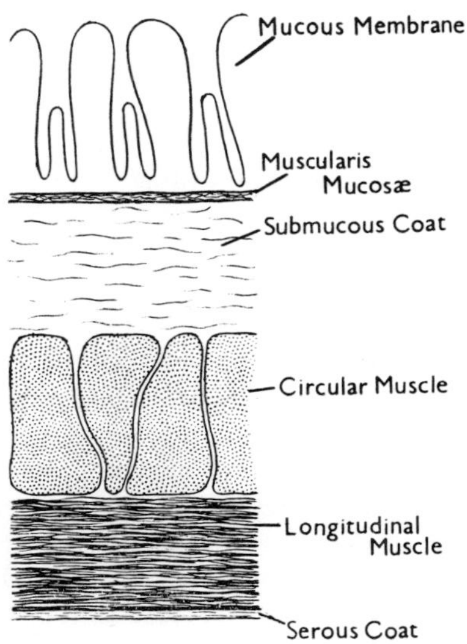

Fig. 8.8. **T.S. stomach. (Diagrammatic.)**

(From Wallis—Practical Biology.)

the amount of food inside it but the left-hand end, the **cardiac portion**, is wider than the outlet end, the **pyloric portion**. The oesophagus enters the cardiac portion by the **cardiac orifice** and the region of the stomach above this is called the **fundus**. The pyloric portion leads into the first part of the small intestine, the duodenum. At this point, the **pylorus**, is a ring of unstriated muscle known as the **pyloric sphincter**. The stomach is partly concealed by the left lobe of the liver. Its wall consists of four coats. The outermost is the **peritoneal** (or **serous**) **coat**, then comes the **muscular coat** composed of three layers of unstriated muscle, longitudinal, circular and oblique, in that order, though the last mentioned is restricted to the cardiac end. The **submucous coat,** composed of areolar tissue and containing blood vessels and nerves follows and the innermost lining is the **mucous coat**. When the stomach is empty this is thrown up into folds called **rugae** but these disappear when the organ is distended with food. The mucous membrane contains branched tubular glands, the **gastric glands**, which secrete *gastric juice.* At the upper ends of these glands are goblet cells which secrete *mucin.* Lower down along the sides are **oxyntic cells** which secrete *hydrochloric acid* and scattered throughout the glands are **peptic cells** which secrete the enzyme *pepsin.* The outer boundary of the mucous coat consists of a thin layer of muscle, longitudinal and circular, known as the **muscularis mucosae**. The food is churned over in the stomach by the action of its muscular wall and its movements can be seen under X-rays after a barium meal has been taken.

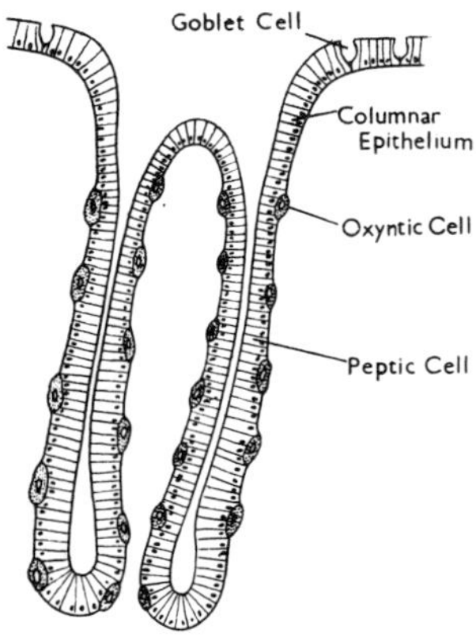

Fig. 8.9. **Gastric Gland.**
(From Wallis—Practical Biology.)

DIGESTION IN THE STOMACH

Gastric juice consists of 99·4 per cent. water, 0·02 to 0·2 per cent. of free *hydrochloric acid,* the enzymes *pepsin, rennin (chymase)* and *gastric lipase* and some *mucin* which protects the tissues against the action of the enzymes. Its flow is stimulated by the hormone **gastrin**, secreted by the stomach. The **hydrochloric acid** provides a medium in which the enzymes act, serves as an antiseptic by killing any bacteria in the food and hydrolyses sucrose to a mixture of glucose and fructose (*invert sugar).* **Rennin** (or **chymase**) curdles milk, converting the soluble protein caseinogen into insoluble casein which is thereby precipitated and then attacked by pepsin, a process of great importance to babies and young children. Rennet, used for making junket, is obtained from the stomach of calves. **Pepsin** hydrolyses proteins to **proteoses** and **peptones. Gastric lipase,** present only in small amounts, digests a limited amount of fat. This is, in fact, done mostly in the duodenum. It has been suggested that gastric lipase is regurgitated duodenal lipase. There is no starch-splitting enzyme in gastric juice but the action of ptyalin continues as long as the food is not in contact with hydrochloric acid and ceases completely when it is saturated with gastric juice, usually after about half an hour. The food acquires the consistency of porridge and is then known as **chyme** and, as the result of a series of peristaltic contractions this is passed into the duodenum a little at a time for about four to six hours by which time the stomach is emptied. A limited amount of absorption takes place in the stomach but only as such substances as glucose and alcohol which accounts for the speed with which the intoxicating effect of the latter takes place.

III. THE SMALL INTESTINE

The **small intestine** in which digestion is completed and in which absorption takes place is some 675 cm. in length and consists of three parts, the duodenum, the jejunum and the ileum. It is supported from the abdominal wall by the **mesentery**. The **duodenum** is a single loop about 25 cm. long and is the shortest, widest and most fixed part of the intestine. It lies in the right

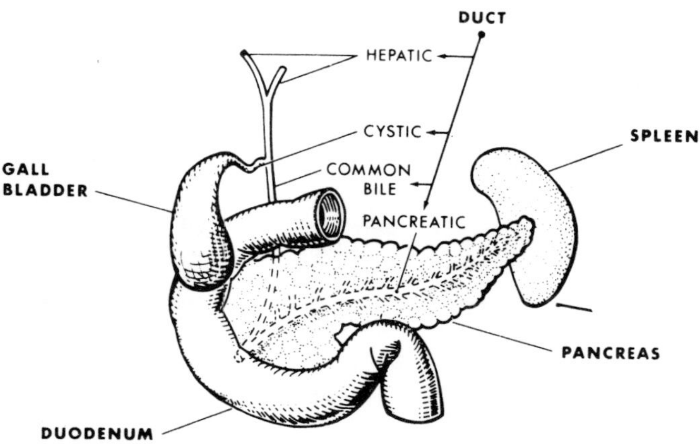

Fig. 8.10. **Duodenum, pancreas, gall bladder showing ducts. (The spleen is not part of the alimentary system.)**

hypochondriac, epigastric and umbilical regions of the abdomen. Into it open the **common bile duct** from the liver and the **pancreatic duct** from the pancreas, a gland which lies partly within its loop. The wall of the small intestine consists of the same four coats as the stomach—peritoneal, muscular, sub-mucous and mucous but the muscular coat lacks oblique fibres.The inner surface of the mucous coat is projected into microscopic finger-like processes called **villi** and into transverse folds, **valvulae conniventes,** which encircle the

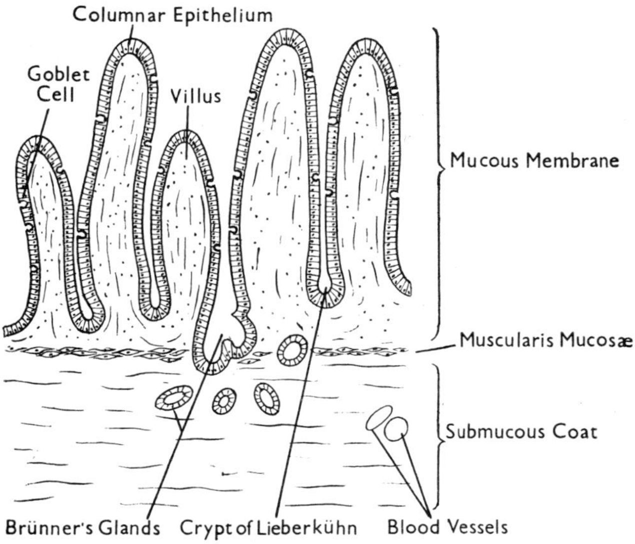

Fig. 8.11. **T.S. duodenum. Mucous and sub-mucous coats.**
(From Wallis—Practical Biology.)

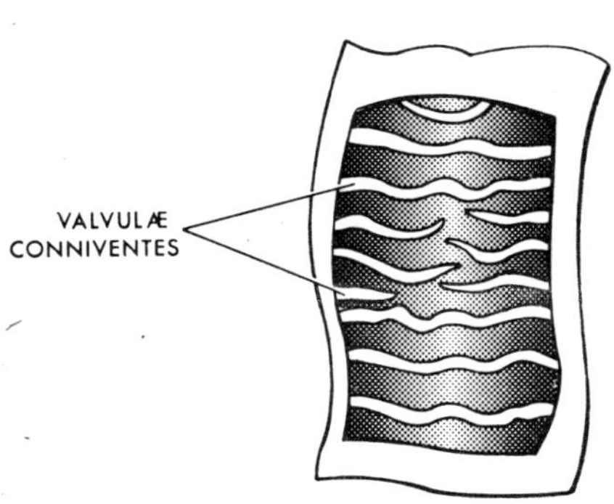

Fig. 8.12. **Part of small intestine dissected open to show the valvulae conniventes.**

intestine. In this way the absorptive area is considerably increased. Between the villi are the **crypts of Lieberkühn,** simple tubular glands which secrete the *intestinal juice (*or *succus entericus)* which contains digestive enzymes. The villi and crypts of Lieberkühn are lined by columnar epithelium in which mucus-secreting goblet cells are scattered and each villus contains a network of blood capillaries and a lymphatic vessel known as a **lacteal**; it also contains muscle fibres. In the upper part of the sub-mucous coat are small racemose glands called **Brünner's glands** and these secrete an alkaline fluid and mucus, their ducts opening either between or into the crypts of Lieberkühn.

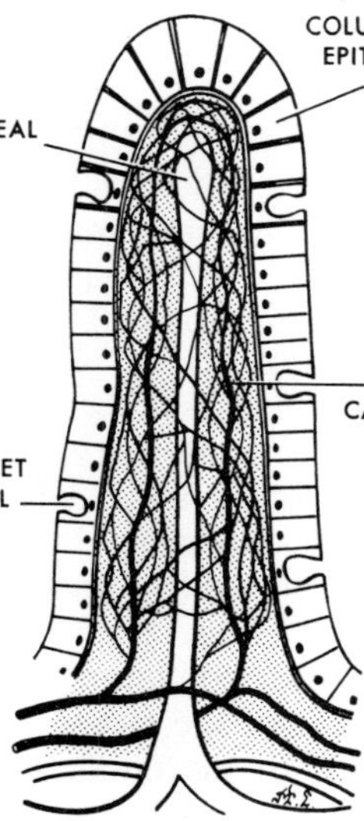

Fig. 8.13. **Structure of a villus.**
(From Wallis—Practical Biology.)

The thick-walled **jejunum,** usually empty after death, a fact from which its name is derived, is around 270 cm. in length and constitutes about 2 /5 of the small intestine. It is narrower than the duodenum and lies mainly in the umbilical region. The remaining three fifths of the small intestine, the **ileum**, around 360 or more cm. long, occupies the umbilical and hypogastric regions. Being of such length it is very much coiled; it is also thin-walled and its villi are longer than those in the duodenum. In the right iliac region it enters the large intestine. Although the small intestine is divided into these three parts there are no definite points or indications of demarcation. Scattered over the surface of the mucous membrane are areas of lymphoid tissue known as **Peyer's patches**.

IV. THE PANCREAS

This soft and somewhat flattened organ, popularly known as the "sweetbread" is bluntly triangular in shape and 12 to 15 cm. in length. It has its wider end, the head, in the loop of the duodenum and the tail extends behind the stomach as far as the spleen Fig. 8.10. It is a racemose gland and its main duct joins the common bile duct from the liver just before the latter enters the duodenum. The microscopical structure of the gland is similar to that of the salivary glands but the alveoli are rather longer. In addition there are isolated groups of cells known as the **islets of Langerhans** which secrete the hormone **insulin** which regulates the sugar content of the blood. The pancreatic cells secrete pancreatic juice and this contains three enzymes.

V. THE LIVER

Situated in the right hypochondriac and epigastric regions of the abdomen, with its convex surface immediately beneath the diaphragm and reaching down to the arch formed by the lower ribs, is the liver, the largest gland in the body. It is about 28 cm. in length, some 16 cm. in width and nearly 9 cm. at its greatest thickness. It is wedge-shaped with its wider end to the right. The inner surface lies over part of the stomach, duodenum and other organs. It is supported from the diaphragm by the **peritoneal ligaments** and is attached to the stomach and the first part of the duodenum by the **lesser omentum**. The organ is dark red in colour and is divided into *right* and *left lobes*, the former being considerably larger than the latter and sub-divided into three smaller lobes, visible on the under surface. An **hepatic duct** leaves each lobe and the two join to form the **common bile duct** which runs down to the duodenum. It receives the **cystic duct** from the **gall bladder**, a pearshaped muscular sac about 8·75 cm. long lying under the right lobe. It is a storage sac for bile and its capacity is about 50 ml. The lesser omentum lies in the cleft between the two lobes and through this cleft the **hepatic artery** and **portal vein** enter the liver. It is here, too, that the bile duct leaves it.

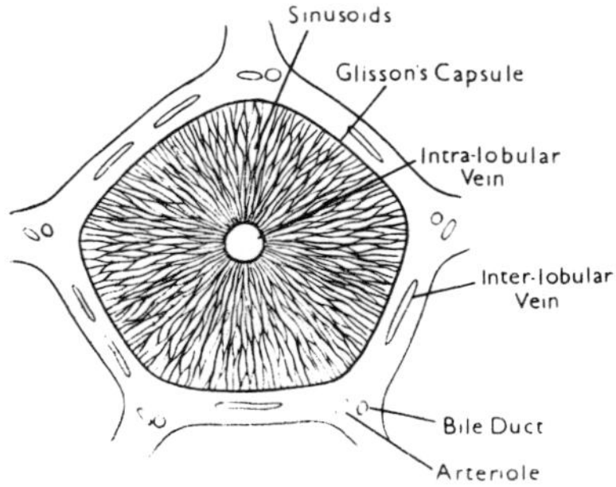

Fig. 8.14. **T.S. liver showing one lobule.**
(From Wallis—Practical Biology.)

A microscopical examination of the liver reveals that each lobe is composed of small many-sided **lobules** enclosed in a connective tissue sheath known as **Glisson's capsule**. In the centre is an **intra-lobular vein** and radiating out from this are minute vessels called

sinusoids which lead into **interlobular veins** in the capsule which originate from the portal vein. The intralobular veins ultimately join up to form the **hepatic veins** which take the blood away from the liver. Between the sinusoids are **hepatic cells** which secrete

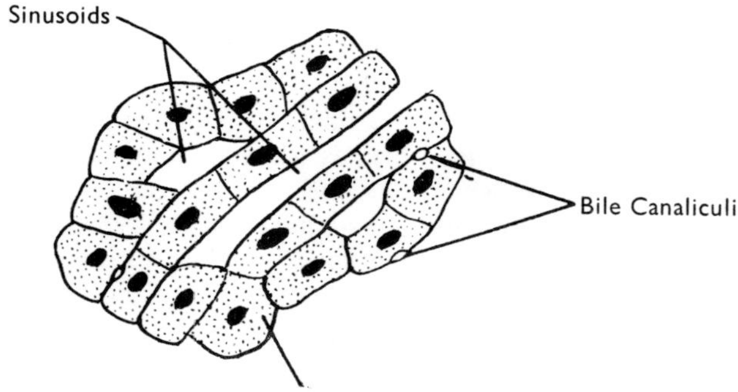

Fig. 8.15. **Part of liver lobule under high power.**

(From Wallis—Practical Biology.)

bile. Minute **bile canaliculi** lie between them and these form small bile ducts outside the lobules which join up to form the hepatic duct of the lobe.

DIGESTION IN THE SMALL INTESTINE

It will be recollected that the chyme leaving the stomach is acid in reaction. In the duodenum digestion is continued by the enzymes of the pancreatic juice and this is alkaline, due to the presence of sodium bicarbonate, since these enzymes cannot act in an acid medium. Further alkalinity is provided by the **bile** which enters the duodenum by the bile duct. But bile performs another important function. It emulsifies the droplets of fat which have been set free by gastric lipase, thus giving it a greater surface area.

Bile is a yellowish-green fluid with a very bitter taste and is composed of the bile salts *sodium glycocholate* and *sodium taurocholate*, two pigments, *bilirubin* (reddish-brown) and *biliverdin* (green), *calcium carbonate* and *phosphate, cholesterol* formed by the breakdown of fats and other constituents. It is alkaline in reaction. The liberation of bile from the gall bladder is thought to be under the control of a hormone produced in the duodenum in the presence of the acid chyme.

Pancreatic juice contains three enzymes, **lipase**, **amylase** and one inactive form, **trypsinogen**. This is activated by conversion into **trypsin** by a co-enzyme called **enterokinase** in the intestinal juice.

Lipase hydrolyses the emulsified fats to fatty acids and glycerol. **Amylase** hydrolyses any remaining starch to maltose, and **trypsin** converts peptones and proteoses to **polypeptides** and **amino-acids**. The secretion of pancreatic juice is stimulated by the hormone **secretin** produced by the duodenum.

The mucous membrane of the small intestine secretes the **intestinal juice** or **succus entericus**. This, as already seen, contains the co-enzyme **enterokinase** which activates trypsinogen. It also contains the following four enzymes:— **maltase** which hydrolyses maltose to glucose, **sucrase** which hydrolyses sucrose to invert sugar and **lactase** which converts lactose to glucose and galactose, and **erepsin**, an enzyme complex, which completes the digestion of proteins to amino-acids.

ABSORPTION IN THE ILEUM

The digested foods are absorbed into the villi by diffusion and osmosis, amino-acids and glucose into the blood capillaries and the fatty acids and glycerol into the lacteals. The fatty acids tend to react with the alkaline bile salts to form *soaps* but this is a temporary process to aid passage into the villi. Recombination of fatty acids and glycerol takes place, thus reforming fats. It is this which gives the lymph a milky appearance; hence the name *lacteal* for these lymphatics. The lymphatic vessels transport the fats. These vessels eventually join the main lymphatic vessel known as the thoracic duct which enters the blood system at the junction of two veins near the heart, the left subclavian and left internal jugular, and so the fat is then transported by the blood to the various regions of the body where it is stored until it is required.

FURTHER ACTIVITIES IN THE LIVER

The blood vessels leaving the small intestine unite to form the large portal vein which, as already seen, enters the liver. Such amino-acids as are immediately required by the body continue in the blood stream and leave the liver by the hepatic veins. It must be appreciated, however, that as all protein foods do not contain all the amino-acids required by the body, excess of some is taken in the food in order to obtain an adequate supply of others. The excess is dealt with by the liver in a process known as **deamination**. By removal of amino-groups, this results in the formation of urea $CO(NH_2)_2$. This is an excretory product and eventually reaches the kidneys by the blood stream and is eliminated from the body in the urine.

The liver cells also allow sufficient glucose to continue in the blood stream leaving the liver to maintain the blood sugar content at its normal level of 0·18 per cent. but retain the rest which is converted into **glycogen** for storage. When there is a demand for more sugar, glycogen is hydrolysed by the enzyme **glycogenase**, produced by the liver cells, into glucose. As already stated the hormone insulin secreted by the islets of Langerhans in the pancreas regulates the sugar content of the blood. The glycogen storage capacity of the liver is, however, limited and if excess carbohydrate is taken in the food over and above the body's requirements, some of the glucose is changed not into glycogen but into fat. This is then transported by the blood to other parts of the body for storage. This accounts for the fact that people who eat considerable amounts of starchy foods tend to become fat.

It should be noted here that in addition to the secretion of bile, deamination and the storage of glycogen, the liver has other functions unconnected with digestion. These will be mentioned in their proper context. Two, however, will be stated now. One is the **production of heat** which the blood transfers to other parts of the body and this is the result of the enormous amount of work done by this very large organ. The other function is **detoxication**. Reticulo-endothelial cells lining the sinusoids absorb poisonous materials, particularly such as may be taken in food, and which enter the blood. The cells are able to break down these foreign substances into harmless ones which are eventually eliminated from the body.

VI. THE LARGE INTESTINE

The **large intestine** is composed of the **caecum**, the **colon** and the **rectum**. It is much shorter than the small intestine, being only about 180 cm. in length but it is of much greater calibre. Whereas the small intestine has a diameter around 2·54 to 3 cm., the colon is over 7 cm. The **caecum** is a short, pouch-like sac in the right iliac region which the ileum enters and at the point of entry is an **ileo-caecal valve** which regulates the passage of the contents of the ileum into the caecum. Just below this is the **vermiform appendix,** a blindly ending, worm-like tube about 9 to 10 cm. long in which the cavity of the caecum extends. In herbivorous animals the appendix, like the caecum, which in these animals is a much larger and longer structure, is concerned in the digestion of cellulose by bacterial action but in man it appears to be vestigal and functionless. In fact its removal has none other than beneficial effects when it becomes infected or inflamed in *appendicitis*.

Continuous with the caecum is the **colon** which at first runs upwards as the **ascending colon** in the right lumbar region in front of the right kidney and behind the right lobe of the liver. Here it bends to the left at what is known as the **hepatic flexure** and passes as the **transverse colon** under the stomach in the umbilical region as far as the spleen, in the left hypochondriac region. At this point, the **splenic flexure**, it bends downwards to become the **descending colon**. In the left iliac region it bends towards the centre at the **sigmoid flexure** and enters the pelvic cavity in the hypogastric region where it joins the rectum.

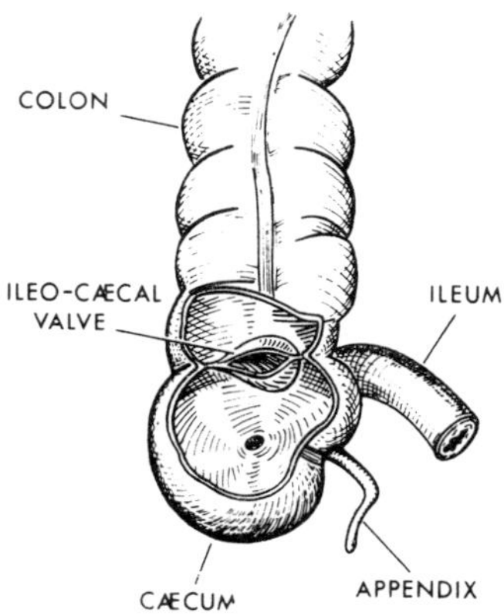

Fig. 8.16. **Caecum cut open to show ileo-caecal valve.**

The **rectum** in spite of its name, bends both sideways and from front to back. It is about 16 cm. in length and continuous with it is the **anal canal**, a short tube about 3·5 cm. long which leads to the exterior at the **anus** which is surrounded by sphincter muscles. The bladder and part of the reproductive organs lie in front of the rectum in the male and in the female the uterus occupies the anterior portion of the pelvic cavity while the vagina and the neck of the uterus lie below it.

Histologically the wall of the large intestine is similar to that of the small in so far as the actual coats are concerned but there are certain marked differences. Except in certain parts the peritoneal coat is incomplete. The longitudinal muscle is collected into three bands and being shorter than the actual intestine, these cause the latter to be sacculated, a formation which retards the flow of the contents and so enabling water to be absorbed. This is the only function of the large intestine for no digestion or other absorption takes place in it. The sacculations are not present in the rectum. The mucous membrane is devoid of villi and valvulae conniventes but it contains numerous tubular glands which secrete mucus. Enzyme-producing glands are, of course, absent.

FUNCTIONS OF THE LARGE INTESTINE

About four hours after a meal the contents of the ileum begin to enter the large intestine. By this time practically all the nutritious substances which are of benefit to the body have already been absorbed. The only substances which are absorbed by the large intestine are water and some mineral salts. The object of this is to prevent dehydration of the body. Glucose can be absorbed by the colon, however, and this is administered by an enema inserted into the rectum. Salines and drugs can also be injected in this way. Rectal feeding is used when oral feeding is impossible or undesirable under certain conditions of ill-health. As a result of this absorption of water the contents of the intestine become more solid as they travel through the colon, about four-fifths of the water entering the caecum being absorbed here. The solid contents are known as **faeces** and consist of the undigested part of the food such as cellulose (none being digested in man), bacteria, mucus, pigments and epithelial cells from the wall of the intestine. The cellulose content is increased when large quantities of vegetable food are included in the diet. Cellulose has a beneficial effect in digestion, in that it stimulates peristalsis by its bulk; it is known as "roughage". Peristalsis is slow in the large intestine however. The majority of bacteria in the faeces are dead. These bacteria, such as *Bacillus coli communis*, are normal inhabitants of the large intestine. The colour of the faeces is mainly due to the brown pigment *stercobilin* which is formed from the bile pigments by bacterial action. Variation in colour is dependent on the diet. For example a lack of red meat makes the stools paler in colour while excess darkens them. Haemorrhage in the upper part of the alimentary tract due to gastric or duodenal ulcers will make them black. The odour of faeces is caused by colonic decomposition by bacteria and to a substance called *indol.* Gases are produced as the result of the decomposition.

The faeces pass into the rectum, which is normally empty, usually about eighteen hours after a meal. The taking of food produces what is known as a **gastrocolic reflex** which causes the desire to defaecate and this normally occurs after the first meal of the day, for most of the food taken during the day reaches the descending colon by the following morning. The involuntary contraction of the muscles of the rectal wall, accompanied by the voluntary contradiction of the abdominal muscles then expels the faeces through the anus when the sphincter ani muscles relax. Immediately after defaecation these muscles contract and close the anus.

Disorders and Diseases of the Digestive System

Dyspepsia or **indigestion**—can be caused by a faulty diet, irregularity of meal times, over-indulgence or insufficient mastication. One of the most frequently occurring forms is **acid dyspepsia** due to excessive secretion of acid in the stomach. This may cause much discomfort and considerable pain and vomiting may result. Treatment lies in strict attention to diet and the administration of magnesium or aluminium trisilicate which takes up the excess acid. The taking of sodium bicarbonate temporarily relieves the condition by neutralising the acid. A common form of dyspepsia is **nervous dyspepsia**, caused by incorrect nervous control of digestion. It occurs in people, often of a nervous temperament, who are under stress.

Flatulence—which often accompanies dyspepsia, is the result of the fermentation of food by bacteria thereby causing the production of gases.

Gastritis—or inflammation of the mucous membrane of the stomach produces such symptoms as nausea, biliousness, and headache. The cause may be food-poisoning, over-eating or alcoholic indulgence.

Gastric and **duodenal ulcers**—usually develop in the pyloric portion of the stomach and in the duodenum near the entrance from the stomach respectively. Both cause severe pain, the onset of which generally occurs before a meal with a gastric ulcer and about three hours after it in the case of a duodenal ulcer. Vomiting may accompany the pain and if the ulcer penetrates a blood vessel, haemorrhage will occur and blood will appear in the vomit and may be passed from the anus. In some cases the ulcer heals, leaving a scar (visible under X-rays after a barium meal) but in severe cases it may perforate the stomach or duodenal wall thus expelling the contents into the coelom and setting up *peritonitis*. This demands immediate surgery to repair the damage. Gastric ulcers can sometimes be cured by following a strict diet sheet. Failing this, removal of the affected part of the stomach, an operation known as *partial gastrectomy*, is necessary after which no further trouble should occur.

Gastro-enteritis—inflammation of the stomach and intestines causing vomiting and diarrhoea, often caused by infection by bacteria or a virus.

Vomiting—brought about by anti-peristaltic movements effected by the abdominal muscles resulting in the expulsion of the stomach contents through the mouth. Bile may be included in the vomit. Sometimes nothing is expelled and this is known as "retching". Various causes of vomiting have already been given above. It can be induced in order to expel the stomach contents in cases of poisoning by swallowing warm salt water and other substances which are called *emetics*.

Diarrhoea—may be caused by indigestible food, a chill, food-poisoning or by one of the causes given above. It is also caused by an infection of the bowels and may be a symptom of more serious diseases.

Dysentery—mainly a tropical disease, is caused either by a bacterium *Shigella dysenteriae*, or by a protozoon called *entamoeba histolytica* (bacillary or amoebic dysentery). Ulcers and inflammation develop in the lower part of the intestines, fever and severe diarrhoea occur and blood is passed in the faeces. Infection may come from food to which the parasite has been carried either by flies or by human carriers.

Constipation—is a very common complaint and is the result of too great an absorption of water in the colon making the faeces dry and hard. Deficiency of roughage in the food is often a cause but it can be due to organic causes such as a stoppage in the bowel.

Mumps—which cause swellings of the salivary glands is a virus infection.

Obstruction—occurs when the passage of the intestinal contents is prevented by kinking of the wall of the intestine, pressure on it from outside or by hernia.

Haemorrhoids—commonly known as "piles" are really varicose veins in the lower end of the rectum (internal piles) or outside the anus (external piles). Constipation is one of the causes but their presence may be a symptom of other diseases.

Gall stones—are sometimes found in the gall bladder and consist of cholesterol mixed with bile pigment. When they are small they enter the intestine and are expelled with the faeces but if they are large they may become lodged in the cystic duct and cause great pain. Surgical removal of the gall bladder may be necessary.

Jaundice—caused by the obstruction of the flow of bile into the intestine and this causes it to be absorbed into the blood and lymphatics. As the blood and lymph circulate round the body the bile gives a yellow colour to the skin and bile is excreted in the urine. There is also an infective form of the disease.

Hepatitis—or inflammation of the liver is most frequently caused by a virus infection but it can be brought about by a gallstone blocking the bile duct.

Cirrhosis of the liver—in which fibrous tissue develops in the organ replacing the normal cells, is often the result of a deficiency in diet when no other cause can be found. It can give rise to **cancer** of the liver.

Congestion of the liver—often occurs in people suffering from heart disease. As a result of this the liver becomes enlarged.

Liver transplants—The first ever was performed at Addenbrooke's Hospital in Cambridge in the Spring of 1968. Several further operations have since been performed in this country but nearly all the patients have died. Of three known survivals one was still alive in 1974 after four years, the others being for slightly shorter periods.

PRACTICAL WORK

Practical Work on the Muscular System

It was suggested at the end of Chapter VI that the practical work on the muscular system should be deferred until a rabbit was dissected for an examination of the digestive system.

The muscles should be examined on the skinned rabbit as explained in 1, 2 and 3 on pp. 73 seq. before the animal is dissected.

Anatomy of the Digestive System

A rabbit should be dissected, opening up the neck, thorax and abdomen. The various organs which constitute the alimentary system should be identified, first of all those which can be seen *in situ* and afterwards by deflection as necessary to expose those which were hidden. Finally the entire alimentary canal should be removed and unravelled and laid out neatly on another dissecting board. It will be noted that, unlike the human being, the rabbit has a very large caecum. This is because the rabbit is an herbivorous animal and cellulose is digested in its caecum by bacterial action. It also has a very long rectum in which pellets of faeces can be seen. The liver has five lobes whereas in the human subject there are only two.

Histology of the Digestive Organs

Salivary glands

Examine a *T.S. of a salivary gland.*

Identify the alveoli, secretory cells and ducts as shown in Fig. 8.3.

Stomach, duodenum and ileum

Examine *transverse sections of the stomach, duodenum and ileum.*

Using Figs. 8.8, 8.9 and 8.11 identify the four layers and carefully examine the mucous membrane in order to see the various glands mentioned in the text.

Liver

Examine a *T.S. of the liver.*

Note the liver lobules and examine one carefully in order to find the parts shown in Fig. 8.14.

Pancreas

Examine a *T.S. of the pancreas.*

Note the general similarity in structure to the salivary gland already seen but that the alveoli are longer and that there are groups of small cells in the pancreatic tissue. These are the islets of Langerhans.

Large intestine

Examine a *T.S. of the large intestine.*

Again identify the four coats of which the wall is composed and note the absence of villi in the mucous membrane.

The Physiology of Digestion

Perform the following experiments in clean test-tubes, label them, plug them with cotton wool and place them in a thermostat kept at body temperature (38° to 40°C).

Digestion in the Mouth

Saliva

Chew a piece of rubber tubing for a few minutes: this will produce a copious supply of saliva. Transfer some to a test-tube and dilute with an equal volume of distilled water. (It will act more quickly.) Filter a portion.

1. Test the reaction of saliva with a piece of litmus paper.

2. (*a*) To some filtered saliva add a drop of dilute acetic acid. Mucin, a viscous muco-protein, is precipitated. (*b*) Perform the Biuret and Millon's tests to show the presence of the protein, mucin.

The Action of Ptyalin

3. Add some diluted saliva to (*a*) some starch sol, (*b*) a small quantity of albumin or finely chopped lean meat, (*c*) a little fat meat. Label the tubes and place them in the thermostat, kept at 38°–40°C. Shake periodically. Later examine to see whether the food has been digested and test samples of (*a*) with iodine every five minutes on a white tile until no colour change is produced; then test the residue with Fehling's solution.

Digestion in the Stomach

Gastric Juice

In the following experiments use Artificial Gastric Juice (*Liquor peptici*).

4. Find the reaction of gastric juice with litmus.

The Action of Pepsin and Rennin

5. Add a large quantity of gastric juice to a small quantity of (*a*) finely chopped lean meat or albumin, (*b*) starch sol, (*c*) a few particles of fat. Label the tubes and place them in the thermostat. Shake periodically. Examine a few hours later.

Then carry out the Biuret test with the residue in (*a*).

6. Repeat experiment 5 (*a*) with gastric juice which has been made alkaline with a little caustic soda.

7. (*a*) Add a little rennin to some milk, diluted with an equal volume of water. Place in the thermostat and examine a few minutes later. (*b*) Replace the tube in the thermostat, leave it for an hour and then examine again.

8. Repeat experiment 5 (*a*) but first heat the enzyme to about 70°C for a few minutes. Note the effect of this temperature.

Digestion in the Duodenum

Pancreatic Juice

In the following experiments use Artificial Pancreatic Juice (*Liquor pancreatini*).

9. Test the reaction of pancreatic juice with litmus.

The Action of Trypsin, Amylase and Lipase

10. Repeat Experiments 5 (*a*), (*b*) and (*c*), using pancreatic juice.

Test (*b*) with iodine at intervals of five minutes on a white tile. When no colour change is observed, test the residual solution with

Fehling's solution. Test the reaction of (*c*) with litmus solution when digestion appears to be complete.

11. Acidify some pancreatic juice with dilute hydrochloric acid and repeat Experiment 10.

12. The "*Lenham chromatography kit**" will be found to be most helpful in demonstrating enzyme action and the effects of *p*H on it. The kit consists of paper chromatography strips, miniaturised tanks, dyes, reagents and all apparatus necessary for the experiments, together with full instructions for carrying out the work.

The suppliers of this kit will be found in Appendix II.

* *Paper chromatography* is a method of analysis of mixtures of compounds in solution using the different effects and rates at which they move across areas of dyes on a piece of paper which is specially prepared for the process.

Chapter IX

THE CARDIO-VASCULAR SYSTEM—I

THE STRUCTURE AND FUNCTIONS OF BLOOD

The **blood** forms about one fifteenth of our body weight and amounts to some five litres in volume, though this is a variable figure. It is pumped through closed tubes, arteries and veins, by the muscular action of the heart and these, together with the microscopic capillaries in the organs, constitute the cardio-vascular system. Blood is composed of a fluid, the *plasma,* in which are suspended two kinds of cells, *red corpuscles* and *white corpuscles.* In addition there are very much smaller structures known as *blood platelets.*

The **plasma** is a yellowish, alkaline fluid consisting of about 90 per cent. water containing substances known as fibrinogen and prothrombin concerned in clotting, **mineral salts** (chiefly sodium chloride) which maintain the correct alkalinity of the blood and including calcium salts which play their part in clotting, and **heparin** which prevents internal clotting. As the blood system forms the main part of the transport system of the body, the plasma also contains digested food—amino-acids, fats and 0·18 per cent. glucose—excretory substances such as urea and carbon dioxide and the secretions of the endocrine glands known as autacoids.

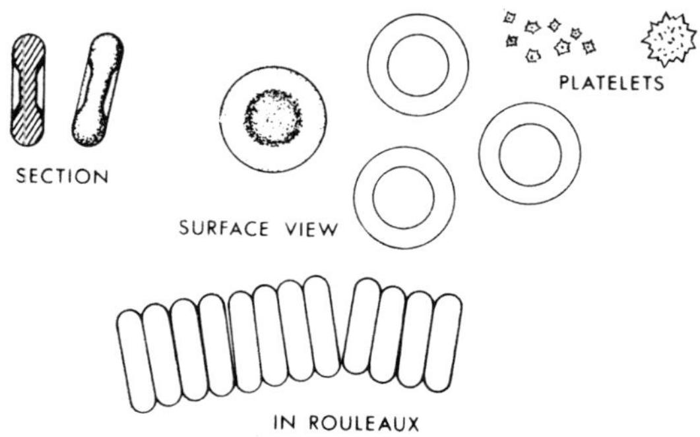

Fig. 9.1. **Blood-erythrocytes and platelets.**

The **red corpuscles**, technically known as **erythrocytes**, are bi-concave discs, $7{\cdot}2\mu$* in diameter, devoid of nuclei and individually more or less straw-coloured though en masse imparting a bright red colour to the blood. They contain a pigment called **haemoglobin**. Their origin is in the red marrow at the end of the long

* μ (pronounced *mew*) = 1 *micron* = 0·001 mm.

bones where they develop from nucleated cells. They have a tendency to collect like a pile of coins when they are said to be *in rouleaux*. There are some 5,000,000 per cu. mm. in man and 4,500,000 in women and they outnumber the white corpuscles by 600 or 700 to 1 but as the latter vary considerably in number in different parts of the body and under differing conditions, this is not a constant ratio. The erythrocytes are said to have a life of about four months and the worn-out corpuscles are destroyed by the reticulo-epithelial cells in the liver and spleen and elsewhere. The bile pigments, bilirubin and biliverdin, are derived from the haemoglobin from these disintegrated cells. Under these and certain other conditions, haemoglobin is set free from the corpuscles and this is called *haemolysis*. It will take place if blood is mixed with water or with saline solutions which are more dilute than the blood plasma. **Haemoglobin** is a compound of iron known as *haem* combined with a protein, *globin*. It has a great affinity for oxygen with which it very readily but very loosely combines to form **oxyhaemoglobin** which gives the blood a scarlet colour. This occurs in the lungs where oxygen passes into the blood. The **reduced haemoglobin** leaving the tissues is darker and more purplish in colour and on reaching the lungs oxyhaemoglobin is again formed. Haemoglobin also has an affinity for carbon monoxide and this gas combines with it more firmly than oxygen, the compound formed being known as **carboxyhaemoglobin**. It gives a cherry-red colour to the blood.

The number of red corpuscles in the blood is therefore of extreme importance and a deficiency in their number or a lowering of the haemoglobin content will reduce the oxygen-carrying capacity of the blood as in the disease anaemia. Blood counts are made by means of an instrument known as a *haemocytometer*, using a microscope.

The **white corpuscles** or **leucocytes**, really almost colourless, are larger than the erythrocytes, varying between 9μ and 20μ. On an average there are 6,000 to 8,000 per cu. mm. of blood but, as already stated, the number varies considerably under different conditions and is increased to a very great extent in localities of infection. Most of them serve as the scavengers of the body and ingest and destroy infective organisms. Leucocytes are capable of amoeboid movement which enable them to reach the parts of the body where they are needed. This is brought about by pseudopodia and the cells are able to ingest infective bacteria and the poisonous toxins they produce. Thus they serve as guardians against infection and as eliminators of it. Under these conditions they are known as **phagocytes**. They have a short life, a few days or at the most little more than a week and many succumb to infective bacteria.

These dead leucocytes constitute what is known as **pus** and an accumulation of pus forms an abscess. New leucocytes are formed in the red marrow of the bones and elsewhere as will be seen below. There are five different kinds of leucocytes:—

1. **Polymorphonuclear leucocytes**, usually referred to simply as polymorphs. Somewhat larger than the erythrocytes, they are up to 12μ in diameter and form 65–75 per cent. of the total number. The nucleus has a characteristic U-shape with the free ends lobose or it may be composed of several lobes. They arise in the spleen, the tonsils and in other lymphoid tissue. Their function is phagocytosis.

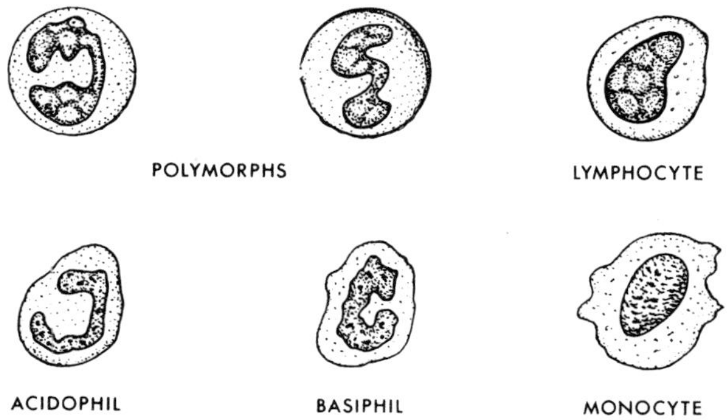

Fig. 9.2. **Blood—types of leucocytes.**

2. **Lymphocytes.** These are smaller than the polymorphs, being about 8μ across, have a large round nucleus, and form 20–25 per cent. of the leucocytes. They originate in lymph nodes, Peyer's patches and other lymphoid tissue. They pass easily through the capillary walls and occur freely in lymph. Their function is not phagocytosis but they produce anti-toxins.

3. **Acidophils (eosinophils).** Forming only about 3 per cent. of the white corpuscles, they are larger than polymorphs, being up to 16μ in diameter and having a nucleus with two or three lobes. These, too, are non-phagocytic but their function is detoxification. They arise in bone marrow.

4. **Basiphils.** Also larger than the polymorphs, being some 16μ in size, these form a mere 0·5 to 1 per cent. of the leucocytes. The nucleus is round or bilobed and, like the acidophils, they arise in the bone marrow. They are not very motile nor are they phagocytic. Their function is in some doubt.

5. **Monocytes.** The largest of the leucocytes, up to 20μ in diameter, they form 1–3 per cent. of the total number. The nucleus is round. These are the most actively motile of all and they are phagocytic.

Blood platelets or **thrombocytes** are minute oval or irregularly-shaped, granular and non-nucleated bodies. Measuring about 2μ in diameter, they are the smallest of the structures in the plasma. Their number varies considerably up to about 400,000 per cu. mm. of blood. They take part in the clotting of the blood.

The Clotting of the Blood

It is a well-known fact that when blood has been escaping from a wound for a short time it "sets" into a jelly or coagulates. This is known as **clotting**. The means by which this comes about is not fully understood but the following account will give some idea of what happens during the process.

Dissolved in the plasma is a protein called **fibrinogen**, a substance known as **prothrombin** and **calcium salts**. The blood platelets contain an enzyme, **thrombokinase (thromboplastin)**. When blood clots the platelets break up and release thrombokinase and this and the calcium ions in the plasma react with the prothrombin and convert it into **thrombin**. The thrombin then changes the soluble fibrinogen into insoluble **fibrin**. This is a stringy substance in the meshes of which the red corpuscles become entangled, thus forming the clot which gradually shrinks due to the contraction of the fibrin and a yellow fluid known as **serum** is exuded.

Internal formation of thrombin in the blood stream is prevented by an *anti-coagulant*, **heparin**, which occurs in the plasma and is formed in the liver. But when a wound occurs and blood escapes, thrombokinase, having been set free from the platelets, is able to form thrombin and the counter-action of heparin is stopped. Clotting prevents undue loss of blood or haemorrhage and by closing the entrance to a wound, the clot prevents the entry of bacteria. There is a disease known as *haemophilia* in which the blood is unable to clot. Clotting can be prevented by adding sodium citrate to blood as this removes calcium ions.

Blood Groups

It is found that if the blood of different individuals is mixed, in some cases there is no observable effect but in others the red corpuscles clump together or **agglutinate** after which *haemolysis* occurs and this can, of course, prove fatal. Investigation has shown that the erythrocytes contain substances called **antigens** (or **agglu tinogens**) and the plasma contains others known as **antibodies** (or **agglutinins**). Some antigens and antibodies from different bloods interact when mixed causing agglutination. As a result of these discoveries blood can be classified into four groups known as **A**, **B**, **AB** and **O.** A person's corpuscles may contain one antigen (**A**) or (**B**), both (**AB**) or neither (**O**) and the same person's plasma may contain one antibody (**b**) or (**a**), neither or both (**ab**). If blood having red corpuscles containing an antigen is mixed with blood which lacks it, then the antibody in the plasma of the latter reacts with the antigen and agglutination results. Thus:—

Agglutination takes place between:—

The plasma of **O** and the corpuscles of **A**, **B** and **AB**
The plasma of **A** and the corpuscles of **B** and **AB**
The plasma of **B** and the corpuscles of **A** and **AB**

and

No agglutination takes place between:—

The plasma of **A** and **AB** and the corpuscles of **A**
The plasma of **B** and **AB** and the corpuscles of **B**
The plasma of **AB** and the corpuscles of **any group**
The plasma of **any group** and the corpuscles of **O**

On account of the fact that the corpuscles of O are not agglutinated by the plasma of any group, individuals in this group are **universal donors** and their blood can be transfused to people in any other group but they cannot safely receive blood from any other than their own group. Also because the plasma of AB cannot agglutinate the corpuscles of any group, individuals in this group are **universal recipients** and people in this group can receive blood from any other group but cannot with safety give blood to any other than those in their own group. The proportion of people in Great Britain in the various Groups is approximately as follows:—

A .. 40 per cent. B .. 9 per cent.
AB .. 3 per cent. and O .. 48 per cent.

It has also been found that, quite regardless of the above groups in 85 per cent. of our population blood is agglutinated by antigens in the serum of rabbits previously injected with the blood of the rhesus monkey. This is due to what is known as the **rhesus factor** (**Rh**) and such blood is said to be **Rh positive**. The remaining 15 per cent. are **Rh negative**.

Both the ABO and Rhesus groups are inheritable. Clearly all these facts are of supreme importance in **blood transfusions** used in surgical operations, childbirth and when severe haemorrhage occurs and it is essential that blood of the correct group is administered. Group tests can be made in hospitals. It is a very wise precaution to ascertain one's blood group and make a note of it in a prominent place in one's diary or elsewhere in case of accident.

A person's Blood Group can be determined by one or the other of the following methods:—

1. (i) A drop of serum of Group **A** and (ii) a drop of serum of Group **B** are placed on a white tile. A little sodium citrate is added to prevent clotting. Then a drop of the patient's blood under investigation is added to each and the specimens examined under the lower power of the microscope. If agglutination takes place it will be seen within a few minutes. No agglutination occurs between the blood of the person in Group A and (i) or Group B and (ii) or Group O and (i) or (ii). Agglutination does occur between the blood of a person in Group A and (ii), in Group B and (i) and in Group AB and (i) or (ii). Thus the correct Blood Group is found.

2. Using *Eldon Cards*. These cards have a test portion coated with a cellulose strip divided into four test panels on which specific dried serum reagents are placed. The panels are labelled *anti–A, anti–B, anti–D* and *Control*. To each of these in turn a drop of water is added by means of the pipette provided. The reagents are then dissolved by means of a plastic strip also provided. A half-sphere of blood, obtained by pricking the finger tip with a sterilised needle, is removed and added to each of the panels with the plastic strip and spread over the entire panel. After a minute's pause the results are examined and the group determined according as to whether agglutination has taken place and, if so, in which panel. A table is provided and the anti–A, and anti–B panels determine the ABO Group and the anti–D whether the blood is Rh positive or negative. No agglutination should take place in the control and if it does this indicates that the test is unreliable.

Full instructions are provided with the cards and these must be implicitly followed if success is to be obtained.

The suppliers of Eldon Cards will be found in Appendix II.

Blood and serum are stored in what are known as **Blood Banks** in hospitals and elsewhere. It is safe in urgent cases to transfuse Rh negative Group O blood into any recipient since people with O Group are universal donors and Rh negative can be received by either group whereas Rh positive cannot be administered to an Rh negative patient. But if the blood group of the patient is determined, then the correct group can be used in the transfusion.

As a person's ABO and Rh blood groups are inherited the fact can be used in medico-legal cases of disputed paternity. The blood group of a child is determined by the blood group or groups of its parents and it cannot therefore have as a parent one who has not contributed to its group. Obviously this can only be used as negative evidence.

The Functions of the Blood

From the foregoing examination of the blood, it will be seen that its functions may be summarised as follows:—

1. The transport of oxygen to the tissues by the haemoglobin in the red corpuscles.
2. The transport of digested food products from the alimentary canal to the tissues in the plasma.
3. The transport of excretory substances in the plasma—carbon-dioxide, mainly as sodium bicarbonate, to the lungs and urea and other waste-products of metabolism to the kidneys.
4. The distribution of heat, generated mainly in the liver and muscles, to all parts of the body, thus helping to maintain an even temperature.
5. The protection of the body against infection by the leucocytes in phagocytosis, the transport of anti-toxins and antibodies, and by clotting.
6. The protection against haemorrhage in body injuries by clotting.

A better understanding of how these functions are performed will be obtained after a study of the cardio-vascular system has been made and this we shall do in the next three chapters.

Disorders and Diseases of the Blood

Anaemia. In this disease there is a deficiency of red corpuscles or of haemoglobin. Blood pressure is low and there is general weakness. It may be due to an insufficiency of iron in the diet or, more frequently, to its being inadequately absorbed. In this case treatment lies in the administration of iron salts. Other diseases can cause anaemia and it will occur after haemorrhage in which case a blood transfusion is given.

Pernicious anaemia is a severe form of the disease in which there is a deficiency of erythrocytes. Treatment is by administration of liver extracts and vitamin B12.

Leukaemia or **cancer of the blood** in which the number of leucocytes is enormously increased has no known cure and the disease is fatal. The cause of this disease is unknown but research has recently indicated that it may possibly be an extremely minute virus. Virus leukaemia in cats is known and can now be prevented by immunisation. Considerable progress has been made in the treatment of child leukaemia with drugs, some children apparently being quite fit after five years of treatment. There still remains the chance of relapse but it is hoped that the result may prove to be permanent. Adult leukaemia is much more difficult to treat and so far, although a fair percentage of cases respond to treatment, they all tend to prove fatal eventually. There are various forms of the disease, one of which does not respond to treatment at all.

Cyanosis or blueness of the blood is due to insufficient oxygenation. It is seen in "blue babies" who suffer from a *hole in the heart.* This hole is in the septum and allows blood to leak from the left side of the heart to the right. Consequently when this mixed arterial and venous blood reaches the lungs it is insufficiently oxygenated, giving it a bluish tinge. This is evident in the lips, at the base of the nails and elsewhere. If this congenital malformation is discovered in time, a baby's life may be saved by surgical treatment.

Haemophilia of which there are thousands of sufferers is a disease in which the blood is unable to clot and wounds of any kind, particularly internal ones, are therefore extremely dangerous and may prove fatal in severe cases when a major operation is necessary. Its cause is lack of the *anti-haemophilic factor, AHF* or *Factor VIII,* present in normal blood and which controls clotting. Factor VIII can be obtained in concentrated form from blood plasma and stored as a powder. An even more concentrated form is *cryoprecipitate* which has to be stored at a temperature of $-30°C$. Thawed out and dissolved in saline it can then be injected into the patient. Transfusions can be given and local applications applied to the wound. An interesting fact concerning this disease is that it is hereditary, that it is only through the females that it is transmitted and then only to males and that the females do not themselves suffer from it. This aspect will be dealt with in Chapter XVIII.

PRACTICAL WORK

Structure of Blood

1. Clean the tip of your left fore-finger with spirit and sterilise a sharp mounted needle in the flame. Prick the tip of the finger. Squeeze a drop of blood on to a coverslip and invert it quickly on a slide or squeeze the blood on to a slide and draw the

edge of another slide across it to make a film. Examine under the high power.

2. Make another preparation as in (1) but dilute with physiological saline (0·6 per cent.) and cover.

The red corpuscles will be more easily examined.

3. Prepare a smear of human blood, fix at once by waving rapidly in the air, and stain with Leishman's stain.

The erythrocytes will be stained pink and the nuclei of leucocytes blue.

Search for the various types of leucocytes. You are certain to find polymorphs, one or two monocytes, perhaps, and probably one or two lymphocytes, but you may be unable to identify other types.

4. Examine a prepared slide, stained to show platelets.

5. Examine a prepared slide of human blood.

Haemoglobin

Fresh blood should be used in the following experiments. Add sodium citrate to prevent clotting.

1. Bubble oxygen through the blood and note that the colour is bright red due to the presence of oxyhaemoglobin.

2. To the oxyhaemoglobin from above add some ammonium sulphate, which is a reducing agent. Observe the colour change as reduced haemoglobin is formed. Then bubble oxygen through the blood and note the return of the bright red colour due to the restoration of oxyhaemoglobin.

3. Bubble coal-gas, which contains carbon monoxide, through a sample and note the cherry-red colour produced owing to the formation of carboxyhaemoglobin.

Clotting of Blood

1. Prick the finger with a sterilised needle and place a drop of blood on a microscopical slide. Examine later under the microscope.

Note that the blood has clotted.

2. Put a few drops of 10 per cent. sodium citrate solution on a microscopical slide. Introduce a drop of fresh blood from the finger. Put aside and examine later.

The sodium citrate, by removal of the calcium salts, will have prevented clotting from taking place.

3. To the "salted" blood from (2) add a few drops of calcium chloride solution.

Observe that the blood clots, showing that clotting depends on the presence of calcium salts.

4. Place some freshly drawn blood in a boiling tube and examine about twenty minutes later.

Note that the blood has clotted, a reddish jelly-like mass of red corpuscles being suspended in a yellowish fluid called serum.

5. Whip up some freshly drawn blood with twigs for a few minutes.

Note that a stringy substance collects on the twigs. This is fibrin.

Leave the blood for about twenty minutes and examine again.

It will be seen that this *defibrinated blood* has not clotted.

Haemolysis

Prepare a blood film on a coverslip, invert on a slide and irrigate with distilled water.

Note the effect on the erythrocytes which swell up by absorption of the water and either burst or become so distended that the haemoglobin escapes into the plasma.

Blood Groups

1. Mix the blood of various pairs of individuals on a slide by means of a mounted needle and examine under the low power to see whether agglutination takes place or not.

2. If Group A and Group B sera are available, place a drop of each on a white tile. Add a drop of a student's blood to each and determine his Blood Group by reference to the text on p. 111.

3. Determine the blood group of a fellow student by means of *Eldon Cards*, carefully following the instructions provided with the cards.

Chapter X

THE CARDIO-VASCULAR SYSTEM—II

THE HEART AND THE STRUCTURE OF THE BLOOD VESSELS

In order that the blood may carry out the functions given at the end of the last chapter, it is necessary for it to be circulated round the body. This is made possible by a pumping organ, the **heart**, and a series of closed vessels. Those taking blood away from the heart are called **arteries**. These branch into smaller **arterioles** and give rise in the organs to a network of microscopic thin-walled **capillaries**. These, in turn form small **venules** which join up with others to form **veins**, by means of which blood is taken back to the heart. The heart, arteries, capillaries and veins and the contained blood constitute the **cardio-vascular system** associated with which is a further arrangement of vessels containing the fluid **lymph** and forming the **lymphatic system**.

THE HEART

The heart is a bluntly conical organ situated in the thorax roughly in the mid-line but extending about 2·5 cm. more to the left than to the right of the sternum. The **base** of the heart is directed upwards and this is level with the third costal cartilages, the **apex** lying between the fifth and sixth ribs. It is partly covered by the lungs which are situated on each side of it. Its size is about that of the clenched fist, being, on an average, 12·7 cm. long, nearly 9 cm. broad and around 6·5 cm. in thickness. The blood vessels entering and leaving the base of the heart hold it in position, the rest of the organ being free. Protection is given by a double serous membrane, the **pericardium**, the inner layer of which covers the surface of the heart and between this and the outer membrane is the **pericardial fluid** which lubricates the surface, reducing friction during the movement of the organ.

The heart is divided into four chambers, the **right** and **left atria** at the basal end and the **right** and **left ventricles** below. A septum prevents any direct communication between the two sides. It is interesting to note that during development in the uterus, the two atria are in communication and the locality of this opening is indicated in the fully developed heart by a closed depression in the inter-atrial septum called the **fossa ovalis**. The cavities are lined by a serous membrane, the **endocardium**, and the wall of the heart is composed of the striated but involuntary *cardiac muscle*, the

myocardium. The walls of the ventricles are thicker than those of the atria and the left ventricle has a thicker wall than the right. The reasons for these differences are firstly that the atria have only to pump blood to the nearby ventricles and whereas the right ventricle has only to pump it to the lungs lying alongside the heart, the left is responsible for pumping it through the entire body. Blood enters the heart by the atria and leaves it by the ventricles.

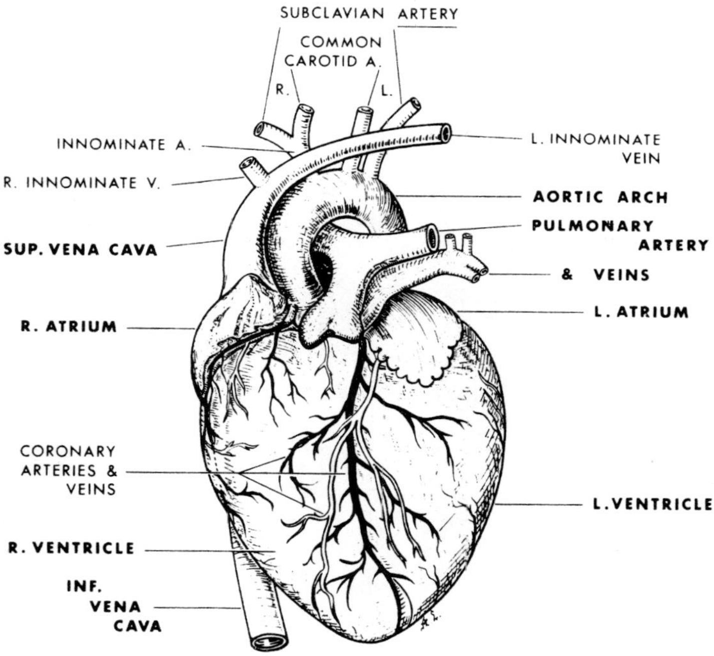

Fig. 10.1. **The heart and great vessels. Ventral view.**

Entering the right atrium are the **superior vena cava** returning blood from the head and neck, the arms and the upper part of the wall of the thorax, the **inferior vena cava** bringing blood back from the rest of the body and the **coronary sinus** returning blood from the heart itself. There are muscles around the entrances of the venae cavae. Leaving the right ventricle is the **pulmonary artery** carrying blood to the lungs. At its entrance are three **semi-lunar (pulmonary) valves** to prevent regurgitation of blood into the ventricle. Entering the left atrium are four **pulmonary veins** returning the blood from the lungs, two veins from each lung. Finally, leaving the left ventricle is the **aorta** with three **semi-lunar valves (aortic**

valve) at its entrance. This large vessel gives rise to the arteries which carry blood to the whole of the body.

The heart, the hardest worked organ in the body, has its own blood supply in the **coronary arteries** which arise from the aorta and this blood is returned to the right atrium mainly by the **coronary sinus**. The coronary arteries and veins can be seen on the surface of the heart. The orifices between the atria and the corresponding ventricles can be closed by **atrio-ventricular valves** which prevent the return of the blood into the atrium when the ventricle contracts.

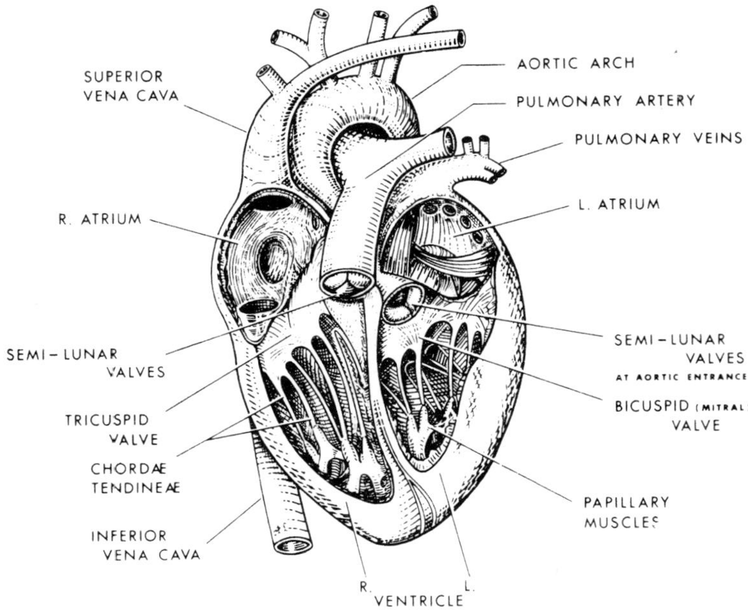

Fig. 10.2. **The heart. Internal structure. Ventral view.**

That between the right atrium and ventricle has three flaps or cusps and is known as the **tricuspid valve** whereas the valve between the left atrium and ventricle has only two and is therefore called the **bicuspid valve**. Owing to its resemblance to a bishop's mitre it is also known as the **mitral valve**. The cusps are supported by strong cords of tendon known as **chordae tendineae**, which are connected to **papillary muscles** in the ventricular wall because during ventricular contraction considerable pressure is produced. There is a collection of specialised muscle fibres in the right atrium close to the entrance of the superior vena cava known as the **sinu-atrial node** which carries rhythmic impulses to the wall of the atrium, causing it to contract. The wave of contraction continues to a further collection of specialised muscle where the atrium and ventricle join and this is called the **atrio-ventricular node**. As the tissue between the atria and the ventricles is non-conducting another highly specialised region of muscle known as the **bundle of His** is situated in this locality and it passes through the septum to both ventricles. This results in the ventricles contracting at the same rate as the atria. Thus the sinu-atrial node controls the rate of heart beat and it is therefore referred to as the *pacemaker* of the heart. It must be understood that the contraction and relaxation of the heart is automatic and is entirely due to its muscle but the *rate* can be increased or decreased according to the body's requirements and this is under nervous control.

THE PHYSIOLOGY OF HEART BEAT

The so-called heart "beat" is really alternative contraction and relaxation of the heart and the two atria contract together and the two ventricles contract together. Blood enters the right atrium from the venae cavae and the left atrium from the pulmonary veins when the walls of the atria are relaxed. As the atria become filled the bicuspid and tricuspid valves open and the blood continues into the ventricles, the walls of which are relaxed. This period of relaxation, known as **diastole**, is followed by contraction or **systole**, first of the atria and then of the ventricles. Atrial systole forces more blood into the ventricles, return into the venae cavae and pulmonary veins being prevented by the muscles around their openings. Immediately afterwards ventricular systole begins and the increase of pressure closes the bicuspid and tricuspid valves. Exit from the ventricles is avoided by the fact that the semilunar valves at the entrance to the pulmonary artery and the aorta are closed until pressure in the ventricles is such that they are forced open and then the blood enters these vessels. Directly ventricular diastole begins the pressure begins to fall below that in the pulmonary artery and aorta and the pressure in the vessels closes the semilunar valves and prevents the return of blood into the ventricles. This whole series of events constitutes the **cardiac cycle**. Atrial systole takes 0·1 sec., ventricular systole 0·3 sec. and diastole of the whole heart 0·4 sec.; so the entire cardiac cycle occupies 0·8 sec. and there are about seventy-five heart beats per min. However, this varies somewhat in different individuals, with age and under different conditions. Immediately after birth, for example, it is about 140 per min., at seven 90–85, at fourteen 85–80, in adult age 80–70 and in old age 70–60 per min. The rate is increased with vigorous exercise, when it may reach as much as 140 per min., during nervous excitement, as a result of the passage into the blood of the hormone adrenalin, and when fever occurs. It will be noticed that the period of rest is brief, 0·4 sec., and remember that the heart is at work throughout the twenty-four hours of the day. The cardiac output is in the region of 4·5 litres of blood per min.

The heart sounds are usually described as "lubb-dup" and are mainly caused by the closing of the valves. The "lubb" is a long, rather low-pitched sound caused by the atrio-ventricular valves together with the contraction of the ventricular muscle and the short, sharp "dup" by the semilunar valves.

As with skeletal muscle, the contraction of the cardiac muscle is partly due to changes in electrical potential and these can be recorded graphically by an instrument known as an **electrocardiograph**. Leads from the thorax or limbs of a patient are connected to the instrument and the vibrations set up in it are recorded on a moving photographic plate. The **electrocardiogram** thus produced indicates whether or not the heart is behaving normally and helps in the diagnosis of heart diseases.

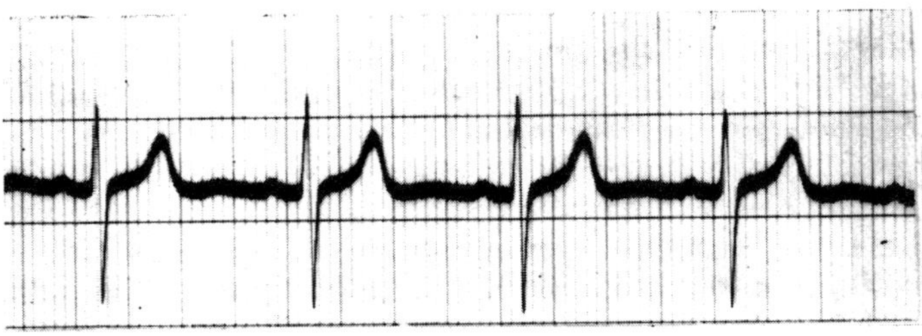

NORMAL

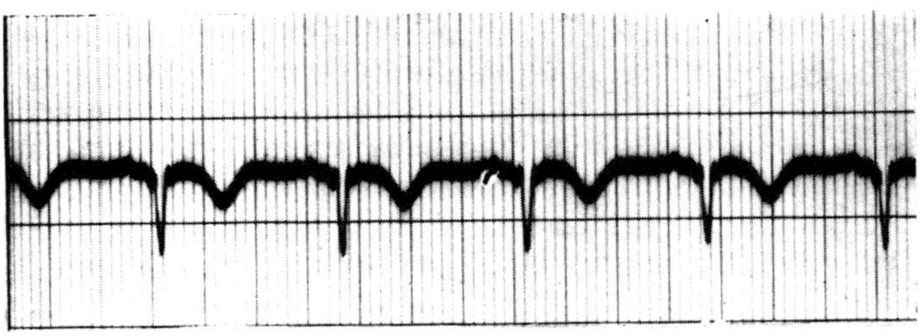

MYOCARDIAL INFARCTION (CORONARY THROMBOSIS)

Fig. 10.3. **Electrocardiograms.**

The long traces show the ventricular systole and the short thick ones the atrial systole.

(By courtesy of the National Heart Hospital, London.)

ARTERIES, VEINS AND CAPILLARIES

Arteries, having to withstand considerable pressure from within, have thick walls which consist of three coats. The outer coat, or **adventitia** is composed of areolar tissue, the **tunica media** or middle coat of unstriated muscle fibres and, particularly in the case of the larger arteries, of elastic tissue, and the inner coat, **the tunica interna**, consists of endothelium. In arteries, blood always flows

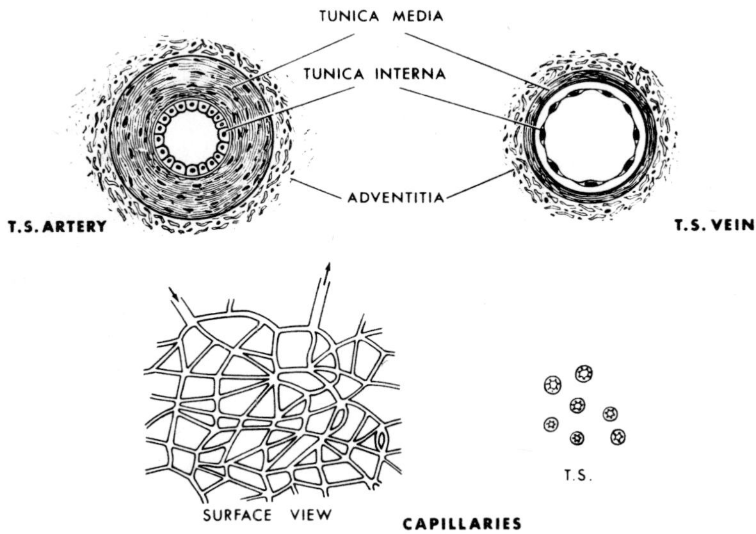

Fig. 10.4. **Structure of arteries, veins and capillaries.**

away from the heart and, with the exception of the pulmonary artery, the blood is always oxygenated. The pressure of the blood can be felt in more superficially placed arteries such as the radial artery in the wrist.

Veins are thinner-walled than arteries. They have the same three coats but there is much less muscular and elastic tissue. It is less necessary because by the time the blood has reached the veins the pressure has fallen very considerably and the tougher texture found in the arteries is not required. At intervals infoldings of the inner coat form **valves** which prevent the blood from flowing in the wrong direction. In veins blood must always flow towards the heart and with the exception of the four pulmonary veins, the contained blood is always deoxygenated.

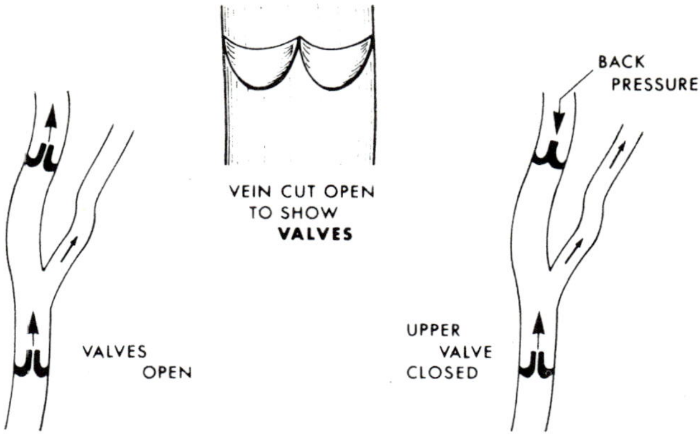

Fig. 10.5. **Action of valves in veins.**

Capillaries are very minute, microscopic vessels, all having the same calibre and their walls, composed of endothelium, are only one layer in thickness. The thinness of the walls allows diffusion to take place through them.

HEART SURGERY

Many successful **heart operations** have been performed by heart surgeons including one in which three of the four heart valves were replaced in a six-hour operation during which the heart was stopped for three hours, circulation being maintained artificially. Valves from pigs' hearts have proved to be the most successful for these replacements but plastic and other compositions have been used.

The **heart-lung machine** is used during these operations, putting the heart and lungs temporarily out of action and taking over their functions so that the brain and other organs receive their normal oxygen supply. The surgeon can work unimpeded by the normal movement of the heart and lungs. In fact, without this machine much heart surgery would be impossible.

The **cardiac pacemaker** is an instrument which is used when the heart's natural pacemaker, the sinu-atrial node, functions improperly. Electrodes from the minute unit are embedded in the heart muscle and supply it with impulses, the energy for which is provided by a tiny battery embedded under the skin of the thorax. This has to be replaced at lengthy intervals. The impulses control the rate of heart beat.

In 1975 a 13-month-old boy who had undergone a heart operation was kept alive for sixteen hours by connecting his circulatory system

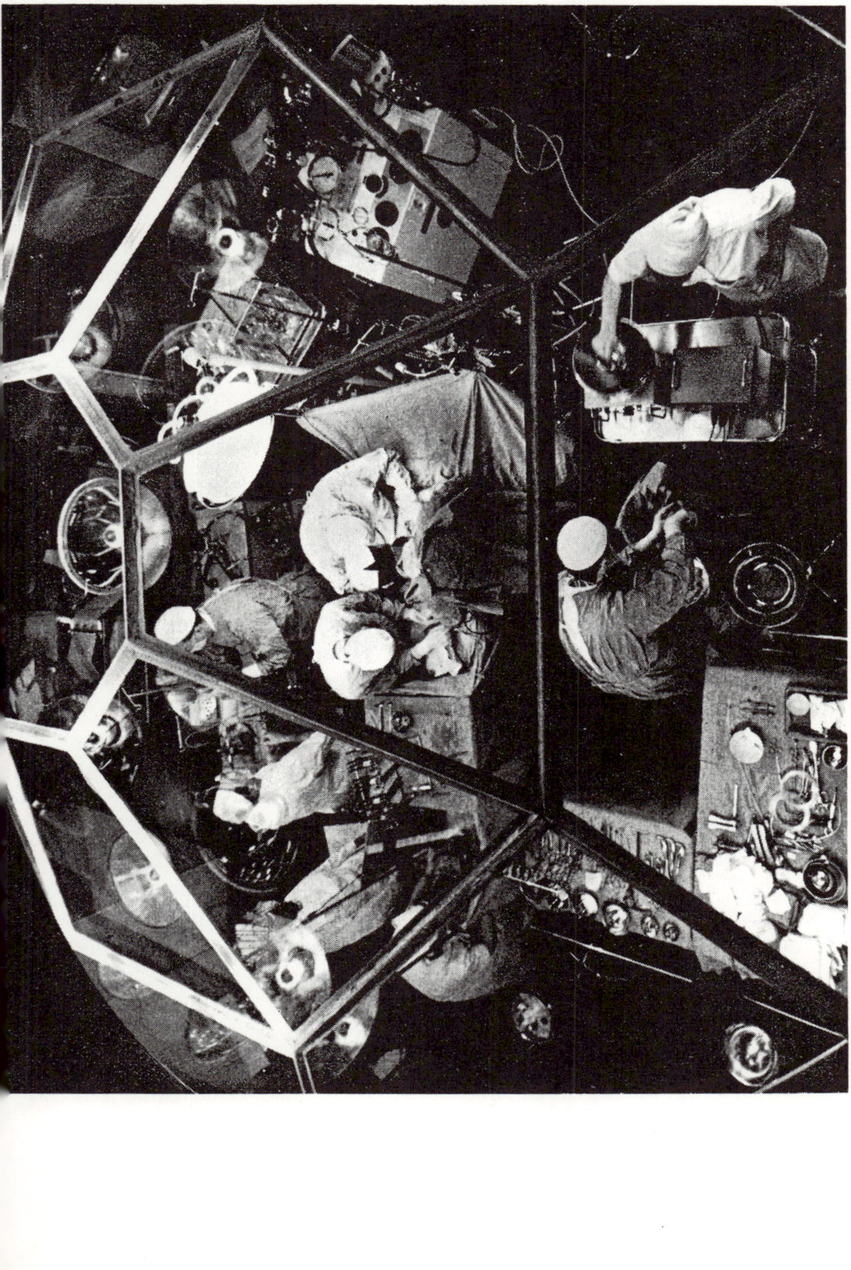

Fig. 10.6. A heart operation in progress.
In the *centre* is the chief surgeon with two assistants. On the *left* is the instrument table. On the *right* is an Armström respirator. In the *left background* is the pump operator and his assistant and in the *right background* is a television screen recording arterial and venous pressures, and electrocardiogram and electroencephalogram tracings.

(*By Courtesy of the World Health Organisation, Geneva.*)

to the heart, lungs and kidneys of a baboon. This operation was done as a last resort two days after the heart operation as the child's heart and kidneys had failed and the heart-lung machine kept him alive. All went well for fifteen hours, the boy's heart and lungs showing signs of recovery but then the high toxicity of the child's blood affected and eventually killed the baboon and the child died.

Heart Transplants

A large number of heart transplants from human subjects, removed immediately after death of the donor, have been performed in various countries. One of the chief obstacles to be overcome in any transplant operation is the rejection of foreign tissues by the body and it is essential that a matched donor be found. A further risk is the danger of secondary infections developing afterwards.

A long-surviving patient was Dr. Philip Blaiberg, aged 59, on whom a heart transplant was performed by the South African surgeon Dr. Christian Barnard on January 2nd, 1968. This was Dr. Barnard's second heart transplant and the world's third. The patient left hospital when he had recovered from the operation but had to return later owing to the development of liver inflammation and a lung complaint. He was later able to return home and a year later his condition was excellent and he was said to be hale and hearty but he died nineteen months after the operation In 1974 only some 15 per cent. of people who had received a heart transplant were still alive, thirty-eight patients having survived for a year, twenty-three for more than two years, twelve for over three years, eleven for more than five years and one patient was well into his sixth year. The first heart transplant to be done in this country was at the National Heart Hospital in London. The patient survived the operation but eventually contracted what was described as an overwhelming infection, to which he succumbed, forty-seven days after the operation. It is unlikely that heart transplants are still being performed in this country and certainly not at the National Heart Hospital in London because the rejection phenomenon and the need for frequent return to hospital for treatment by immuno-suppressive drugs has made surgeons hesitant to carry out further transplants.

Transplants of the heart and lungs have also been done but in all cases so far recorded the patients have eventually succumbed.

Chapter XI

THE CARDIO-VASCULAR SYSTEM—III

THE CIRCULATION OF THE BLOOD

The blood is pumped by the ventricles into the arteries and so a circulation is set up. The circulation of the blood was first discovered and demonstrated by William Harvey in 1628. The lesser circulation between the right ventricle, the lungs and the left atrium via the pulmonary artery and vein respectively is known as the

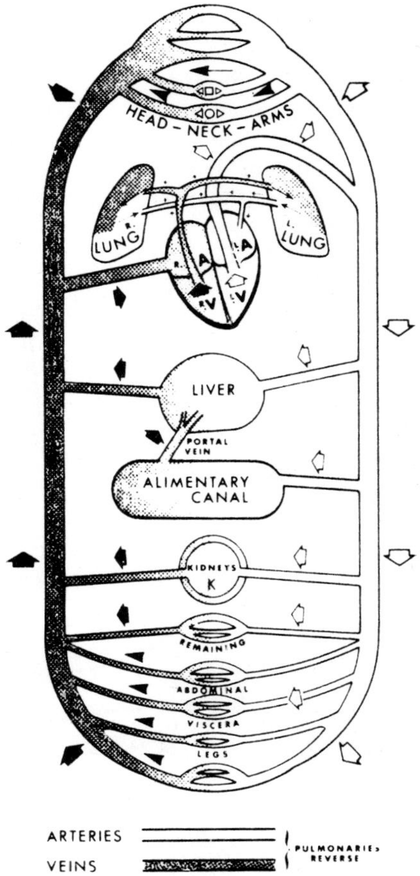

Fig. 11.1. **Diagram to show the course of the circulation.**

pulmonary circulation. That from the left ventricle through the body and back to the right atrium by the venae cavae is the **systemic circulation**.

The ventricular systole forces the blood in a succession of spurts into the arteries and the pressure of the blood against the walls of the blood vessels, naturally greatest in the arteries close to the heart, is known as **blood pressure**. It is necessary in order to maintain the circulation but it must not be excessive or haemorrhage may result. Although this pressure is set up by the pumping action (systole) of the ventricle, the blood continues to flow during diastole and this is where the elasticity of the arteries comes in for it keeps up the pressure so that the flow is continuous and avoids too great a difference in pressure between systole and diastole. The sudden distention of the arterial walls due to systole is what is known as **pulse** and this is the rate of the heart beat.

The systolic pressure in the large arteries reaches up to 140 mm. of mercury and in the medium arteries it is 110 mm. In the capillaries it drops to about 20 mm. and in the small veins it may be as high as 20 mm. but drops in the larger veins to 10 mm. while in the veins of the neck and thorax it falls further to between 0 and -8 mm.

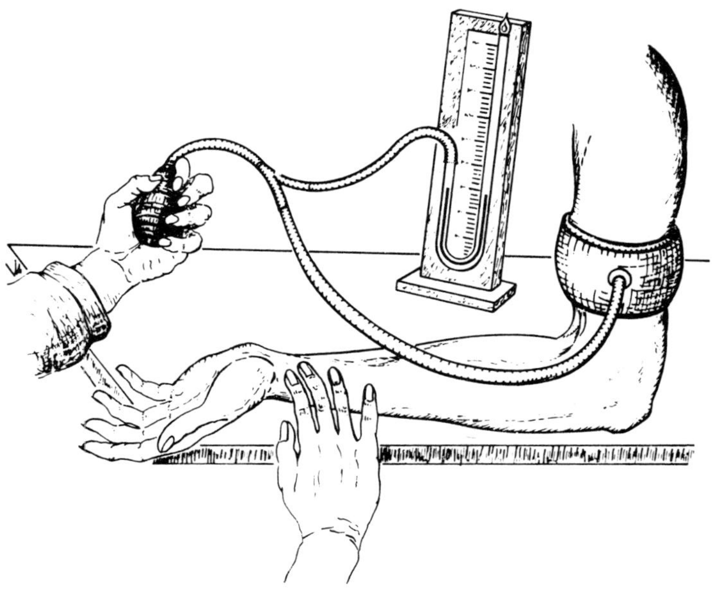

Fig. 11.2. **Sphygmomanometer.**

Blood pressure is measured by an instrument called a **sphygmomanometer**. This consists of an inflatable non-elastic bag tied round the upper arm where the pressure in the brachial artery can be measured. The bag is connected by rubber tubing to a bulb so that air can be pumped into it and to a mercury manometer which records the pressure. As air is pumped into the bag the pressure of the whole armlet compresses the blood vessels in the arm. This is done until the pulse, taken at the wrist, disappears. This, the systolic pressure, is noted. The bag is then deflated slightly until the pulse returns and the pressure is again noted. The mean of the two readings is the blood pressure.

When the blood enters the capillaries the pressure becomes lower and evens down. It is here that the digested foods and oxygen enter the tissues from the blood and carbon dioxide and other excretory products pass from them into the blood. Part of the plasma diffuses through the capillary walls and this is lymph or tissue fluid. For this reason lymph is often referred to as "the middleman" between the blood and the tissues. Some of it passes back into the capillaries after the exchange has taken place and the balance moves into lymph capillaries which ultimately give rise to lymph vessels whence it is returned to the blood as will be explained in a later chapter.

The capillaries contract as the result of cold and dilate with heat. They also become dilated when a tissue is damaged. This is due to a substance called **histamine** which is formed under these conditions. Dilation also occurs as the result of shock and blood may pass into them to such an extent that blood pressure falls. The blood from the capillaries passes as a steady stream into the venules and veins and so eventually enters the left and right atria of the heart by the pulmonary veins and the venae cavae respectively. Thus the pulmonary and systemic circulations are completed.

Chapter XII

THE CARDIO-VASCULAR SYSTEM—IV

THE CHIEF BLOOD VESSELS OF THE BODY

THE PULMONARY SYSTEM

The **pulmonary artery** arising from the base of the right ventricle on the left side of it divides into two trunks, one to each lung. It carries deoxygenated blood to the lungs which it enters between the pulmonary veins and the main air tube or bronchus. In the lungs the artery branches considerably and finally breaks up into capillaries which line the air-sacs. From these capillaries arise branches which form the **pulmonary veins**, two from each lung. Devoid of valves and smaller than the arteries, these four veins enter the left auricle.

THE SYSTEMIC SYSTEM

The **aorta**, after leaving the left ventricle ascends for a short distance as the **ascending aorta** and then arches over to the left in a backward direction over the root of the left lung. From this point it bends over as the **aortic arch** whence it continues downwards on the dorsal side of the body as the **dorsal aorta**. From this vessel arise the chief arteries of the body.

First, arising from the ascending aorta are the **left** and **right coronary arteries** which supply the heart and then from the aortic arch comes the **innominate artery** which divides into a **right subclavian artery** and a **right common carotid artery**. Further down the arch gives rise to the **left common carotid artery** and then the **left subclavian artery**. It will be seen that these two arteries arise directly from the aorta and not from a common root as is the case on the right side.

THE ARTERIES

Arteries of the Head, Neck, Thorax and Upper Limbs

Each **common carotid artery** divides into an internal carotid and an external carotid. The **internal carotid artery** runs upwards to enter a foramen, the *carotid canal*, in the temporal bone of the skull and so into the cranium to supply the brain. En route it gives off a branch, the **ophthalmic artery**, to supply the eye. The main branch divides into the **anterior** and **median cerebral arteries** which supply the greater part of the cerebral hemispheres of the brain, the hinder part being supplied by the **posterior cerebral arteries** arising from the basilar artery (see below).

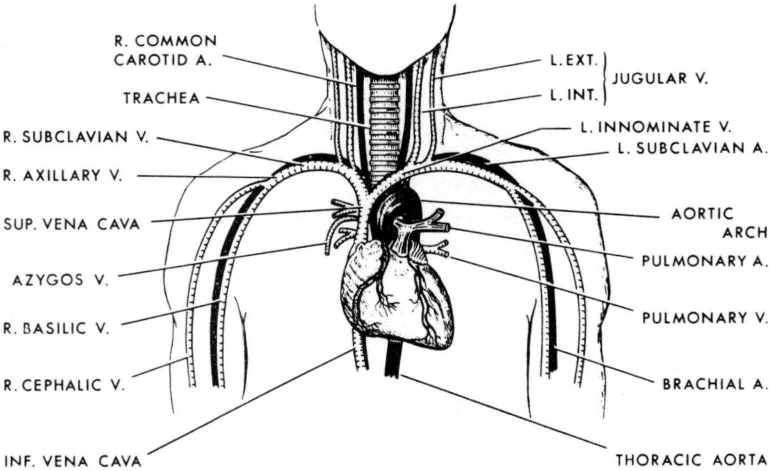

Fig. 12.1. **Chief arteries and veins of the neck, thorax and upper arms.**

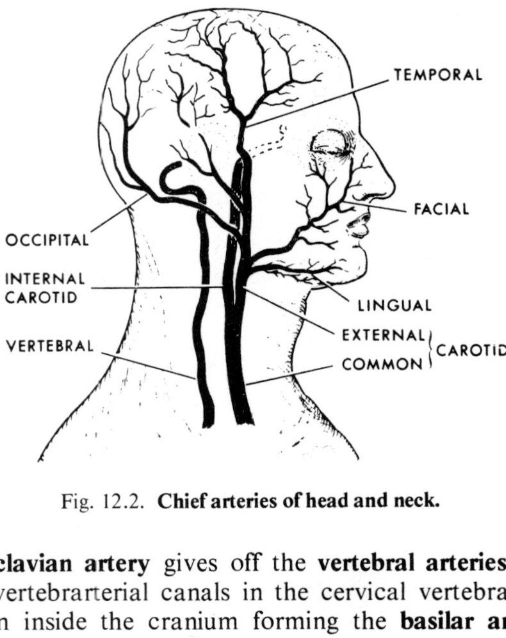

Fig. 12.2. **Chief arteries of head and neck.**

The **subclavian artery** gives off the **vertebral arteries** which run inside the vertebrarterial canals in the cervical vertebrae. The two arteries join inside the cranium forming the **basilar artery** which divides into **left** and **right posterior cerebral arteries.** These are joined by branches from the internal carotids. An anastomosis of arteries known as the **Circle of Willis** lies at the base of the brain. It is formed by the union of the anterior and posterior cerebral arteries mentioned above and from it arise several branches which supply the brain.

The **external carotid artery** divides into three branches. The first is the **facial artery** which runs to the angle of the jaw where it forms many branches. The second branch of the external carotid is the **occipital artery** which goes to the occipital region of the head and the third branch is the **temporal artery** which runs upwards in front of the ear to the temporal region of the head. The pulse can be felt in this artery against the skull as is the case with other superficially placed arteries elsewhere in the body. Their positions will be found in books on First Aid. In the case of accidents in which arterial haemorrhage occurs, exertion of pressure at the point nearest the wound on the heart side, pressing the artery against the underlying bone, will arrest the haemorrhage. It is, of course, only a temporary or first aid expedient. A knowledge of the location of the pressure points may therefore prove useful in case of accident.

The **subclavian arteries** run under the clavicles to the axilla on each side where it is known as the **axillary artery**. At the lower

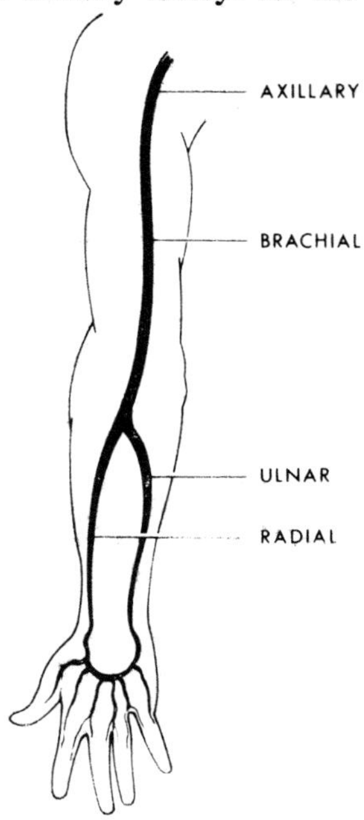

Fig. 12.3. **Chief arteries of arm.**

edge it becomes the **brachial artery** which passes down the inner side of the arm. At the elbow it divides into the radial and ulnar arteries. The **radial artery** continues along the radial side of the forearm to the wrist where it enters the palm of the hand and provides branches to the fingers. The **ulnar artery** runs along the ulnar side of the forearm and this, too, provides branches to the palm and fingers. A further branch of the subclavian, the **mammary artery**, supplies the muscles of the wall of the thorax. The **thoracic aorta** provides **intercostal arteries** to the intercostal muscles, **superior phrenic arteries** to the upper surface of the diaphragm and **bronchial arteries** to the tissue of the lungs.

Arteries of the Abdomen and Lower Limbs

On passing through the diaphragm and entering the abdomen, the **abdominal aorta**, as it is now called, gives off an **inferior phrenic artery** to the lower surface of the diaphragm. Almost immediately beneath this vessel is the unpaired **coeliac artery** (or **axis**). This divides into the **hepatic artery** to the liver, the **gastric**

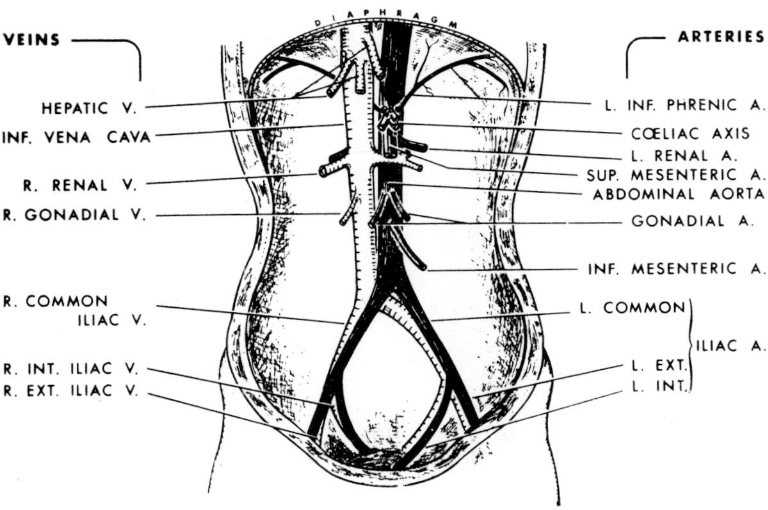

Fig. 12.4. **Main arteries and veins of abdomen.**

artery to the stomach and the **splenic artery** to the spleen. About 1 cm. behind it is another single artery, the **superior mesenteric artery**, which runs in the mesentery and supplies the whole of the intestine from the centre of the duodenum to the caecum and the ascending and transverse colons. The next arteries are paired.

the **supra-renal arteries** to the supra-renal glands above the kidneys, the **renal arteries** to the kidneys and the **gonadial arteries (spermatic** in the male and **ovarian** in the female) to the reproductive organs. Then comes another single artery, the **inferior mesenteric artery**, which goes to the descending colon and rectum.

Level with the fourth lumbar vertebra the aorta bifurcates, the two vessels thus formed being the **common iliac arteries** and each divides into an **internal iliac artery** supplying the pelvic viscera and the muscles of the buttock and an **external iliac artery** which passes into the thigh where it is known as the **femoral artery**.

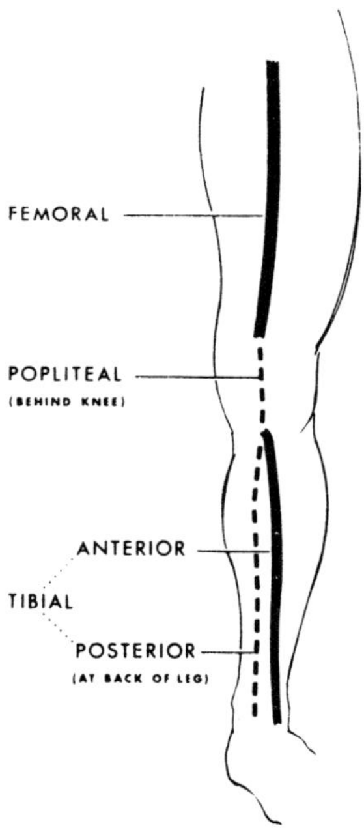

Fig. 12.5. **Chief arteries of leg.**

This runs down the inside of the thigh as far as the knee and continues behind it as the **popliteal artery** after which it divides into an **anterior tibial artery** which runs down the front of the leg to the ankle and a **posterior tibial artery** which passes down the back of the leg. The anterior tibial continues into the foot supplying the dorsal side of the organ, the sole and the toes. The posterial tibial also supplies the foot and toes.

THE VEINS

The veins are classified as *superficial veins* which lie just beneath the skin and *deep veins* which accompany the corresponding arteries and usually bear the same names.

Veins of the Head, Neck, Thorax and Upper Limbs

The venous blood from the brain collects in wide channels formed by separation of layers of its outermost protecting membrane, the dura mater, and known as **venous sinuses** and the blood from these sinuses passes into the two **internal jugular veins**. Branches of these veins bring blood from the face, tongue and pharynx. The two internal jugular veins pass down the neck, one on each side, and join the **subclavian veins** from the arms, level with the clavicles, thus forming the **innominate veins** in the thorax.

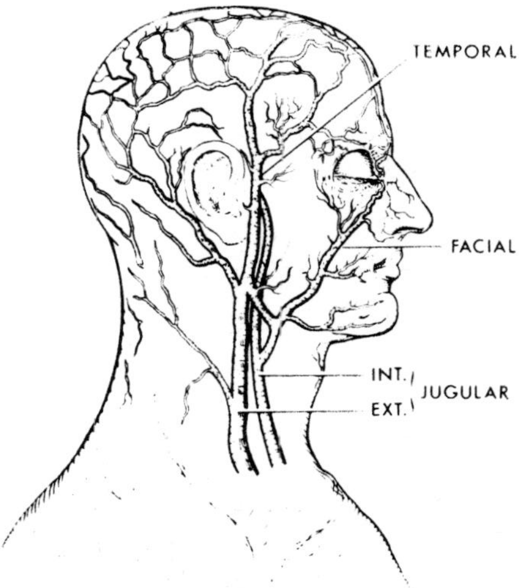

Fig. 12.6. **Chief veins of head and neck.**

Most of the deep veins of the upper limb correspond with the arteries and are in pairs, one on each side of the artery. They are known as **companion veins** or **venae comites**. The veins from the fingers and the palm of the hand form the two **radial** and two **ulnar veins** which run alongside the **artery**. These unite in front of the elbow forming two **companion veins of the brachial artery** which, in turn, join one another to form the **axillary vein**. This becomes the subclavian vein which, as already seen, is joined in the thorax by the internal jugular vein to form the innominate vein. The veins in the hand also give rise to a **median vein** running in front of the forearm and at elbow level this divides into two branches, one joining the **basilic vein** and the other the cephalic. As already seen the companion veins of the brachial artery also join the basilic and thus form the axillary vein. On the side of

Fig. 12.7. **Chief veins of arm.**

the arm is the **cephalic vein** and in the region of the shoulder this, too, enters the axillary vein.

The two innominate veins, after receiving **internal thoracic veins** from the anterior wall of the thorax, unite to form the **superior vena cava** which enters the right atrium. This receives the **azygos vein** from the posterior thoracic wall. Also entering the right atrium is the **inferior vena cava** returning blood from the abdomen and lower limbs.

Veins of the Lower Limbs and Abdomen

Superficial veins bringing blood from the toes form a **venous plantar arch** and other veins from the foot and branches from this arch give rise to two **anterior** and two **posterior tibial companion veins** which accompany the corresponding arteries. A **short saphenous vein** carries blood from the outside of the foot, passes up the back of the leg and is joined by the tibial veins to form the **popliteal vein** which runs up the thigh. Blood from the inside of the foot is carried by the **long saphenous vein,** the longest vein in the body, which runs up the inside of the leg and, higher up in the thigh, is joined by the popliteal vein and the **femoral**

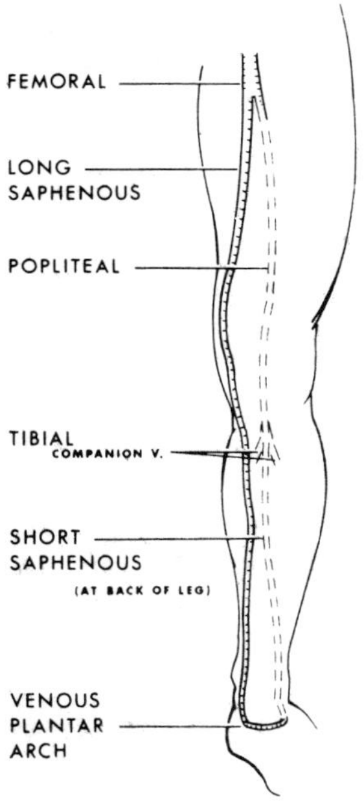

Fig. 12.8.
Chief veins in the leg.

vein is formed. This runs alongside its artery and higher up this is known as the **external iliac vein**. An **internal iliac vein** joins it and thus the **common iliac vein** is formed. At the level of the fifth lumbar vertebra the two common iliacs unite to form the **inferior vena cava**. This large vein runs along the right side of the aorta in the abdomen, pierces the diaphragm and passes into the thorax where it enters the right atrium of the heart as already stated above. On its journey upwards it receives the **renal veins, suprarenal veins, gonadial veins** and **phrenic veins** corresponding to the paired arteries of these names. Immediately below the liver it also receives the large **hepatic veins**, four or more, from that organ.

It will be noticed that so far no mention has been made of veins taking blood from the alimentary canal. These vessels, the **superior mesentric vein,** the **splenic vein,** the **inferior mesenteric vein** and the **gastric veins** give rise to a large vein which takes the blood to the liver. This is known as the **portal vein** which enters the liver where it breaks up into capillaries. The **portal system**, as it is called, is thus a by-pass system taking blood from the capillaries of the alimentary canal direct to the liver. It will be remembered that it is in this organ that glycogen is formed and stored and deamination takes place and the purpose of this system will become obvious. Thus the liver has a double blood supply, oxygenated blood by the hepatic artery and venous blood by the

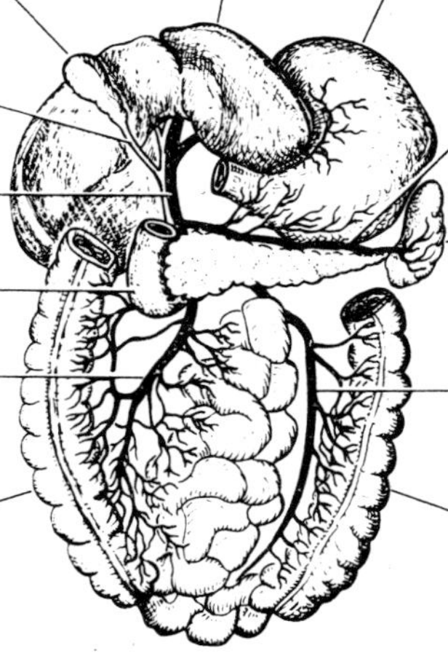

Fig. 12.9. **The portal vein and its tributaries.**
The transverse colon and part of the duodenum have been removed and the liver deflected upwards to give complete exposure.

portal vein. When you studied the microscopic structure of the liver you came across interlobular veins which it was stated originated from the portal vein. The sinusoids, formed from the portal capillaries after union with capillaries from the hepatic artery enter the intralobular vein in the centre of the lobules and the intralobular veins ultimately give rise to the **hepatic veins** which carry the blood to the inferior vena cava.

FUNCTIONS OF THE CARDIO-VASCULAR SYSTEM

The functions of the cardio-vascular system, which have already been seen in the foregoing chapters, may be summarised as follows:—

Transport of oxygen from the lungs to all the tissues:—
carbon dioxide from the tissues to the lungs;
digested food from the small intestine to all tissues;
excretory substances from the liver and elsewhere to the kidneys;
hormones from the endocrine glands to all tissues;
heat, particularly from the muscles and liver, to all tissues.

Protection of the body against infection and disease.

Clotting to prevent haemorrhage and to protect the body from invasion of infective agents.

Disorders and Diseases of the Cardio-Vascular System

Diseases of the Cardio-Vascular System accounted for 33·5 per cent. of the total deaths from natural causes in this country in 1974.

Diseases of the blood itself, anaemia, leukaemia and haemophilia, have already been described at the end of Chapter IX.

Palpitation—when one becomes aware of the normally unnoticed heart beat. Occurs after vigorous exercise and during nervous excitement and it can be caused by excessive smoking and alcoholic indulgence. The very forcible heart beats make it apparent.

Angina pectoris—due to a deficiency in blood supply to the heart muscle, is indicated by sudden violent and severe pain in the region of the organ.

Embolism—caused by the blocking of blood vessels by a clot or other obstruction and which, by preventing blood from reaching a particular part, may result in the destruction of the tissues which the vessels supply.

Thrombosis—or formation of a clot inside the blood vessels is usually brought about by damage to those vessels and it may occur in arteries or veins. **Coronary thrombosis (myocardial infarction)** occurs in the coronary artery and this, of course, is a very serious condition in which the blood supply is cut off from the heart muscle which therefore dies. These "heart attacks" as they are popularly called, come on suddenly and cause severe pain in the chest. Recovery is possible with prolonged rest and care but they may recur and ultimately death may result. Deaths in 1974 amounted to 153,227 and a further 42,649 died from other heart diseases. **Cerebral thrombosis** occurs in one of the branches of the cerebral artery and this causes paralysis, usually on one side of the body (**hemiplegia**), which is what is commonly known as a "stroke".

Apoplexy—usually happens only late in life and results from the rupture or blocking from thrombosis or embolism of cerebral blood vessels causing damage to the brain. This produces a sudden loss of sensibility and the brain tissue degenerates owing to lack of blood.

Aneurism—or distention of arteries due to a weakening of their walls which may, under the pressure of the blood within them, be in danger of bursting and this can prove fatal.

Varicose veins—are veins which become stretched and thereby weakened. They frequently occur in the legs, particularly in people whose work entails long periods of standing. This produces a dull, aching pain in the affected part and, unfortunately, the condition usually worsens. The wearing of elastic stockings, which give support, gives relief and treatment by injection is often beneficial. **Haemorrhoids**, as explained in Chapter X are varicose veins in the rectum.

Phlebitis—inflammation of the veins. It may cause swelling and pain.

High blood pressure—not at all uncommon and quite frequently associated with some disease of the arteries. There is always a danger of apoplexy.

Valvular disease of the heart—in which the bicuspid or aortic valves lose their power to close their apertures or when the edges of the valves become joined together from disease thereby restricting the size of the openings, a condition known as **stenosis**; obviously interferes with the proper functioning of the heart.

Hypertrophy of the heart—when the heart becomes enormously enlarged; occurs under certain diseased conditions.

Fatty degeneration of the heart—a disease in which some of the cardiac muscle changes into fat, thereby impairing its efficiency. It is usually restricted to old age.

Arteriosclerosis—or hardening of the the middle coat of the arteries is a condition which develops with age. Serious effects will be produced if this takes place in the coronary arteries when a coronary thrombosis may result, and its occurence in the cerebral arteries may result in apoplexy. There is evidence which shows that a high cholesterol level in the blood resulting from animal fats in the diet can be a cause of this disease.

Septicaemia—or general blood poisoning occurs when pathogenic bacteria multiply considerably in the blood which circulates them round the body. It is, of course, a very serious condition and can be caused by a wound.

PRACTICAL WORK

The Heart

Structure

1. Examine the external structure of a sheep's heart which has the great vessels attached. Identify the two atria, the two ventricles and the vessels entering and leaving them, and note the coronary arteries and veins on the surface.
2. Make a longitudinal section of the heart, wash out the contained blood and study its internal structure as described in the chapter.
3. Examine a dissectable model of the human heart (if available).

Heart Beat

Listen to the heart beat of a seated fellow student using a stethoscope placed in the appropriate position on the bare chest. Identify the sounds "lubb" and "dup" and count the number of beats over a quarter of a minute. Record the number per minute. Then get the student to run around quickly for a couple of minutes or so and on his return record the number of heart beats per minute. Compare the two readings.

The Blood Vessels

1. Examine prepared slides showing **transverse sections of arteries, veins** and **capillaries**.
2. Examine a complete ventral dissection of a rabbit and with the aid of suitable diagrams identify the chief arteries and veins. An injected specimen in which the arteries are coloured red and the veins blue is most instructive if available.

The Circulation

1. Lightly anaesthetise a frog with ether or use a pithed frog (in which the brain has been destroyed) and stretch and secure the web of one of the hind feet on a frog plate. Place this on the stage of a microscope and examine under the low power. Note the network of blood capillaries in the web of the foot in which the blood will be seen circulating.
2. Demonstrate the flow of the blood in the veins of a human subject by tying a bandage tightly over the lower end of the upper arm and clenching the fist. This hinders the flow of blood upwards towards the heart and the veins in the forearm become swollen and prominent. If they are stroked with the finger towards the wrist blood will collect beyond the valves in the veins and their

location can be identified by the nodular appearance which develops where they are situated.

3. If the veins on the back of the hand are at all prominent, stroke one of the larger ones with the finger away from the wrist, and exert slight pressure at the distal end with the finger. The portion of the vein between the valves will empty. Now release the pressure of the finger and the blood will shoot forwards along the vein in the direction of the heart and refill the emptied portion.

Pulse

Take the pulse of a fellow student by placing the first finger lightly over the radial artery at the wrist. Count the number of pulsations over a quarter of a minute. This should be done with the student (1) sitting at rest (2) after walking a short distance (3) after walking steadily up a flight of stairs (4) after running up the same flight of stairs and (5) after running around outside for two or three minutes. In each case record the pulse rate per minute and compare results.

Chapter XIII

THE LYMPHATIC SYSTEM AND THE SPLEEN

It was mentioned in Chapter X that associated with the cardiovascular system is another known as the lymphatic system. This consists of very extensively distributed **lymph capillaries** which are found in all the tissues of the body which at their origin are minute blindly-ending sacs and which give rise to **lymph vessels**. These vessels are transparent, have a structure similar to that of veins and contain valves. Those in the villi of the small intestine, known as lacteals, have already been seen. The contained fluid, **lymph**, flows in one direction only and is collected into a large **thoracic duct** carrying the lymph from all parts of the body except the right side of the head, the right arm and the right side of the thorax which discharge it into a short **right lymphatic duct**. The former, which begins in the abdomen by a sac called the **cisterna chyli,** runs upwards and enters the blood system at the junction of the left subclavian and left jugular veins. The latter enters it in a corresponding position on the right.

Lymph is a colourless fluid, alkaline in reaction, having a salty taste and capable of clotting. It is similar in composition to blood plasma from which, in fact, it is derived. This **tissue fluid** contains **lymphocytes** and passes through minute openings in the walls of the blood capillaries when they are distended and thus bathes the tissues. Some of the fluid returns to capillaries direct but the greater part of it is carried by the lymphatics. There is no direct and intimate contact between the blood and the tissues as they are bathed in lymph and this fluid therefore acts as a "middleman" between the two. By means of it digested foodstuffs and oxygen are passed from the blood to the tissues and carbon dioxide and other excretory substances from the tissues into the blood. The lymph carries fats which would be harmful to the red corpuscles so that a safe, regulated amount can be passed into the blood.

Along the course of the lymphatic vessels in the neck, the axilla, the groin and elsewhere, is a very large number of what are known as **lymph nodes**. These serve by reticulo-endothelial cells within them to remove foreign matter such as bacteria and dead red corpuscles from the lymph to prevent their entering the blood and in these nodes lymphocytes arise. Swelling in the arm sometimes follows an infected finger and this is because the lymphocytes in the lymph nodes in the arm are endeavouring to cope with the infective bacteria which have reached them from the finger.

Furthermore the lymphocytes are produced in enormous numbers when infection occurs. Each lymph node is more or less kidney-shaped and is enclosed in a **capsule** from which ingrowths called **trabeculae** divide it up into separate cavities or **lymph spaces.** These

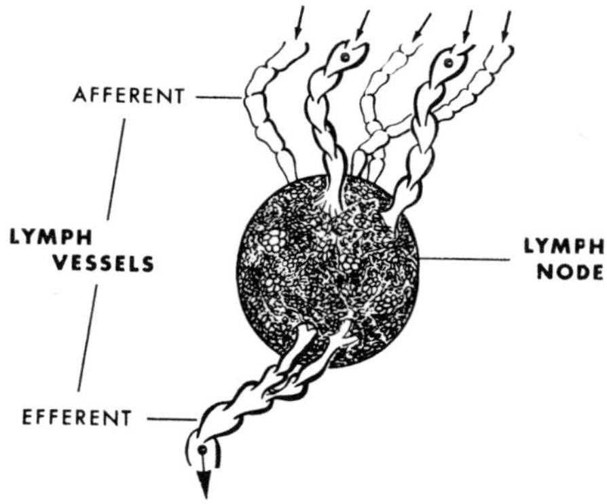

Fig. 13.1. **Lymph node.**

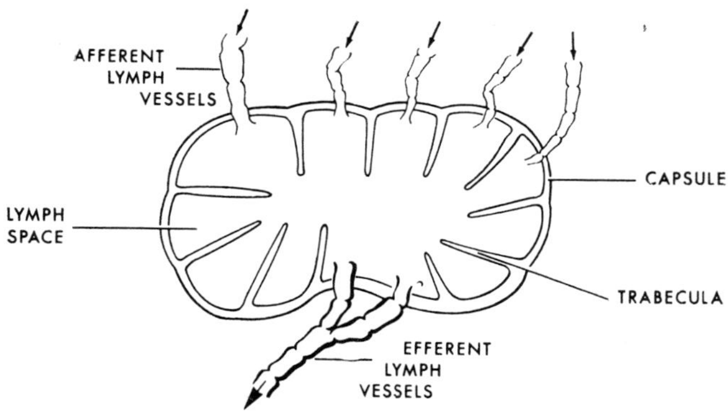

Fig. 13.2. **T.S. Lymph node.**

contain the reticulo-endothelial cells referred to above. Lymph enters the node by several **afferent lymph vessels,** passes through the lymph spaces and leaves by **efferent lymph vessels.**

It is clearly necessary for the lymph to be circulated through the body but in this system there is no pumping organ. Movement of the lymph is effected by pressure of fluids on the outside of the vessels and by the general muscular movements of the body. It will be remembered that the Peyer's patches in the intestine and the tonsils consist of lymphoid tissue.

THE SPLEEN

This organ has already been seen lying in the left hypochondriac region of the abdomen, just below the diaphragm and to the left of the stomach and to which the tail of the pancreas reaches. It is 10 to 12 cm. long, about 7·75 cm. wide and 3 cm. thick. Like the lymph nodes it is enclosed in a capsule from which trabeculae arise internally and in the spaces between them is **splenic pulp**. This contains red corpuscles and lymphocytes as well as reticulo-endothelial cells. In the spleen worn out red corpuscles are destroyed and haemoglobin is broken down and, leaving by the splenic vein, part of this gives rise to the bile pigments in the liver but the iron part is used in the spleen for the formation of more haemoglobin. Lymphocytes are also formed in the spleen. Finally, it serves as a storage organ for blood which reaches it by the splenic artery and which can be passed into the circulation when the capsule contracts.

Disorders and Diseases of the Lymphatic System and Spleen

Oedema—in which considerable swelling occurs, is caused by leakage of lymph through the walls of the lymphatics. This may be in internal organs, or as in the case of **dropsy**, beneath the skin.

Inflammation and other disorders of the **lymph nodes** and **lymphatic leukaemia** can occur.

Hodgkin's disease—in which the lymph nodes become enormously enlarged has always been considered as incurable but recent treatment with high voltage radiation has produced some very satisfying results.

The **spleen**, too, can become very much enlarged in leukaemia and malaria and other less serious diseases. If it ruptures, it may bleed profusely, with fatal results. It can, however, be removed surgically without ill effect on the body.

Chapter XIV

BREATHING, RESPIRATION AND VOICE

Respiration is the process whereby the cells of the body obtain the energy they require for the performance of their functions. It takes place within the cells and is consequently known as **internal** or **tissue respiration**. In the process sugar is broken down to carbon dioxide and water in order that the energy may be released and to enable this to take place, oxygen is necessary and must therefore be taken into the body. Furthermore the katabolically produced and excretory carbon dioxide must be expelled. This intake of oxygen and expulsion of carbon dioxide is the process of **breathing** and the actual gaseous exchange between the air in the lungs and the blood is called **external respiration**. The structures concerned in these processes constitute the respiratory system. Associated with it, as we have already seen, is the blood which is responsible for the transport of the oxygen and carbon dioxide between this system and the tissues.

THE RESPIRATORY SYSTEM

The expanded entrances to the nasal passages, the **external nares** or **nostrils** are provided with a number of hairs which act as filters of the ingoing air and the two nasal passages which are within the nose are separated by a **nasal septum**, the upper part of which is made of bone and the lower part of elastic cartilage. The lacrimal ducts from the eyes open into the nasal cavities, the upper parts of which are sensory to smell. A mucous membrane covered by ciliated epithelium lines the nasal cavities which are thus kept moist and free from obstruction as the cilia move the mucus, in which dust and micro-organisms become entrapped, towards the external nostrils. The **turbinated bones**, covered with this membrane, increase its area and the membrane continues into the sinuses which are in communication with the nasal passages. The upper ends of the cavities are lined by columnar epithelium and in it are the **olfactory cells** concerned with the sense of smell. Their method of functioning and the study of smell will be studied in Chapter XXI. The nasal passages, which are narrowed at their upper ends, open by the **internal nostrils** or **posterior nares** into the **naso-pharynx** which lies above the soft palate.

The respiratory passage continues by entry through the **glottis** into the windpipe or **trachea**, at the upper end of which is the sound box or **larynx**, popularly known as the "Adam's apple".

It will be remembered that the entrance to the glottis is closed during the act of swallowing by the **epiglottis** in order to prevent entry of food so that breathing temporarily ceases during this process. The larynx and the production of sound and speech will be considered at the end of the chapter.

The trachea which lies ventral to the oesophagus is narrower than the larynx and is about 10·5 to 12·5 cm. long and 2·54 cm. in diameter. It is kept permanently open by rings of cartilage which are, however, incomplete on the dorsal side to allow room for the swelling of the oesophagus during swallowing. Its walls are composed of muscular and connective tissue and it is lined by a mucous membrane covered with ciliated epithelium. At the

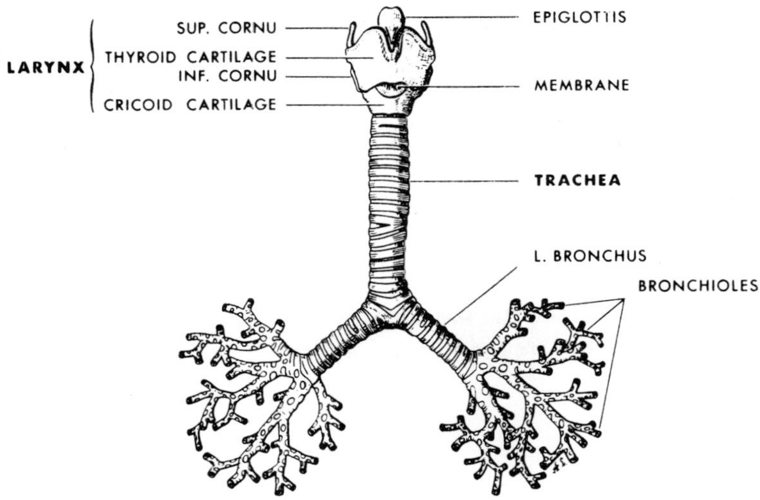

Fig. 14.1. **The larynx, trachea, bronchi and bronchioles.**

level of the fifth thoracic vertebra the trachea divides into two **bronchi** of similar structure to that of the trachea, the left bronchus being a little longer than the right. These pass downwards and outwards to the lungs in which they branch repeatedly into smaller **bronchioles**, the cartilage in the walls of which is plate-like and completes the circumference of the tubes. The bronchioles terminate in **air sacs**.

The **lungs**, spongy in texture and bluntly cone-shaped, occupy the greater part of the cavity of the thorax, the heart lying between them and partly covered by them. Their bases rest on the diaphragm and their apices lie just above the clavicle. The left lung consists of two **lobes** and is longer and narrower than the right lung which has three lobes. The bronchi together with the pulmonary arteries and veins accompanied by lymphatic vessels and nerves enter the sides of the lungs and this point is known as the **root of the lung**. Each lobe is composed of a very large number of **lobules**,

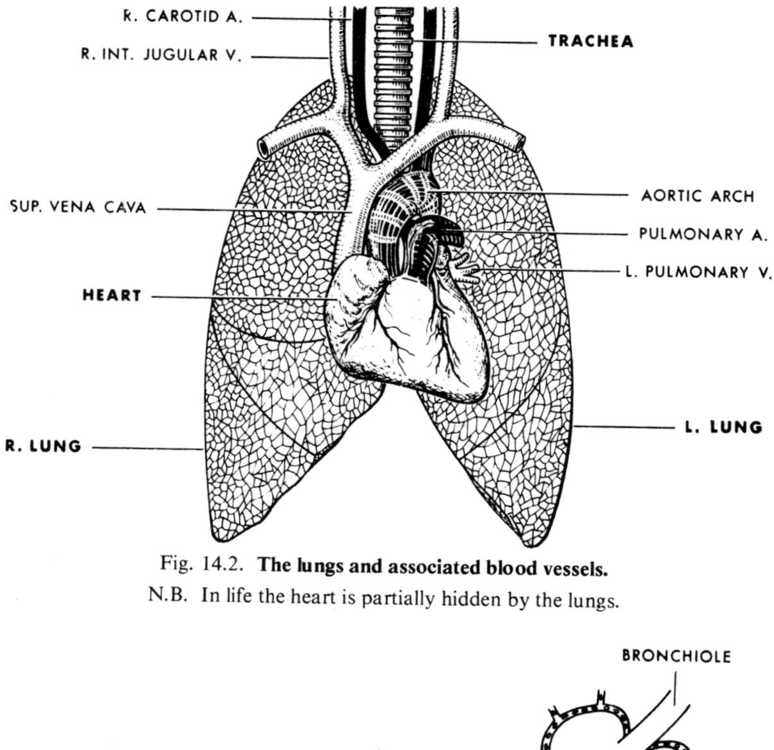

Fig. 14.2. **The lungs and associated blood vessels.**
N.B. In life the heart is partially hidden by the lungs.

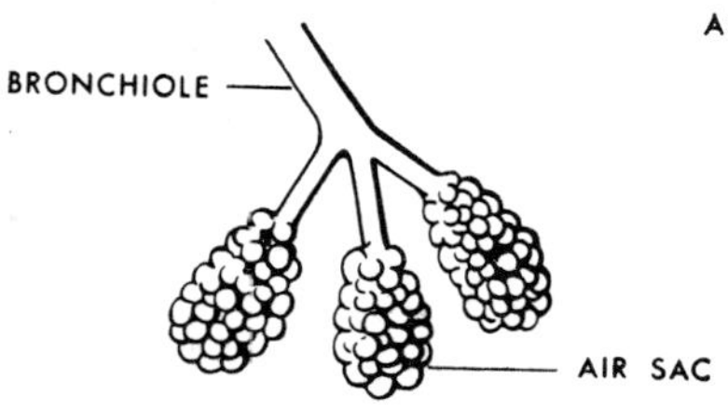

Fig. 14.3. **Termination of bronchiole in lung.**

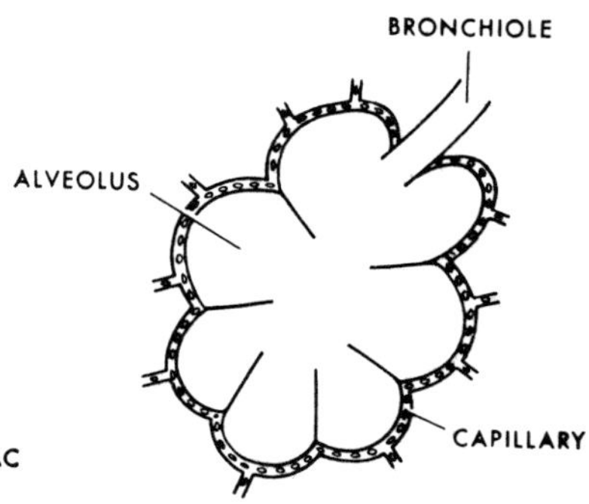

Fig. 14.4. **Structure of air sac.**

each consisting of the terminus of a bronchiole continuing into an air sac which is itself composed of many projecting **alveoli**. These are lined by a single layer of pavement epithelium. Branches of the pulmonary arteries reach the lobules and give rise to capillaries which form a plexus over the surface of the alveoli and run between them. These ultimately give rise to the pulmonary veins. Thus gaseous exchanges between the blood and the air sacs is facilitated. The lobules are held together by connective tissue. The lungs themselves contain a large amount of elastic tissue which is important in the process of breathing.

At this stage it might be as well if we consider the thorax as a whole. Its boundaries have already been seen but nothing will be lost by repeating them here. The sternum and costal cartilages form its ventral wall, the vertebral portions of the ribs and the vertebral column its dorsal wall, the diaphragm its floor and the lower end of the neck its roof. In it, apart from blood vessels, lymphatics and nerves, are the oesophagus, trachea, lungs and heart. The cavity is lined by a serous membrane, the **parietal pleura**, and continuous with this at the root of each lung is the **visceral pleura** which completely invests the lung so that each one is in

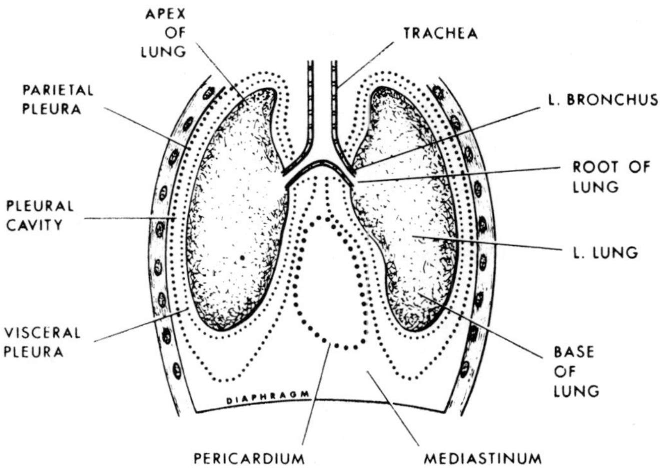

Fig. 14.5. **Diagram of thorax to show the pleurae.**
N.B. The parietal and visceral pleurae are in contact in life.

its own **pleural sac**. There is a narrow **pleural cavity** between the parietal and visceral pleura. The central portion of the thoracic cavity, known as the **mediastinum**, consists of the heart in its pericardium with the great vessels, oesophagus, trachea, nerves and so on. The mucous secretion of the pleurae enables them to glide over one another enabling smooth movement of the lungs to take place without friction. In young children the lungs are pinkish in colour but in adults they become a slate-grey and, in those living in smoky industrial districts, almost black.

BREATHING AND EXTERNAL RESPIRATION

There are, of course, two stages in breathing, inspiration in which air is taken into the lungs and expiration in which it is expelled from them. This is to make possible gaseous interchange between the air sacs in the lungs and the blood in the capillaries lining them. **Inspiration** is a muscular act in which the volume of the thoracic cavity is increased. The external intercostal and other respiratory muscles raise the rib cage upwards and forwards and the diaphragm, domed when at rest, becomes more flattened. This causes a decrease of pressure in the air in the lungs and air from outside therefore rushes in through the nostrils to fill them.

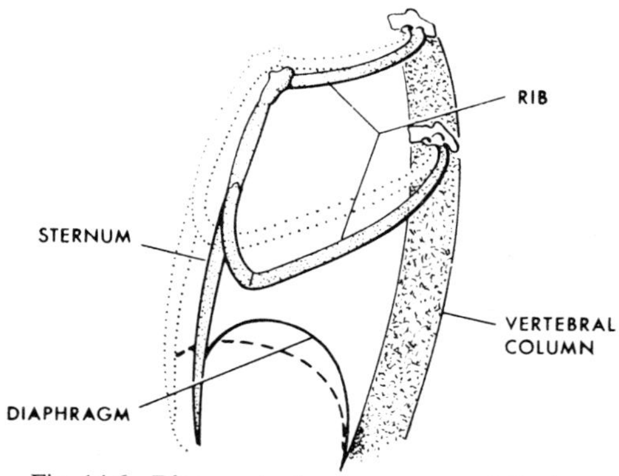

Fig. 14.6. **Diagram to show movement of diaphragm and rib cage during breathing.**

Expiration is not due to muscular contraction. The intercostal muscles relax and the rib cage is moved downwards and backwards while the diaphragm muscles relax causing it to become more domed. The effect of this, aided by the elasticity of the lungs, is to expel the air from them.

During inspiration the chest expands about six to eight cms. in adults and the normal number of inspirations and expirations per minute is 14 to 18, which is one to every four or five heart beats. This rate varies with age and is increased considerably during exercise when the muscles require more oxygen per minute. At birth the rate is 40–44 per minute. Breathing is entirely automatic and although the rate can be altered at will, the breath can be held for only a short time after which the rate automatically increases. This also results from oxygen deficiency and excess of carbon dioxide. Similarly forced breathing produces a retardation in the rate. Oxygen deficiency has little serious effect until the percentage in the air falls from its normal 21 per cent. to about

13 per cent. when loss of consciousness results and may lead to death by asphyxiation due to the tissues being deprived of oxygen. This is what happens in carbon monoxide poisoning when carboxyhaemoglobin is formed instead of oxyhaemoglobin, as already explained in Chapter IX. Excess of carbon dioxide in the air has little effect beyond increasing the rate of breathing until the percentage increases from its normal 0·1–0·3 per cent. to 5 or 6 per cent. when a general "muzziness" occurs and at around 10 per cent. loss of consciousness results. Anaesthetics, of course, produce a similar effect.

Depth of breathing varies according to the needs of the body. Under vigorous exercise it increases greatly and during sleep it is much less than it is during the day. Failure to breathe (*apnoea*) as the result of drowning, suffocation and other causes calls for immediate measures to restore it in order to maintain life. There are several methods of **artificial respiration**. In the well-known "Kiss of Life", the patient is placed on his back, his head pushed back, the chin held down to prevent the tongue from obstructing the passage and the nose pinched to put it out of action. Then the administrator takes a deep breath and places his lips firmly on the mouth of the patient and blows air into his lungs. This is repeated until normal breathing is restored or death is confirmed by a doctor. Other methods involve alternate pressure on the thorax to force air out of the lungs and the release of this pressure to allow air to enter. This is done either by the hands of the administrator or by his moving the arms of the patient. Detailed instructions for these procedures will be found in books on First-Aid.

Changes in the pressure of the air will also affect breathing. At high altitudes, where the pressure of the air is much lower than it is at sea level, the depth of breathing increases and every movement calls for greater exertion. Although the percentage of oxygen is the same as it is at sea level, the pressure is considerably less. At the top of Mount Everest, for example, it is only 250 mm. of mercury compared with 760 mm. at sea level. Anyone who has been up in high mountains abroad will have experienced this, particularly if he has been carried up by a cable car or mountain railway. With more gradual ascent, the effects are less marked. Headache and mountain sickness may follow unless he has taken the wise precaution of resting for a while before doing anything which calls for exertion. People living at high altitudes become acclimatised to the altered conditions, however. These symptoms are due to a deficiency of oxygen to the tissues. High altitude climbing and flying call for additional oxygen supplies. Increase of pressure will also have adverse affects unless suitable precautions are taken. Men working in caissons (diving bells) do so in a chamber from which entry of water is prevented by pumping air into the chamber at a higher pressure, seldom above four atmospheres. On leaving the chamber there is considerable risk, for the blood has been saturated with oxygen and nitrogen at this pressure and if this is suddenly reduced, bubbles of nitrogen collect in the blood and serious damage to tissues results, a disease known as "bends". Slow release from this high pressure in a decompression chamber avoids this danger. Short periods of work in the chamber are also necessary as this prevents the blood from becoming saturated with nitrogen.

Shortness of breath, known as *dyspnoea*, happens during exercise because, as the result of the extra activity, the carbon dioxide and lactic acid content of the blood increase. It is often noticed that continuation of the exercise produces what is called "second wind". This is because the blood circulation has by then had time to adapt itself to the changed conditions and the muscles are working more efficiently so that less carbon dioxide is produced. During severe exercise more oxygen is required than one can inhale in a given time and an oxygen debt is set up with an accumulation of lactic acid in the blood, as we saw when we were studying the physiology of muscular contraction. When resting after the exercise, this debt is recovered by increase of oxygen intake and the lactic acid is oxidised.

It is interesting to compare the oxygen requirement per minute under different conditions as shown in the following table:–

Sitting still	250 ml.
Standing still	330 ml.
Walking slowly ..	780 ml.
Walking at 3 m.p.h. ..	1065 ml.
Running	4175 ml.

The volume of air normally changed in each act of breathing is about 500 ml. and is known as **tidal air** but this is insufficient to completely fill the lungs. The volume over and above this which can be drawn into the lungs by the deepest possible inspiration is around 1,600 ml. and is called **complemental air**. Similarly the most forceful expiration will expel a further 1,600 ml. over and above the tidal air and this is referred to as **supplemental air**. Even so, some remains in the lungs and it is impossible to expel all the air from them, though some of it may be forced out when a man receives an abdominal blow. This **residual air** amounts to 1,000 ml. or more. Finally if the deepest possible inspiration is followed by the most forceful expiration possible, it will be found that 3,500–4,000 ml. of air are expelled. This sum total of the tidal, complemental and supplemental air is called the **vital capacity**. The 4,000 ml. mark has been known to be considerably exceeded in certain cases and a high vital capacity is required of civil and military air crews. As already stated during vigorous exercise depth of breathing increases considerably and the tidal air may be doubled or trebled. Again, during sleep the depth of breathing decreases and the tidal air may be reduced to as little as a third.

The various measurements given above are obtained by means of an instrument called a spirometer, and it should be noted that the quantities vary according to physical development in different individuals while vital capacity, of course, also varies with exercise, and trained athletes manage to increase it very considerably.

A **spirometer** consists of a cylindrical tank inverted in a similarly shaped tank of water and balanced by a weight to which it is attached by a cord passing over pulleys. A mouthpiece is joined to a tube sealed through the side near the bottom of the outer tank and which leads into the air space above the water in the inner tank. A scale indicates the position of the inner tank and records the volume of air. By making appropriate respiratory move ments through the mouth-piece recordings can be made of the various physiological respiratory measurements described above. A portable form is made for use in hospital wards.

Breath sounds can be heard by placing the ear against someone's chest, or, better, by using a **stethoscope**. In healthy breathing characteristic sounds are made as the air enters and leaves the lungs and these vary in different regions. The character of the sounds is altered in diseases of the lungs and by placing the stethoscope in different positions on the chest, the part of the lung affected can be located.

Now let us consider **external respiration**, the gaseous interchange between the air in the alveoli of the lungs and the blood in the capillaries which line them. The total surface area of these alveoli is enormous, far greater in fact, than the complete area of the skin covering the body, and it is over this surface that the gaseous exchange takes place. The concentration of oxygen in the alveoli is higher than that in the blood which has reached the lungs in the pulmonary artery and a concentration gradient is thus set up so that oxygen diffuses into the blood. This oxygen combines with the haemoglobin in the red corpuscles to form oxyhaemoglobin. Similarly the concentration of carbon dioxide is higher in the blood reaching the lungs than it is in the alveolar air and as a result of the concentration gradient thus produced, carbon dioxide diffuses out of the blood into the alveoli. The blood leaving the lungs by the pulmonary vein therefore contains a higher percentage of oxygen and a lower percentage of carbon dioxide and the rate of breathing is adjusted so that the percentage of carbon dioxide in the alveolar air is kept at a level of 5–6 per cent. If this increases, as in vigorous exercise, the rate of breathing is raised and so the oxygen intake is increased to meet the demand of the tissues. Similarly the forceful breathing which follows such exercise removes large quantities of carbon dioxide from the alveoli and the carbon dioxide content of the blood is lowered. The rate of breathing is then reduced until the alveolar air has once again reached its normal level. The volume of nitrogen in the alveolar air remains constant. The following table shows the differences in composition between inspired and expired air:—

	Inspired Air	Expired Air
Oxygen	20·96 ml.	16·40 ml.
Carbon dioxide	0·04 ml.	4·10 ml.
Nitrogen	79·00 ml.	79·50 ml.

Thus 4·56 ml. of oxygen are removed from every 100 ml. of air inspired and 4·06 ml. of carbon dioxide are added to it.

The average composition of alveolar air is oxygen 14·1 per cent. and carbon dioxide 5·5 per cent. but as gaseous exchange is going on continuously the percentage of oxygen will be at its lowest at the beginning of inspiration and highest at the end while the percentage of carbon dioxide will be the reverse. The ratio of the volume of carbon dioxide expired to the volume of oxygen inspired is known as the **respiratory quotient** and taking the figures given above this works out as follows:—

$$\frac{CO_2}{O_2} = \frac{4{\cdot}06 \text{ ml.}}{4{\cdot}56 \text{ ml.}} = 0{\cdot}89 \text{ (approx.)}$$

However, the figure, which is always less than unity, does vary, depending on the extent to which the carbohydrate and fat content of the food is used in respiration.

Expired air is saturated with water vapour whereas the inspired air normally has a much lower content. Furthermore the air which leaves the lungs is at body temperature whereas inspired air is, of course, at the temperature of the atmosphere which, fortunately, is normally much lower. The question of temperature maintenance of the body will be dealt with in a later chapter.

TISSUE RESPIRATION

As we already know the blood in the arteries takes oxygen to the tissues and the carbon dioxide is carried away from them in the veins, the oxygen being carried as oxyhaemoglobin in the red corpuscles and the carbon dioxide in the plasma, mostly in the form of bicarbonates. These bicarbonates decompose liberating the carbon dioxide on reaching the lungs. The following comparison of the percentage gaseous content of arterial and venous blood gives adequate support to this statement.

	Arterial Blood	Venous Blood	Change in Vol.
Oxygen	18·5 ml.	12 ml.	– 6·5 ml.
Carbon dioxide	50·0 ml.	56·0 ml.	+ 6 ml.
Nitrogen	2·0 ml.	2·0 ml.	nil.

When the blood reaches the tissues a concentration gradient is set up since the concentration of oxygen in the tissues is lower than that in the blood. Consequently oxygen diffuses from the blood into the tissues. In a similar way, the concentration of carbon dioxide in the tissues is higher than that in the blood and again a concentration gradient is set up so that the carbon dioxide diffuses from the tissues into the blood. It will be remembered that lymph is the intervening medium for this exchange. In the tissues the oxygen oxidises glucose to carbon dioxide and water, thus releasing energy to the cells. The process is a complicated one and is dealt with in some detail in Chapter XIX (p. 225 seq.) which is concerned with molecular biology. It may, however, be summarised by the equation

$$C_6H_{12}O_6 + 6O_2 = 6CO_2 + 6H_2O + \text{Energy}$$

Special Respiratory Acts

Coughing is a reflex action in which a short, deep inspiration is followed by a forcible obstructed expiration, the glottis being momentarily closed, and the noise is made by the vocal cords. The force of the expiration serves the purpose of expelling mucus, phlegm and other obstructive matter from the air passages. **Sneezing** differs from coughing in that the air is expelled mainly through the nose. The noise, produced by **snoring**, due to sleeping with the mouth open as a rule, is caused by vibration of the soft palate. **Yawning** is a deep inspiratory act, often combined with stretching of the limbs, and is nature's way of rectifying the condition of the blood due to inactivity. **Sighing**, which often has emotional origins, is similar and there is usually an audible exhalation. **Sobbing** consists of a series of inspirations during which the glottis is partially closed. **Hiccoughing** results from gastric irritation and is caused by a sudden, uncontrollable contraction of the diaphragm which results in an inspiration which is suddenly stopped by the glottis being closed. **Laughing** consists of a deep inhalation succeeded by short, rapid exhalations with vibrations of the vocal cords. There is also vigorous movement of the diaphragm when the laughter is loud and hearty. **Crying**, the result of emotional disturbance, is a long deep inhalation followed by short, sharp exhalations accompanied by facial distortions and tears, though of course, tears alone may be produced by the inhalation of irritant odours such as that from onions.

THE LARYNX, VOICE AND SPEECH

The **larynx** is broader above than it is below where it is more rounded in shape. It is composed of a shield-shaped **thyroid cartilage** formed by two curved wings which unite in front. From the back of each wing two cartilages arise, a **superior cornu** at the upper end which runs upwards and an **inferior cornu** which hinges on a ring-shaped **cricoid cartilage** at the base on which the thyroid cartilage can move upwards and downwards. It is narrow in front and there is a space between it and the thyroid cartilage which is covered by a fibrous membrane. At the back of the cricoid cartilage are the **arytenoid cartilages** to which the two internally placed vocal cords and the muscles which open and close the glottis are joined. The other ends of the vocal cords are connected to the thyroid cartilage. Thus they run across the larynx. These cords are composed of yellow elastic tissues below and are called the **true vocal cords** to distinguish them from the white fibrous ridge which forms the **false vocal cords**. The entire larynx is lined by mucous membrane covered with ciliated epithelium except over the epiglottis and vocal cords where it is stratified.

When the true vocal cords vibrate, they produce sounds and by variation of their tension and variation in the size of the glottis, notes of different pitch can be produced. When the tension in the cords is high and they are consequently tightly stretched and the glottis is narrow, notes of high pitch are produced and when the tension is low, resulting in the cords being slacker and the glottis is wide, the notes are of low pitch. The variation in tension of the cords and the size of the glottis is effected by voluntary muscles. The vibration is set up by air expired from the lungs. Resonance or reinforcement of sound is provided by the cavities of the larynx, the pharynx, the mouth and the nasal passages. This influences the timbre or quality of the sounds and so the quality of voice. The range of the sounds which can be produced is about two and a half octaves and the position of these in the musical scale varies with the sexes and with different individuals. Loudness of the sounds produced depends on the amplitude or extent of displacement of the vocal cords when vibrating and this varies according to the force of the expiratory air from the lungs. The vocal cords of men are longer than those of women and this gives the former a voice of lower pitch. The deepening or "breaking" of the voice of the male at puberty is due to the larynx enlarging at that age.

Speech is produced by altering the size and shape, and therefore the resonance, of the resonating cavities and in this the tongue, teeth and lips all play their part. Whispering occurs when the resonating cavities alone are in action, the vocal cords being out of use.

In February 1969 a larynx transplant was performed on a sixty-two year old Belgian in Ghent University Hospital. The patient was able to speak normally after a few days though with a voice which was different from his former one, of course.

Disorders and Diseases of the Respiratory System

Nasal catarrh—a very common complaint in this country, particularly in cold damp weather. It manifests itself by excessive mucus secreting through the nose and sneezing. It is often the prelude to or accompanies other diseases.

The **Common cold**—usually the result of a virus infection and should be distinguished from a chill, though the latter may provide suitable conditions for the virus to thrive and a cold easily follows a chill. Its extreme infectiveness is due to the viruses being transmitted through the air by sneezing. This is known as droplet infection. Catarrh accompanies the cold and this may pass to the throat and along the Eustachian tubes to the middle ear causing earache. If the infection continues further along the bronchial tubes into the lungs, bronchitis may set in.

Emphysema—literally means the abnormal presence of air and is mainly applied to the lungs when, as a result of the breakdown of the walls of the alveoli and the consequent production of large sacs, the oxygen absorbing surface of the lungs is seriously reduced.

Bronchitis—one of the commonest diseases in this country, and there is a higher mortality rate from chronic bronchitis here than in any other country in the world, is due to inflammation of the mucous membrane of the bronchial tubes. There is a painful cough and breathing is difficult both in acute and chronic bronchitis. It is caused by a bacterium. Deaths in 1974 from bronchitis and emphysema totalled 20,409. Acute bronchitis accounted for 1,688 deaths

Laryngitis and **pharyngitis**—inflammation of the mucous membranes of the larynx and pharynx respectively. They may be preceded by catarrh and both the breathing and the voice are affected in laryngitis while pharyngitis is what we normally call sore throat.

Tonsilitis—inflammation of the tonsils and the pharynx can also be affected. The tonsils become enlarged and pain is experienced when swallowing. The cause is usually bacterial infection.

Whooping cough—a common disease of children caused by bacterial infection. The "whoop" which gives it its name occurs during inspiration.

Diphtheria—a very infectious disease, caused by a bacterium, *Corynebacterium diphtheriae*, which starts with the inflammation of the throat. Other symptoms follow and if anti-diphtheritic serum is not administered, post-diphtheritic paralysis may follow. This is not permanent but it takes time for recovery. The well-known Schick test is used to ascertain whether people are subject to the disease and immunisation against it is now practised.

Croup—occurs when the entrance to the larynx is swollen and constricted and breathing is therefore affected.

Asthma—attacks may be brought on by a number of agents such as pollen from grasses or by house dust, feathers, the fur of cats or dogs' hair to which these sufferers are allergic. Breathing is extremely difficult owing to constriction of the bronchial tubes.

Hay fever—occurs in people whose mucous membranes are extremely sensitive to the pollen of grasses and therefore occurs only during the Summer. Considerable nasal discharge and the usual symptoms of the common cold develop. Antihistamines are often effective in relieving the symptoms. People subject to hay fever should obviously avoid being in a grass field.

Pneumonia—of which there are several forms, is an inflammation of the lung tissues and the commonest form is caused by a bacterium, *Pneumococcus.* High temperature, rapid, shallow breathing accompanied by cough and pain in the chest are symptoms. The coughing results in the expectoration of a thick, brown fluid. Sulphonamide drugs and antibiotic treatment are now used with effect. Pneumonia is now often produced by virus infection.

Pleurisy—inflammation of the pleurae and may result from pneumonia and other more serious diseases such as tuberculosis.

Empyema—a disease in which pus collects in the pleural cavity.

This, too, may follow pneumonia and it may be necessary to resort to surgical means to drain off the pus.

Pneumoconiosis—the term used for disabling lung diseases caused by inhalation of dusts of various kinds. **Asbestosis** occurs amongst workers in factories where asbestos is in constant use in manufacture and who inhale asbestos fibres. These cause considerable damage to lung tissue and may result in lung cancer. **Silicosis** is a hazard of workers in industries in which considerable amounts of silica dust are inhaled as in potteries and other industries where this is unavoidable. **Coal miners' pneumoconiosis** is caused by the inhalation of coal dust and this is a very serious hazard in the industry. In *simple pneumoconiosis* the dust diffuses through the walls of the alveoli, kills the cells and forms clusters but it causes little disability if exposure to coal dust ceases. *Complex pneumoconiosis*, on the other hand, is a progressive form of the disease even after exposure to coal dust ceases and results in premature retirement. The dust clusters enlarge considerably and eventually large parts of the lungs are affected and breathing becomes extremely difficult. The condition is probably initiated by tubercular infection on top of simple pneumoconiosis. In 1974 the number of deaths from pneumoconiosis was 403. Several men working in the zinc industry were seriously ill with arsenic poisoning in the Autumn of 1975, this element being an impurity in crude zinc.

Precautions against all forms of pneumoconiosis must be taken in all the industries affected and regular environmental and biological monitoring is essential.

Pulmonary tuberculosis or **phthisis**—commonly called **consumption**, is caused by *Bacillus tuberculosis (Mycobacterium tuberculosis)*. Once regarded as incurable it inevitably led to death but modern methods of therapy have completely changed the picture. The disease is extremely infectious and may easily be caught from the sputum of an infected person in which enormous numbers of the bacteria live and which get into the air by coughing or into the dust of the air from dried sputum. It may also be acquired from infected milk but the bovine bacillus affects children more than adults. Milk must therefore be adequately protected by pasteurisation. Receptacles used by infected persons must on no account be used by healthy people. The name of the disease is derived from the fact that tubercles develop in the lungs, thus tending to obliterate the alveoli. Later these coalesce and the mass breaks down and the particles are discharged in the sputum. As a result cavities develop in the lungs. The disease is now curable and the mortality rate from it has fallen enormously in the last few years. In 1957, 4,784 people died of pulmonary tuberculosis, but by 1960 the number had dropped to 2,354, a decrease of about 50 per cent. In 1974 this fell still further to 776.

Lung cancer on the other hand like other forms of cancer, has been showing a marked increase in mortality rate in recent years. A **cancer** or **carcinoma** is a malignant **neoplasm***, a non-malignant neoplasm being a **tumour**. Although a cancer may be removed surgically, there is always the danger that the cells may be carried by the blood to other parts of the body where they initiate further cancerous growths. Cancer accounted for 121,711 deaths in 1974 in England and Wales, an increase of 14,563 in fifteen years, and of these 33,028 were due to lung, bronchus and trachea cancer, an increase of 11,545 over the same period. It has been proved that a minute quantity of carcinogenic (cancer producing) agents are present in cigarette smoke. The incidence of lung cancer has been shown to increase in proportion to the number of cigarettes smoked and as most cigarette smokers inhale, these dangerous substances get into the lungs. The death rate amongst cigarette smokers from lung cancer in this country is three or four times that among pipe smokers and about ten times that among cigar smokers. Pipe and cigar smoking is therefore very much safer but in non-smokers the death rate from this disease is only about a thirteenth of those among cigarette smokers. These facts speak for themselves. Couple them with the knowledge that of the 563,863 deaths from natural causes in Great Britain in 1974 over 33,028 were due to lung, bronchus and trachea cancer and you cannot fail to see that the matter calls for serious consideration.

In addition to cancer, cigarette smoking can cause *chronic bronchitis* and *emphysema* in which the walls of the alveoli break down thus reducing the surface area of the lungs and consequently their oxygen absorbing capacity. Unfortunately in this country there are few laws which forbid smoking in public places apart from those made by local authorities and transport bodies, though bus drivers, police, waiters in restaurants and shop assistants selling unpackaged food are forbidden by law to smoke when on duty. Many foodshops and other kinds of shops, too, have "no smoking" notices posted. Smoking should certainly be forbidden in telephone kiosks, doctors' and dentists' waiting rooms and other somewhat confined places. Although it remains to be proved that non-smokers are adversely affected to the extent of contracting disease, by a smoky atmosphere, the fact remains that they find such an atmosphere highly objectionable and they have as much right to consideration as smokers.

* **Neoplasm**, literally "new growth"; refers to useless, generally harmful, proliferation of cells.

The World Health Organisation has urged that smoking should be forbidden by law in certain public places, including hospitals and public transport. It has also stated that the advertising of cigarettes should be curtailed and ultimately forbidden.

Cigarette makers have done a great deal of research into the effect of cigarette smoking and of course, they cannot deny its ill effects in the light of authoritative reports on the subject. A cigarette with a lower tar content and therefore less carcinogenic matter, has been produced by them as a result of these investigations. All cigarette packets carry a Government Health Warning and the Tar Yield Content has now been added. The morals to be drawn from these facts: *do not inhale tobacco smoke into the lungs,* smoke a pipe or cigars in preference to cigarettes or, better still, do not smoke at all.

Cancer is a disease which has been known to man for a very long time but the number of deaths caused by it have increased out of all proportion to the increase in the population. In this disease cells multiply in a most disorderly and uncontrolled fashion forming malignant growths to the detriment of the normal tissue in which they are growing. Furthermore they can give rise to further cancerous growths in other parts of the body. The cause remains unknown at present and though a cancer virus has been discovered in monkeys, there is no evidence as yet that such is the case in man. Research is being carried out with a new drug, *metho-dichlorophen,* in the treatment of terminal malignant disease. Out of five cases who responded, four were suffering from lung cancer. Apart from surgery, **radiotherapy** with **radium** and **X-rays** have been used in the treatment of cancer but **cobalt-ray** treatment now largely supersedes them. Radium was first isolated by **M. and Mdme. Curie** in 1898 and radioactivity was discovered by **Antoine Becquerel** in 1890. A German physicist, **Wilhelm Röntgen**, was the discoverer of X-rays and these are not only used in treatment but also, of course, in diagnosis of many diseases.

An entirely new and different method of diagnosis, much safer than X-rays known as **ultra-sonography** has, however, come into use in recent years. In this technique the body is bombarded by sound waves of very high frequency, inaudible to the human ear, and the echoes from them are directed on to a screen where a picture of the internal organs can be examined. The method is quite painless and so far there has been no evidence of any ill effects. There are about a hundred of these machines in use in this country alone and thousands of patients are examined every year. There is still much to be learned about any improvements to be made in the machines and research is being made in several countries. Recently a machine called the **body scanner** has been invented by Mr. Godfrey Hounsfield, F.R.S. which enables soft tissues to be seen. These, of course, are not visible under ordinary X-rays. Pencil-thin beams of X-rays and a computer are used to produce a series of pictures in very thin slices. The picture is taken in 20 seconds and a few minutes later the computer has done its work. It is hoped that internal ulcers, cysts, tumours, cancers, gallstones and so on will become visible by this method and enable their presence to be known without surgical investigation.

PRACTICAL WORK

Anatomy and Histology of the Respiratory System

1. Examine a longitudinal section through the head and neck of a rabbit or a model of a similar section of the human being (if available) and identify the oro-pharynx, naso-pharynx, nasal passages, epiglottis, larynx and trachea.
2. Examine a dissection of the neck and thorax of a rabbit and examine the larynx, trachea, bronchi, lungs, pulmonary artery and vein and the diaphragm. Insert a blowpipe into the trachea and gently blow up the lungs. Then release the pressure and note their collapse.
3. Examine a microscopical slide of a **T.S. of the lung**. Look for the alveoli and for sections through bronchioles. You should also see the blood capillaries.
4. Examine a microscopical slide of a **T.S. of a lung infected**

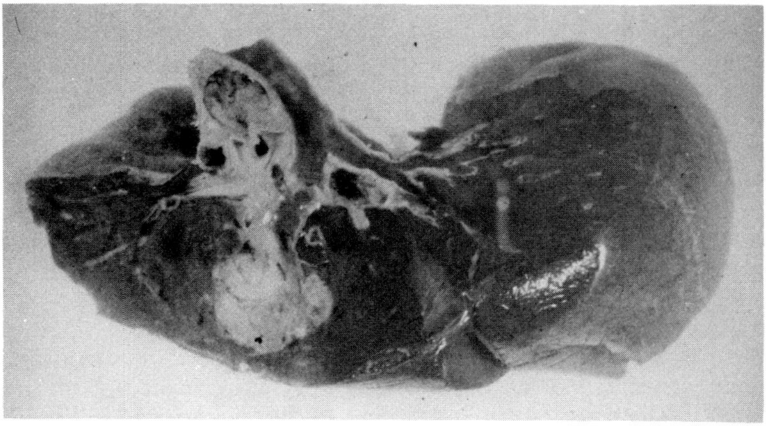

Fig. 14.7. **Photograph of left lung showing a cancer growing in the centre of the upper lobe and its migration in a lymph node at the hilus.**

(By courtesy of Dr. B. E. Heard M.D., F.R.C.P., F.R.C.Path., Consultant Pathologist, Brompton Hospital, London and the Royal Postgraduate Medical School, Hammersmith Hospital, London.)

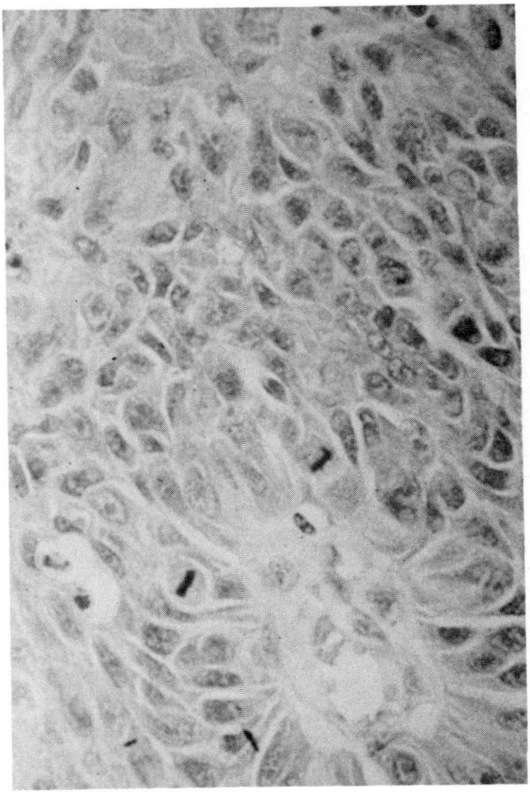

Fig. 14.8. **Micrograph of a cancer of a bronchus.** Mitotic figures can be seen in the cancer cells.
(By courtesy of Dr. B. E. Heard M.D., F.R.C.P., F.R.C.Path., Consultant Pathologist, Brompton Hospital, London.)

with lung cancer. Note the thickening of the epithelium and the loss of cilia and the cancer cells growing down into the tissue or which has already done so, obliterating all the alveoli.

5. The **Bioset "Smoking and Health"** shows up very clearly the structure of a normal healthy lung and increasing stages in the development of the lung cancer.

The Physiology of Breathing

Breathing Rates

1. Count the number of respirations and take the pulse of a fellow student over a quarter of a minute (i) at rest (ii) after running round for a couple of minutes (iii) after holding the breath for as long as possible. Compare these rates and work out the ratio of respirations to heart beats.

2. Measure the expansion of the chest of a fellow-student breathing normally by means of a tape-measure placed just beneath

the armpits (i) when at rest (ii) after exercise (iii) after a very deep inspiration.

Movements of the Diaphragm

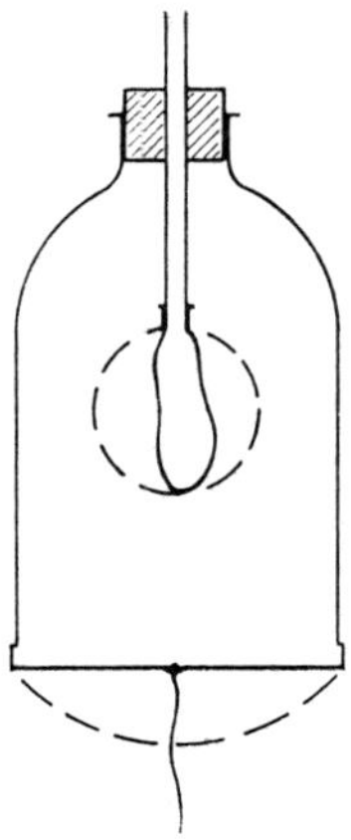

Fig. 14.9. **Experiment to illustrate the movements of the diaphragm.**

The movement of the diaphragm during respiration may be demonstrated by the following experiment.

Fit the neck of a bell jar with a one-holed cork through which is passed a piece of glass tubing, on the lower end of which a small balloon has been securely tied. Cover the wide mouth of the bell-jar with a piece of sheet rubber and tie in on securely so that there is no air leak. Fix a piece of string or tape to the centre of the rubber sheet. The bell-jar represents the wall of the thorax (immovable in this case), the rubber sheet the diaphragm and the balloon the lungs. Lower the "diaphragm" by pulling the string attached to it and note the effect on the "lungs". Now release the string and again observe the effect on the "lungs".

Breathing Sounds

Use a stethoscope placed on the chest and back of the thorax of a colleague and listen to the sounds of breathing during inspiration and expiration.

To Show that Carbon Dioxide is given out in Breathing

Pour some lime water into two conical flasks, A and B, and fit each flask with a two-holed cork through which pass a long and a short right angle tube. Connect the short tube of A and the long tube of B to the cross pieces of a T-tube and fit clips, a and b, on the rubber tubing. To the other limb of the T-tube fit a short piece of glass tubing to serve as a mouthpiece.

With clip a open and clip b closed, draw air through flask A by means of the mouthpiece. Then close clip a and open clip b and blow through the mouthpiece so that the expired air passes through flask B. Repeat several times and observe the effect on the lime water in the two flasks.

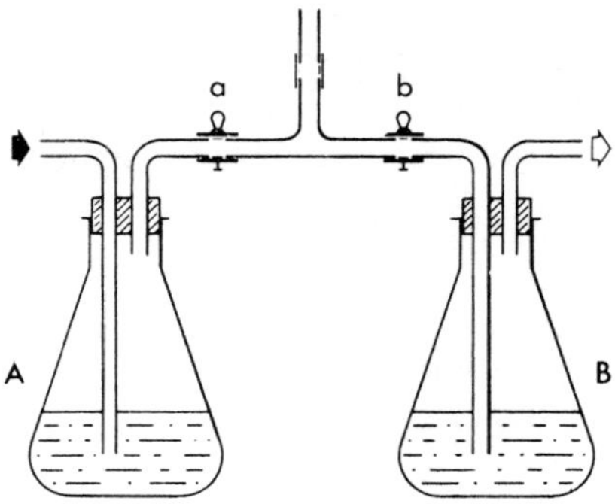

Fig. 14.10. **Experiment to show that carbon dioxide is expired.**

To Demonstrate that Water Vapour is Contained in Exhaled Air

Breath on to the surface of a mirror and note that water condenses on it.

To Measure the Tidal Air, Complemental Air, Supplemental Air and Vital Capacity

If a spirometer is not available these measurements can be found by the methods described below.

1. To Measure the Tidal Air and the Complemental Air:

Take a long piece of rubber tubing and push one end of it well down into a Winchester bottle. Then invert the bottle in water in the sink or a deep trough and breathe in air from the vessel

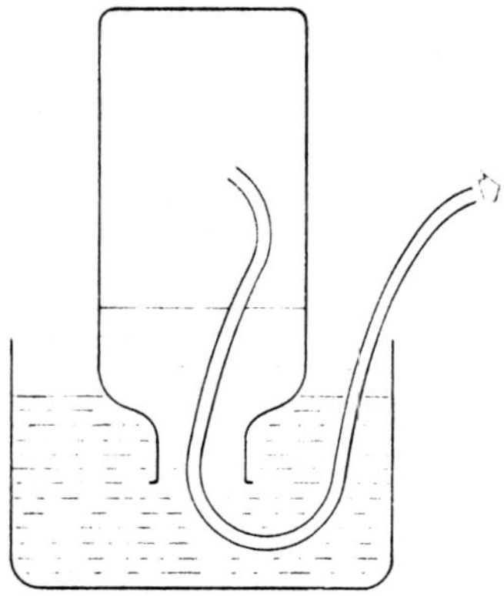

Fig. 14.11. **Apparatus for measuring tidal and complemental air.**

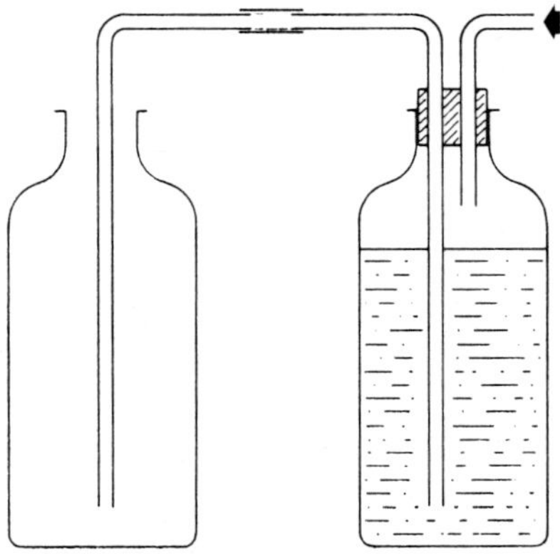

Fig. 14.12. **Apparatus for measuring vital capacity.**

through the rubber tubing. This will cause water to rise in the bottle. Now breathe out through the tube and the air entering the bottle will expel the water again. Do this several times and after the last inspiration remove the rubber tubing from the bottle, insert a cork and lift the bottle out of the water. Then pour the contents of the bottle into a measuring jar. The volume of water is equal to the *tidal air*.

Repeat the experiment with the deepest possible inspiration to find the *complemental air*. Record these readings.

2. To Measure the Vital Capacity and to calculate the Supplemental Air:

Fit up a large Winchester bottle as a siphon bottle and almost completely fill it with water. The short right-angled tube will be used as a mouthpiece and a long piece of glass tubing should be fitted to the long right-angled tube and placed in another Winchester. After practising deep inspirations and expirations a few times, fill the lungs as completely as possible and force this air to the fullest extent of your capacity through the mouthpiece of the siphon bottle. Then measure the volume of this water expelled in a measuring jar. Repeat the experiment at least twice and record the highest reading. This represents the *Vital Capacity*.

Add together the tidal air and the complemental air from experiment 1 and subtract the total from the vital capacity. This will give the *supplemental air*.

Chapter XV

EXCRETION AND THE URINARY SYSTEM

Excretion is the process whereby the waste-products of metabolism are eliminated from the body. We have already seen that the skin is excretory to a certain extent but the organs chiefly concerned in this process are the kidneys which, with the structures associated with them, constitute the **urinary system.**

The two **kidneys** are dark red, bean-shaped organs, their ventral surfaces being convex and each about 11·5 cm. in length, 6·3 cm. in width and 2·54 cm. in thickness and enclosed in a mass of fat. They are situated, one on each side of the vertebral column on the dorsal body wall alongside the first three lumbar vertebrae, the left kidney being slightly higher than the right. They are covered by peritoneum and therefore lie outside the coelom. The concave edge of each kidney is directed inwards and in this region is a fissure, the **hilum,** where the renal artery enters the organ and the renal vein leaves it. Behind these blood vessels the duct from the kidney, the **ureter,** leaves it and passes downwards and inwards. The two ureters, which are about 45·7 cm. in length lead to the **urinary bladder** which they enter obliquely from below. Each consists of an outer fibrous coat then an outer layer of circular muscle and an inner layer of longitudinal muscle, both unstriated, and an innermost coat composed of transitional epithelium. The **urinary**

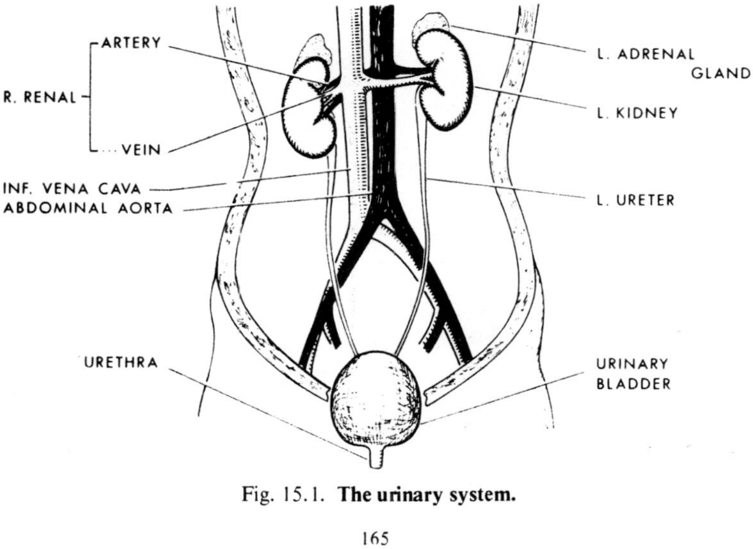

Fig. 15.1. **The urinary system.**

bladder, partly covered by peritoneum, lies in front of the rectum in the pelvic cavity but it expands considerably when filled with urine. Its wall consists of the coats similar to those of which the ureters are composed but the muscular layer is thicker, particularly the part containing the circular fibres. Leading from what is called the **neck** of the bladder to the exterior is the **urethra.** This is longer in the male than it is in the female, being around 18 cm. in the former and only about 3·8 cm. in the latter. In the male it traverses the penis and, for part of its length, also serves as a genital duct whereas in the female it is purely urinary in function, the genital duct or vagina being separate. In the female it opens in front of the vaginal opening in the vulva by the **meatus urinatus.** The urethra consists of three coats, an external muscular coat composed of circular and longitudinal fibres, a submucous coat of areolar tissue and a mucous membrane.

We must now examine the internal structure of the kidneys. If a longitudinal section through the centre of the organ is examined, it will be seen to be enclosed in a fibrous **capsule.** Inside this is the **cortex,** 1·27 cm. in thickness and containing a very large number of minute red specks. These are known as **Malpighian corpuscles.** The

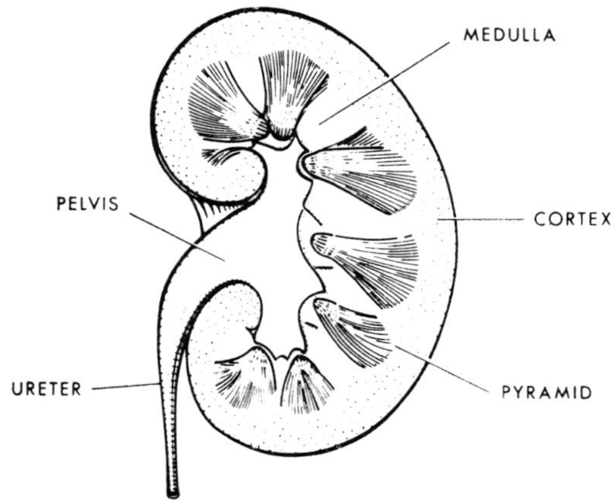

Fig. 15.2. **L.S. kidney.**

centre of the kidney is occupied by the **medulla** composed of a mass of minute **uriniferous tubules,** mostly lined by cubical epithelium (best seen in a transverse section) and bunched together in conical masses called **pyramids,** the tubules opening by **renal papillae** on their apices into a funnel shaped extension of the ureter into the kidney known as the **pelvis**. There is about a dozen of these pyramids. The Malpighian corpuscles and tubules, which are continuous with them, are known

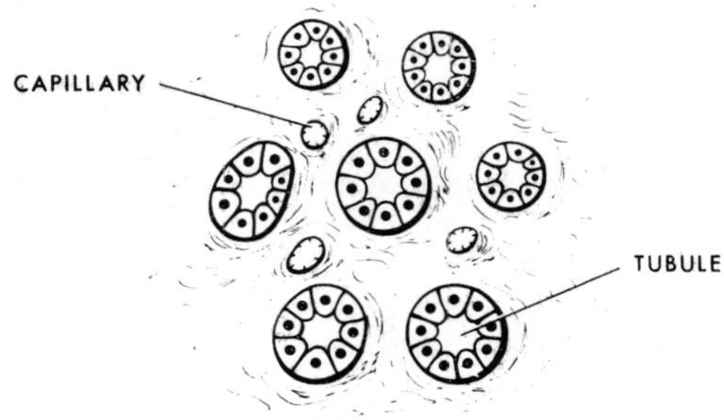

Fig. 15.3. **Microscopical transverse section through medulla of kidney showing uriniferous tubules.**

as **nephrons**, and it is estimated that there are about a million of them in the kidney.

The **Malpighian body** (or **corpuscle**) is a blindly ending extension of the tubule and consists of a spherical **Bowman's capsule** which encloses a tuft of blood capillaries known as a **glomerulus**, the

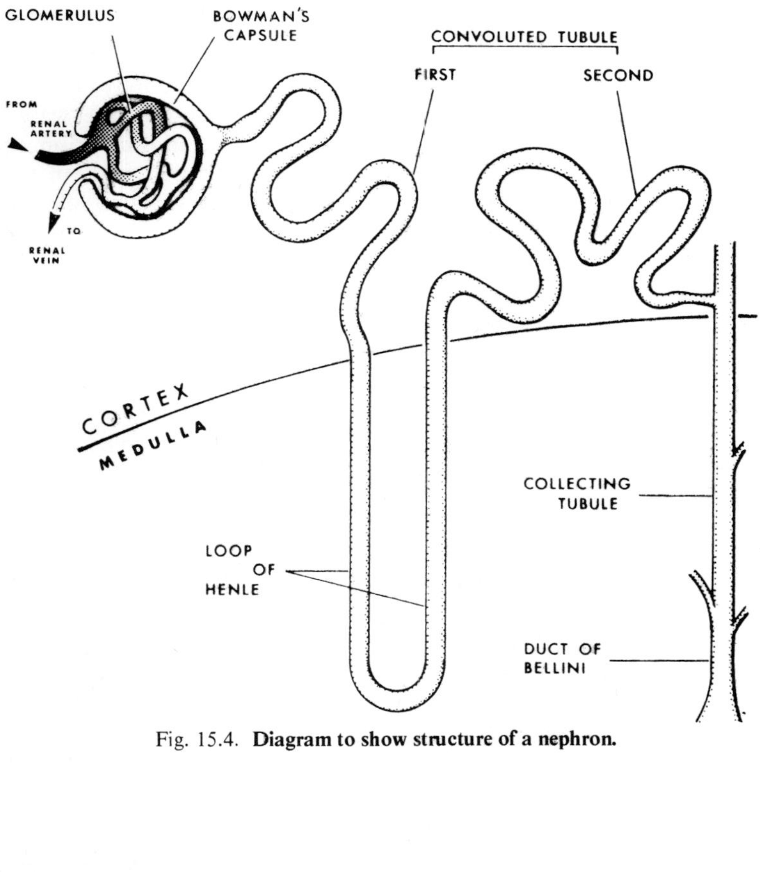

Fig. 15.4. **Diagram to show structure of a nephron.**

afferent vessels into which arise from branches of the renal artery. On leaving the capsule, the tubule is very convoluted, the **first convoluted tubule,** in the cortex. It then passes as a straight tube into the medulla where it loops back like a letter U, this portion being called the **loop of Henle.** After this the tubule re-enters the cortex and forms the **second convoluted tubule** after which it straightens again and is subsequently joined by similar tubules from other nephrons forming a **collecting tubule.** This runs to the apex of the pyramid where it forms the **duct of Bellini** and opens by the renal papilla as already explained. On leaving the glomerulus, the efferent vessels give rise to capillaries which supply the venules from which the renal vein is ultimately formed.

The blood entering the kidney by the renal artery and therefore the glomerulus by afferent vessels contains in addition to the normal constituents of its plasma, **urea, uric acid, hippuric acid, creatinine** and some **ammonium salts,** all of which are excretory products. These diffuse out of the blood in the glomerular capillaries into the capsule together with water, glucose and mineral salts. The protein and haemoglobin molecules are too large to do so. The diffusion is brought about by the fact that the blood pressure in the afferent vessels of the glomeruli is higher than it is when it leaves them. Substances which are beneficial to the body, namely all the glucose, some of the mineral salts and most of the water, pass back into the blood from the tubules. This **ultrafiltration** or **selective re-absorption,** is controlled according to the needs of the body, while the excretory substances pass on in an aqueous solution called **urine.** This passes into the pelvis of the kidney and so by the ureters in the bladder for storage.

Urine is a pale yellowliquid with a specific gravity of 1·015–1·025. Its colour is due to the pigment **urobilin** which has its origin as a breakdown product from worn-out red corpuscles. The specific gravity varies with the amount of water loss by sweating and other means. Urine contains 4 per cent. of solids of which about 2·0 per cent. is urea. This, it will be remembered, is formed in the liver in the process of deamination. The other organic substances, amounting to around 0·8 per cent., are uric acid, hippuric acid (both in the form of their salts, urates and hippurates), and creatinine. The mineral salts are chlorides and sulphates of sodium, potassium, calcium and magnesium. The most abundant constituents are urea and sodium chloride, the latter forming about 0·25 per cent. of the solids, and, of course, water of which there is 95 per cent. A small amount of ammonia is also present, some of which reaches the kidney direct but the bulk of it is derived from the breakdown of urea in the kidney. The uric acid, hippuric acid and creatinine are also nitrogenous compounds. The first two are present only in small quantities but the creatinine is, after urea, the most abundant nitrogeneous waste in the urine. It is derived from muscular metabolism. The reaction of urine is normally acid owing to the presence of acid salts, chiefly acid sodium phosphate but if the diet contains large quantities of vegetables and fruit it may become alkaline. It should be noted that there is no glucose in normal urine. It is always present, a condition known as *glucosuria*, in the disease *diabetes mellitus* but it does not necessarily follow that the cause of glucosuria is diabetes.

The quantity of urine passed by an adult on a normal diet is about 1,500 ml. in twenty-four hours but this varies with age, according to the amount of fluid taken in and the amount of water lost by other means. When sweating increases, for example in hot weather, less urine is produced, the reverse being the case when it is reduced. Furthermore the amount of water lost in expiration and defaecation varies and this, too, affects the volume of the urine. Less, more concentrated urine is produced during sleep as metabolism is then at its minimum. **Diuresis** is the term applied to excessive production of urine and substances such as caffeine in tea and coffee, alcohol and certain drugs which bring this about are known as **diuretics**. Water in large quantities has a similar effect.

The kidneys are responsible for the removal from the body of the nitrogenous waste-products of metabolism. They also excrete acid and alkaline salts and toxic substances from the body. Thus it will be seen that the function of the kidneys is to maintain the normal composition, reaction and volume of the blood. A man can live with one kidney but the removal of both would be fatal. If one is taken away for reasons of disease, the other enlarges and takes on the function of both. The writer knows someone who, fortunately or unfortunately, has three kidneys!

As the quantity of urine in the bladder increases, the muscles of the latter cause it to expand and it extends up into the abdominal cavity. Eventually the pressure is such that the desire to pass urine arises and if this is not done, further increase of pressure makes it imperative to do so. The act of **micturition** or passing of urine is a voluntary act in the adult and is effected by the relaxation of the external sphincter muscle of the urethra and by the contraction of the bladder which forces the urine into the urethra. In babies and very young children, of course, the act is involuntary.

Disorders and Diseases of the Urinary System

Incontinence—The inability to retain urine in the bladder is natural in babies but in adults is usually caused by disease such as paralysis of the sphincter muscles of the urethra or failure of the controlling mechanism.

Retention of urine—or the inability to pass it is usually due to obstruction as, for example, when the prostate gland in the male becomes enlarged resulting in constriction of the urethra. Insertion of a sterile catheter into the urethra enables the urine to be drained off.

Albuminaria—or the presence of albumin in the urine often indicates a disease of the kidneys such as Bright's disease.

Gravel and **stones**—occur in urine when such normal constituents as uric acid are present in excess. A stone, which may attain a considerable size, may become lodged in the urethra and this causes very considerable pain, known as *renal colic.* However, the stone may eventually be passed.

Cystitis—inflammation of the bladder. It may be caused by bacterial infection or by other diseases or disturbances of the urinary system.

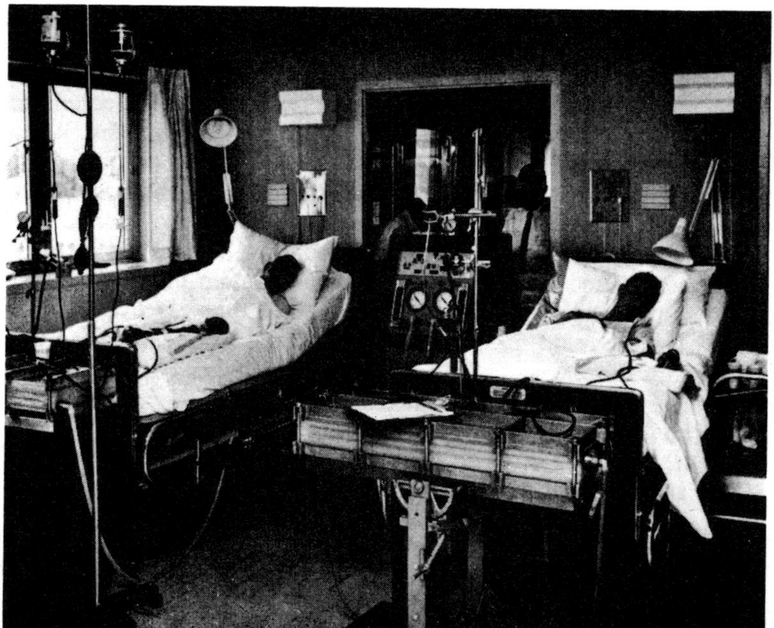

Fig. 15.5. **Artificial kidney unit in operation.**
(By courtesy of the World Health Organisation, Geneva.)

Uraemia—a disease in which the kidneys fail to extract excretory substances from the blood in which they therefore remain and cause a general poisoning of the body.

The **Artificial kidney** is a large machine placed outside the body through which the patient's blood is made to circulate, during which process the excretory products are removed from the blood and thus the normal condition of the blood is maintained. The instrument is used when the kidneys fail to function. The unit can serve up to ten patients simultaneously with only nursing staff in attendance. Patients go to the hospital twice a week and have treatment for up to twelve hours at each visit. The rest of the time they are able to lead normal lives.

Kidney transplants have been successfully made. The risk with this operation, as with all transplants, is tissue rejection but four hundred and eighty kidney transplants were done in the United Kingdom in 1974 and 60–70 per cent. were successful when the donor was not related to the patient. When the donor was a relative, however, 90 per cent. were successful. In the world at large nearly 17,500 have been performed. The survival rate is so high that the operation is performed without hesitation by surgeons all over the world whenever a patient needing a kidney transplant can be matched with a suitable kidney available for transplant. How much it will be possible to apply this branch of surgery to other organs and how permanently successful they will be only the future can decide. Certainly the performances by these pioneers in transplant surgery have been a remarkable achievement as far as the kidney is concerned. The rejection of foreign tissues may, in fact, prove to be a blessing in disguise in one respect for there is some reason to believe that the body is probably capable of recognising cancer cells and endeavours to reject them as it rejects other foreign tissues. This opens up a valuable field in cancer research.

PRACTICAL WORK

Anatomy of the Urinary System

1. Examine a dissection of the abdomen of a rabbit from which the alimentary canal has been removed. Find and examine the kidneys (in the rabbit the right kidney is higher than the left), the renal artery and vein, the ureters and the bladder. The pelvic girdle should be cut through at the symphisis pubis and the two halves pulled apart in order to expose the urethra, which in the male traverses the penis (which note) and, in the rabbit, unites with the vagina in the female to form a common urino-genital duct called the vestibule which opens in the slit-like vulva. (It will

be remembered that the urinary duct or urethra and the genital duct or vagina remain separate in the human subject.)

Structure of the Kidney

1. Remove one of the rabbit's kidneys and cut through it longitudinally right through the centre. (A sheep's kidney, being larger, is more suitable, if available.) Note the outer rind or cortex and the inner medulla in which the pyramids will be seen. Note also the pelvis and, if still attached, the beginning of the ureter and the renal artery and vein.

2. Examine a **L.S. of the kidney** under the microscope. You will see a number of uriniferous tubules, lined by cubical epithelium. If the section passes through the cortex, a number of capsules with their glomeruli will be visible.

Structure of the Urinary Bladder

Examine a **T.S. of the wall of the urinary bladder** under the microscope. Inside the outer fibrous coat are layers of circular and longitudinal muscle, both composed of unstriated fibres. The mucous membrane inside the muscular coat is lined by transitional epithelium.

Urine

Perform the following tests with a sample of normal urine.

1. Observe the *colour, transparency* and *aromatic odour.*

2. Find the *specific gravity* using a hydrometer. It should lie between 1·015 and 1·025.

3. Test the sample with blue litmus paper and note that the *reaction* is acid. If the sample has been taken after a meal, test with red litmus paper to see whether it is alkaline.

4. *Urea.* This test should be performed with a solution of urea in the first instance in order to observe the reaction and then repeated with urine to show the presence of urea. Add a few crystals of sodium hyperbromite. Violent effervescence, due to the liberation of nitrogen, occurs.

5. *Uric acid.* Allow some acid urine (test with blue litmus paper) to stand in a conical vessel or a flask for about three hours. A dark reddish-brown crystalline deposit, often called the "cayenne pepper" deposit from its appearance, may be seen. This consists of crystals of uric acid.

6. *Chlorides.* Add a few drops of concentrated nitric acid to a few ml. of urine and then add a few drops of silver nitrate solution. A white precipitate is obtained. Add some ammonium hydroxide and the precipitate dissolves.

7. *Sulphates.* Add a few ml. of concentrated hydrochloric acid to a small quantity of urine. Then add a few drops of barium chloride. A white precipitate is obtained.

8. *Phosphates.* Acidify some urine by adding half its volume of concentrated nitric acid, followed by a few ml. of ammonium molybdate. Boil the mixture and a yellow precipitate appears.

9. Examine the urine which has been standing for some hours again. A cloudiness will be seen which may have settled to the bottom of the vessel. This is *mucus.*

10. *Glucose.* Add about 8 drops of Benedict's Qualitative Reagent to 5 ml. of non-diabetic urine and boil for a minute or so. No change should occur, showing the absence of glucose. Now add a few crystals of glucose to the urine and repeat the test. This time a yellow precipitate, with a green turbidity in the liquid, should be obtained.

11. The most modern method of urine testing is by means of reagent strips such as "Multistix" and it is strongly recommended that tests with these should be performed. Each is a plastic strip on which separate reagent areas are firmly fixed. The reagent areas are completely immersed in urine (which must be fresh and well mixed) and immediately removed. Colour charts are provided and the colour produced on the reagent areas compared with those on the chart. *Glucose*, for example, produces a blue colour, *protein* green-yellow to green-blue, *blood* blue and *p*H a range of colours. Other reagent strips such as "Clinitix" or "Quantan CI" give a rapid semi-quantitative estimation of the percentage of glucose and "Quantan DT" of protein. In all cases it is *essential* that the instructions provided by the makers are meticulously followed.

If samples of abnormal urine are not available, small quantities of glucose, albumen or blood can be added to a normal sample before performing the tests.

The suppliers of these strips will be found in Appendix II.

Chapter XVI

THE MAINTENANCE OF BODY TEMPERATURE

The maintenance of the constancy of the internal environment of the body is known as **homeostasis**. It involves the control of the transport of the oxygen and carbon dioxide content of the blood, the regulation of breathing under varying conditions, deamination and carbohydrate storage in the liver, the elimination of the waste products of metabolism by the kidneys and the maintenance of body temperature. All of these things have already been studied in their proper context with the exception of the last mentioned which we shall now investigate.

Heat is constantly being produced by various activities in and by the body and is constantly being lost from the body in various ways but, under conditions of normal health, the temperature remains practically constant at 37° Centigrade* or 98·4° Fahrenheit. This **normal temperature**, as it is called, is an average temperature for it varies slightly above or below this figure in different people. It does, in fact, vary a little in different parts of the body and in fever and disease it may rise to as much as 41°C or 106°F or more or fall below 35°C or 95°F. Above and below these extremes the risk to life is considerable. Apart from disease conditions the normal temperature of the blood does not vary with that of the atmosphere and man is therefore said to be **homoiothermic** or "warm-blooded". This is a character which he shares with all other mammals and also with birds. In order that this temperature may be maintained there must be a delicate balance between heat production and heat loss in the body.

Heat Production

Heat is produced by the body in various ways. Firstly, of course, is tissue respiration when sugar is oxidised to carbon dioxide and water in order to liberate energy and some of this energy is set free as heat. When the various glands in the body are functioning, particularly in the case of very large glands like the liver, heat is generated and the liver produces more than any. Secondly muscle contraction liberates heat as we know only too well when we indulge in strenuous exercise. Heat is generated by the heart's contraction and this heat is circulated through the body in the blood stream. In fact, the muscles produce most of our body heat. Again, heat

*Or Celsius.

can be absorbed by the body from outside as we are well aware in hot weather or when we take a hot bath. If outside conditions are cold we shiver, that is our muscles undergo a series of rapid contractions in order to generate heat and we are more physically active in cold weather for the same reason. Food is our source of body energy and in cold weather we eat more food to provide this energy. On the other hand in hot weather we are much less active and we eat less food so that less heat is generated.

Heat Loss

Sweating is a process of evaporation and evaporation causes cooling. Heat is therefore lost from the body by this process and the more we sweat the greater is the amount of heat lost. That is why we sweat more in hot weather when the rate of sweat production may far exceed the rate of evaporation and sweating then becomes visible. In cold weather we sweat to a much lesser extent in order to reduce heat loss. The humidity of the atmosphere will also influence the rate of sweating and that is why a damp climate is unhealthy, whether it is a hot or a cold one. Heat is lost by any body which is at a higher temperature than its surroundings and this applies to the surface of the human body as much as an inanimate one. Contact with cold bodies reduce the temperature locally as we know when, for example, we are making snowballs. A draught or a cold wind increases heat loss and may cause an increase in the rate of sweating which, of course, has a similar effect and this is one way in which we "catch a chill". When the temperature of the body falls, the blood vessels in the skin become constricted and the skin becomes pallid and when the temperature rises these vessels dilate and the skin becomes pink or red.

Heat is also lost in expiration, the amount depending on the humidity and temperature of the atmosphere. Some heat is also lost in the urine and faeces.

The Regulation of Body Temperature

The amount of heat produced by the body and the amount lost from it are obviously nicely balanced so that a constant temperature is maintained. This is made possible by a **heat regulating centre** in the brain which is brought into play by the temperature of the blood passing over it and which exerts a control over the sweat glands, the blood vessels in the skin and the muscles. Subcutaneous fat has an insulating effect but hair plays little part in this way in man. The wearing of clothes made of suitable materials according to the outside temperature form a good substitute for the hair, however.

Chapter XVII

THE GENITAL SYSTEM AND REPRODUCTION

It is a characteristic feature of all living organisms that they are able to reproduce their kind in order that the life of the species may be maintained and whereas in some lower forms of life this may take place asexually by fission or budding, in many animals, including mammals, only sexual reproduction takes place. This essentially involves the fusion of two specialised cells called **gametes**, the resulting cell being termed a **zygote**. These gametes are of two kinds, the male gamete or **spermatozoon** being motile and very much smaller than the female gamete or **ovum** which is static. The organs in which these develop, the **gonads**, are the **testes** and **ovaries** respectively and when they develop in separate organisms there are two separate sexes. *Hermaphroditism,* in which both male and female organs occur in the same animals, is found only in some of the lower forms of life. In mammals **fertilisation**, or fusion of the gametes, is internal to increase the chances of its taking place and the zygote develops into the new offspring in a special organ, the **uterus**, within the female's body. This not only protects the developing **embryo** but also serves as a means of providing it with oxygen and nutrition by the mother. Furthermore, after birth the new organism is fed by milk secreted by the **mammary glands** of the mother and is protected by both parents until it is sufficiently developed to fend for itself. In the human subject the new-born baby is more helpless than is the case with lower mammals. Consequently the period of infancy is prolonged and the child is dependent on its parents for a much longer time. This is an enormous advantage to the child, of course, in making adequate provision for its future life.

We saw in Chapter II that the chromosomes in the nuclei of the cells of an organism are responsible for its body characters, that the number of chromosomes is constant for any one species, that the gametes contain half the normal number (haploid), one from each pair of chromosomes, and that at fertilisation the normal (diploid) number is restored, half being derived from one parent and half from the other. Thus a child bears characters derived from both parents. In the next chapter we shall study this question of heredity and see how the inheritable characters of the two parents affect the characters of the children. We must now turn our attention to the anatomy and histology of the male and female reproductive systems in order that we may gain an insight into their physiology and to the whole process of reproduction.

MALE ORGANS

The two **testes** (*testicles*) are ovoid structures about 3·8 cm. in length lying in an external pouch known as the **scrotum** and in them the spermatozoa develop. The testes develop in the abdomen of the embryo just below the kidneys but they begin to descend about the fifth or six month of pregnancy and eventually pass through the inguinal canal into the scrotum by the end of the

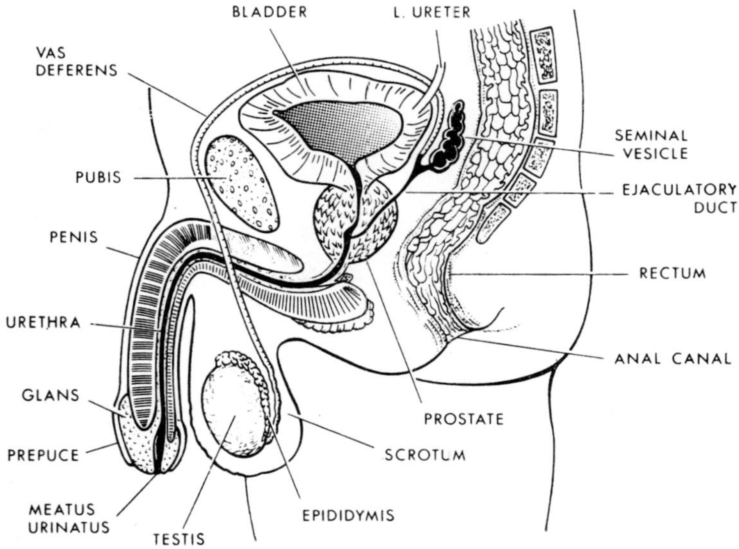

Fig. 17.1. **Male reproductive organs. Section through pelvis.**

seventh month. The left testis is slightly lower than the right. The accepted reason for this descent is that the temperature in the abdomen is too high for the development of the spermatozoa. Each testis is suspended through the inguinal canal by a **spermatic cord** composed of the testis duct or **vas deferens**, spermatic artery and vein. lymphatics and nerves, bound together by connective tissue. The scrotum is divided internally into two separate sacs in each of which lies a testis. Externally it is covered by skin and this is lined by a layer of unstriated muscle.

Each testis is enclosed in a fibrous coat and is divided internally into lobules in which lie very convoluted **seminiferous tubules**.

Between the tubules is connective tissue in which are hormone-producing cells known as **interstitial cells. Germinal epithelium** lines the tubules and it is from this tissue by a series of divisions that the spermatozoa are formed. The convoluted tubules form straight tubules on leaving the lobules and then give rise to a network, the **rete testis**, from which small ducts known as **vasa efferentia** arise. These join up to form a very coiled up body, the **epididymis**,

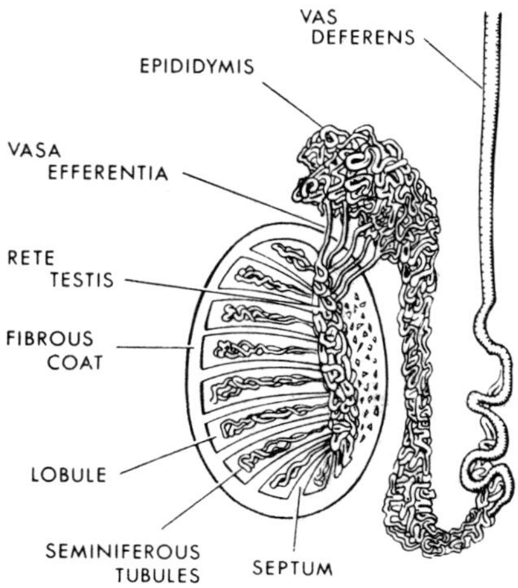

Fig. 17.2. **L.S. testis with epididymis and vas deferens.**

on the side of the testis and from the lower end of this arises the straight vas deferens which, as already seen, enters the abdomen as part of the spermatic cord. At the upper end of the inguinal canal it leaves the cord, loops round the ureter and runs to the upper surface of the **prostate gland** which surrounds the neck of the bladder and the upper part of the urethra. Here it is joined by a duct from a structure called the **seminal vesicle** lying on its outer side, thus forming an **ejaculatory duct** which opens into the prostatic urethra by a large number of ducts. The seminal vesicles are irregular in shape and lobulated. They secrete a mucous fluid and act as storehouses of the seminal fluid. The prostate is both muscular and glandular, the latter part of it consisting of branched tubules through which its secretion passes into the prostatic urethra. It will be seen that the urethra becomes genital as well as urinary in function. Just beneath the prostate are the small **Cowper's glands**, each about the size of a pea, which secrete a sticky fluid into the urethra.

The **penis** is composed of what is known as **erectile tissue** and consists of two **corpora cavernosa** and a **corpus spongiosum**. It is covered by an elastic skin. The corpora cavernosa occupy the sides and upper part of the organ and diverge at the proximal end of the penis to form two roots which are attached to the pubis. The corpus spongiosum which lies on the lower side of the penis consists of spongy tissue through which the urethra passes.

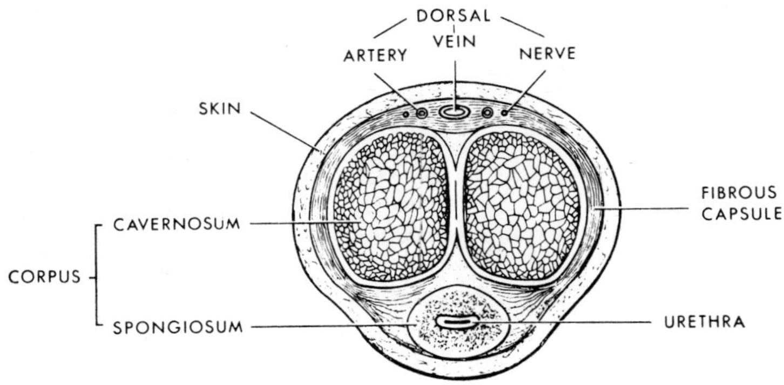

Fig. 17.3. **T.S. penis.**

The free end of the corpus spongiosum, covered with a loose retractable skin forming the **prepuce** or *foreskin*, is expanded to form the **glans penis** and it is on this that the urethral opening or **meatus urinatus** is situated. Both the corpora cavernosa and the corpus spongiosum are well supplied with blood vessels and the glans is provided with sensory nerves. Numerous glands which secrete a thick fluid lie in a deep groove at the base of the glans. *Circumcision* is an operation which has sometimes to be performed on children because the opening of the prepuce is too small.

THE FEMALE ORGANS

The internal reproductive organs of the female consist of the two **ovaries**, the **uterus** or *womb*, the **Fallopian** tubes and the **vagina**. The uterus is a pear-shaped organ about 5 cm. across at its upper, wider end and a little over 7·5 cm. in length. It is situated in the pelvic cavity between the rectum and the bladder and is kept in place by a wide double fold of peritoneum connected to the pelvic girdle known as the **broad ligaments**, and by the **round ligament** which crosses them and is joined to the pelvis which itself give support to the organ. The uppermost part of the uterus is known as the

fundus and beneath this is the greater part of the organ, the **body** which leads into the narrower lower part known as the **neck** or **cervix**. This projects into the vagina into which it opens by the **os uteri**. The wall of the uterus consists of a peritoneal coat, a

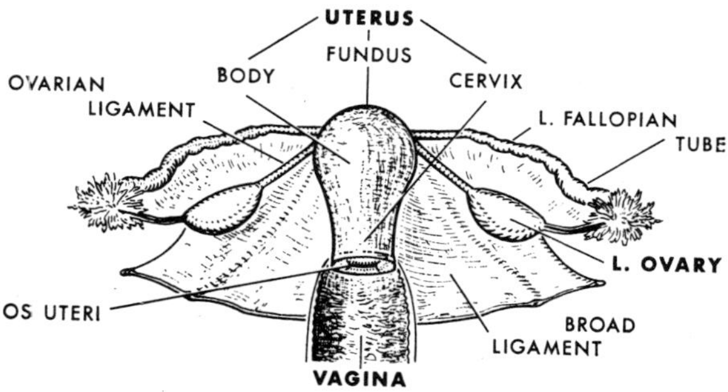

Fig. 17.4. **Internal female reproductive organs viewed from the front.**

muscular coat, unstriated fibres and a mucous membrane called the **endometrium**.

The **ovaries** are almond shaped organs about 2·54 cm. across lying one on each side of the fundus of the uterus at its upper end and to which they are connected by **ovarian ligaments**. The external coat is a modified peritoneum known as **germinal epithelium**.

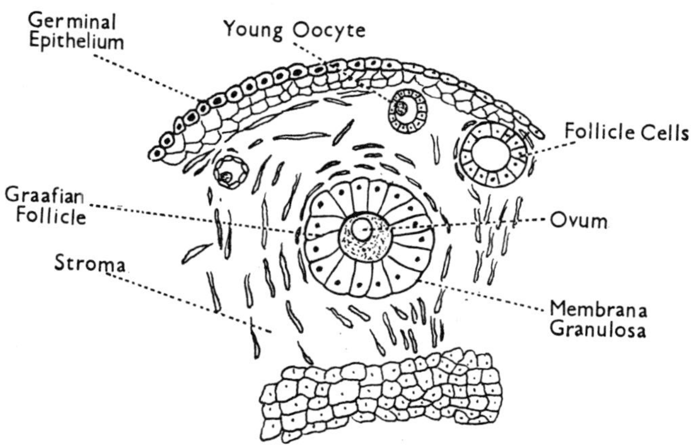

Fig. 17.5. **T.S. ovary.**
(From Wallis—Practical Biology.)

Inside this is a region of fibrous tissue called the **stroma** in which lie a number of spherical sacs known as **Graafian follicles**. Those near the surface are numerous and small but in the deeper layer they are much larger and occur less frequently. It is within these follicles which arise originally from the germinal epithelium that the ova develop. A mature follicle consists of two regions of stratified epithelium, one, the **membrana granulosa**, lining the fluid filled cavity of the follicle and the other, the **discus proligerus**, in which the developing **ovum** lies. Complete maturation of the ovum takes place after it has been liberated by the bursting of the follicle when it passes into the Fallopian tube. When the ovum has been discharged the follicle cells form a yellow body known as the **corpus luteum**. This degenerates later unless fertilisation of the ovum takes place in which case it enlarges.

The **Fallopian tubes** or oviducts are about 10 cm. in length and are attached to the upper end of the fundus of the uterus and at this end they are very narrow. They run outwards across

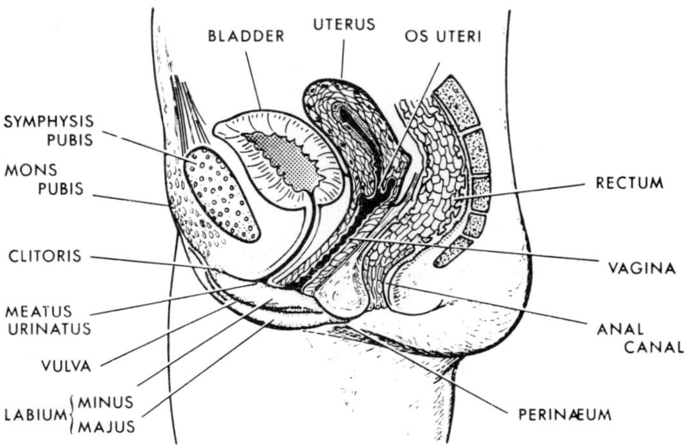

Fig. 17.6. **Section through female pelvis.**

the upper end of the broad ligament and each ends in an expanded process or **ampulla** which has wavy, fringe-like edges called **fimbriae.** The tubes are enclosed in a peritoneal coat, inside which is a muscular coat lined by a mucous membrane which contains ciliated epithelium. Their purpose is to convey ova from the ovaries to the uterus.

The **vagina** into which the neck of the uterus projects leads to the exterior of the body in the vulva. Its anterior wall is up to 7·5 cm. long but its posterior wall is longer by around 2·5 cm. and it runs obliquely forwards as it passes downwards at about 45° with the horizontal. The outer fibrous coat encloses a muscular layer of unstriated fibres mostly running longitudinally but at its outer end they are circular. The mucous membrane, which is covered by fine papillae, is lined with stratified epithelium, the cells on the surface being squamous.

The external genital organs are enclosed by the vulva. Lying in front of the symphysis pubis is a thick layer of fat in which the round ligaments of the uterus end. It is known as the **mons pubis** and beneath this is the slit-like fissure known as the **vulva**.

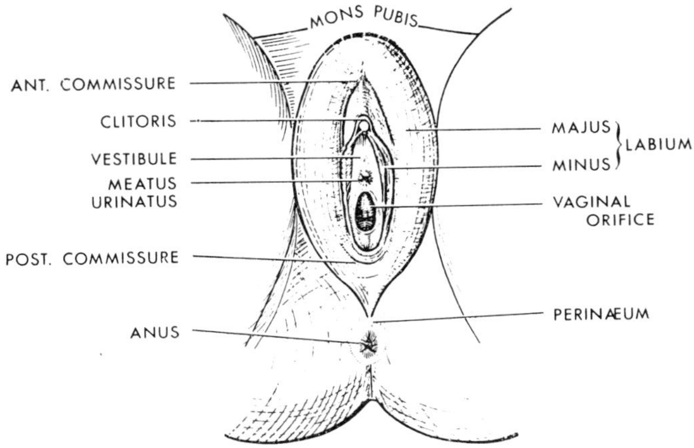

Fig. 17.7. **External female reproductive organs—the vulva.**

Two folds of skin which enclose muscle and fatty tissue form the lateral boundaries of the vulva. They are known as the **labia majora**. Their junction at the upper end is called the **anterior commisure** and that at the lower end the **posterior commisure**. The urethra and the vagina open into the fissure, the former being the anterior of the two. Within the labia majora are two smaller lips or folds of skin composed of erectile tissue and known as the **labia minora**. These become smaller as they approach the vaginal opening. Immediately behind the anterior commisure is the small rod-like **clitoris** which is the homologue of the male penis though not perforated by the urethra. It consists of erectile tissue and terminates in a highly sensitive glans. The anterior ends of the labia minora surround the free end of the clitoris and form its prepuce. The triangular region behind as far as the vaginal opening is called the **vestibule**. In virgins there is a membrane perforated in the centre known as the **hymen** which encloses the entrance of the vagina and on each side of this opening are the **glands of Bartholin** which correspond with the Cowper's glands in the male. Mucous and sebaceous glands are also present. The region between the vulva and the anus is called the **perinaeum**.

The **mammary glands** or *breasts* which are rudimentary before puberty as they are throughout life in man and which enlarge at puberty in the female, are racemose glands lying over the pectoral muscles of the thorax. They lie between the axillae and the sternum and reach from the second to the sixth ribs, though their size

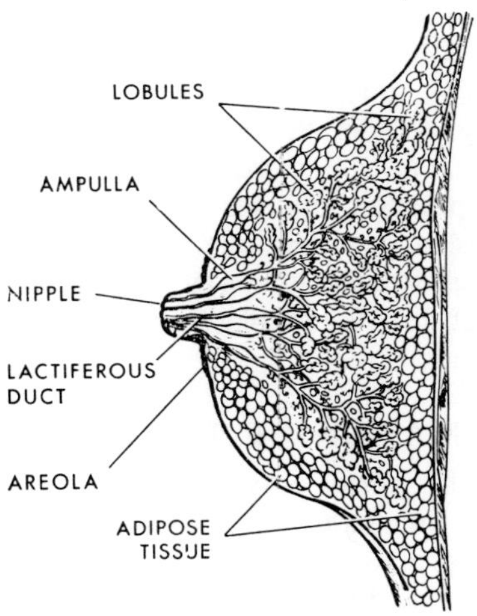

Fig. 17.8. **Section through female breast.**

varies in different women and at different ages. These accessory reproductive organs consist of about twenty **lobes** divided into **lobules** and held together by connective tissue. The lobules are composed of minute tubes lined by epithelium which secretes milk during and after pregnancy and the tubes unite with one another in each lobe to form ducts so that a duct leaves each lobe. These **lacticiferous ducts** converge on the **nipple** which is surrounded by an area of pigmented skin known as the **areola**. Before reaching the nipple the ducts swell out to form **ampullae** which are storage reservoirs for milk. The rounded contours of the breasts are due to masses of adipose tissue lying between the lobes and beneath the skin. The breasts are supplied with blood by branches of the internal mammary, intercostal and axillary arteries.

PUBERTY

The sexual organs do not become functional until what is known as the age of puberty is reached. This normally occurs at fourteen in boys and at twelve or thirteen in girls. The testes then begin to produce spermatozoa and the ovaries to produce ova. This is accompanied by the appearance of what are known as **secondary sexual characters**. In the male the testes enlarge, hair develops on the face, in the axillae, on the chest and in the pubic region and the larynx enlarges causing the voice to "break". These changes are brought about by a hormone called **testosterone**, secreted by the interstitial cells of the testes. In the female the mammary glands enlarge, hair develops in the axillae and in the pubic region, the hips become larger and the body generally assumes the more graceful female form. The hormones responsible are **progesterone** and **oestrin** secreted by cells in the ovaries. They are also concerned with what is known as the menstrual cycle which begins at this time. These developmental changes are accompanied in both sexes by psychological changes resulting in increased interest in the opposite sex. Though fertilisation is, of course, possible once physiological development has taken place, full maturity is not reached until the early twenties.

FORMATION OF THE GAMETES

Maturation of the germ cells or **gametogenesis** is known as **spermatogenesis** in the male and **oogenesis** in the female. Before studying these two processes let us see how the number of chromosomes is halved to the **haploid** number (23) in meiosis to which reference was made in Chapter II.

Meiosis

At one stage in the formation of the gametes in both sexes a reduction division takes place in which the chromosome number is halved. This is essential in order to maintain the constancy of the number in the zygote and this method of nuclear division is meiosis.

At the outset, unlike mitosis, the chromosomes are thin and show no tendency to divide into chromatids. They come together in homologous pairs (one maternal and one paternal in origin) with corresponding genes in juxtaposition and each apparent chromosome is said to be **bivalent**. These bivalent chromosomes become shorter and thicker, split into two chromatids and coil round one another. At this stage interchange of chromatic material may take place between them. (This is the cause of what is known as "crossing over" to which reference will be made in the next chapter.) The two split chromosomes partially separate and ultimately arrange themselves on the equator of the nuclear spindle as in the metaphase

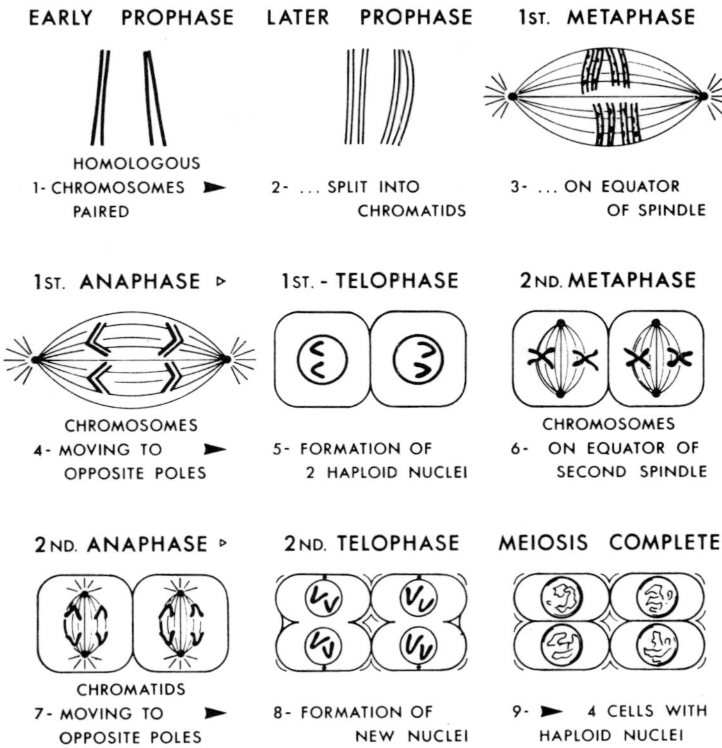

Fig. 17.9. **Meiosis.**

of mitosis. Each is attached at two points, one for each chromosome. Next the two chromosomes, each composed of two attached chromatids, separate and move towards opposite poles as in the anaphase in mitosis. Cytoplasmic division follows and a second nuclear division takes place in which the chromosomes pass quickly to the metaphase stage which is followed by the anaphase and telophase stage as in mitosis.

Thus, in the first of these two divisions half of the chromosomes pass to each pole forming what corresponds with a haploid daughter nucleus and in the second the chromsomes already divided into two chromatids apiece produce four nuclei each having the haploid ($n/2$) number of chromosomes.

Spermatogenesis

Cells of the germinal epithelium in the seminiferous tubules divide mitotically to give rise to **spermatogonia** ($2n$) and after several

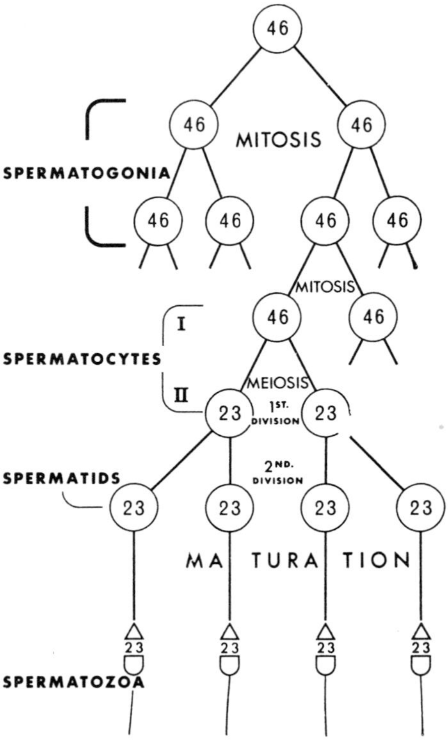

Fig. 17.10. **Spermatogenesis.**

such divisions these divide mitotically to give **primary spermatocytes** (2*n*). The next division into **secondary spermatocytes** is meiotic (*n*). In the subsequent division **spermatids** are formed which develop into spermatozoa, millions of which are produced.

Each **spermatozoon** is about 0·05 mm. in length and consists of a **head**, a **middle piece** or **body** and a **tail** which gives it mobility. The spermatozoa are stored temporarily in the epididymis and afterwards in the seminal vesicles. A fluid is produced to which the secretion of Cowper's glands and the prostate are added. The latter

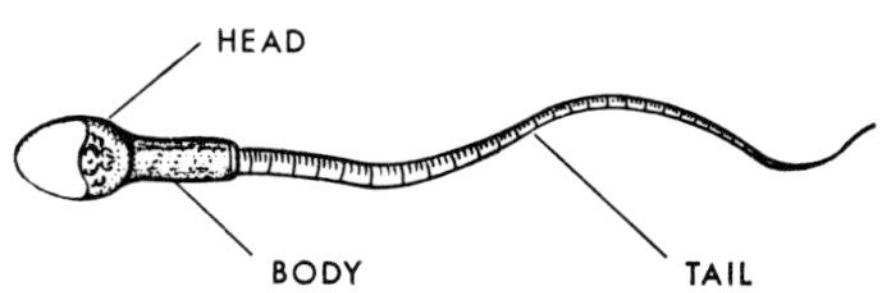

Fig. 17.11. **Spermatozoon.**

is necessary for the activity of the spermatozoa. This fluid containing the gametes is known as **semen** and it contains water, protein, fat and mineral salts.

Oogenesis

This is on similar lines to spermatogenesis but differs from it in certain respects. The **oogonia** ($2n$) formed from the germinal epithelium divide mitotically to form **primary oocytes** ($2n$) and these

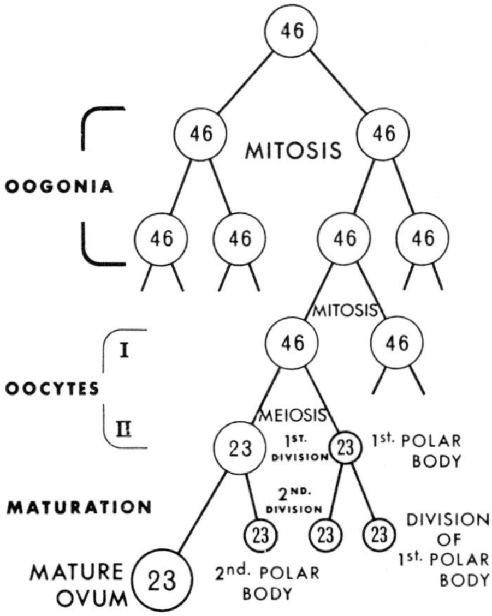

Fig. 17.12. **Oogenesis.**

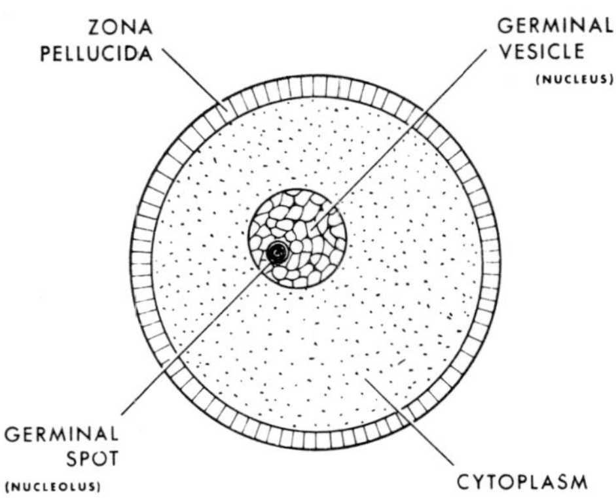

Fig. 17.13. **Ovum.**

divide meiotically to give rise to **secondary oocytes** (n) and what are known as the **first polar bodies**, the division being an unequal one. These divisions take place in the Graafian follicles. In the next division the **secondary oocytes** divide into an **ovum** (n) and a **second polar body**. This takes place after liberation from the follicle and the first polar body also divides. These polar bodies may be regarded as abortive ova, the ovum maturing at their expense, and they eventually degenerate.

The human **ovum** is a spherical cell about 0·2 mm. in diameter, enclosed in a thick transparent membrane called the **zona pellucida**. The nucleus or **germinal vesicle** lies to one side and inside it is a nucleolus or **germinal spot**. Yolk spherules occur in the cytoplasm.

THE PHYSIOLOGY OF REPRODUCTION

The Production of Spermatozoa and Semen

As already explained earlier in the chapter, spermatozoa develop in the seminiferous tubules in the testes and pass along the vas deferens in an albuminous fluid. The secretion it receives from the seminal vesicles is probably nutritive and that from the prostate increases the activity of the spermatozoa. The prostatic secretion is alkaline and this neutralises the acidity due to urine in the urethra. Semen is formed throughout life though in lesser quantity in later life. In fact, a man can remain fertile until he reaches the age of about eighty or more and one well known figure many years ago had a son at the age of eighty! If no coition takes place the semen is discharged during sleep.

The Menstrual Cycle

At intervals of approximately twenty-eight days between the ages of twelve or thirteen and fifty or thereabouts an ovum is liberated by the bursting of a Graafian follicle. This is called **ovulation** and is followed by the formation in the follicle of the **corpus luteum** which secretes a hormone known as **progesterone**. The purpose of this hormone is to prepare the wall of the uterus for the reception of a fertilised ovum. The lining of the uterus, the endometrium, becomes swollen with blood and if fertilisation does not take place the corpus luteum degenerates and its secretion stops. Consequently the endometrium breaks down and the blood, together with epithelial cells from the uterus, is discharged through the vagina. This normally lasts for four or five days or so and is called the **menstrual period**. It is followed by regeneration of the endometrium, a process taking about seven days, after which is a resting period of about fourteen to sixteen days. At the end of this stage ovulation takes place again. **Menstruation** ceases once fertilisation occurs and it remains in abeyance throughout pregnancy. It also ceases at what is known as the **menopause** or **climacteric** after which, of course, child bearing is impossible. This usually occurs at about the age of fifty though it may start somewhat earlier and the process of change often lasts for several years. The hormonal changes accompanying this process produce certain psychological changes also.

Fertilisation

Fertilisation or conception, the fusion of a spermatozoon with an ovum, takes place in the Fallopian tube. When the male becomes sexually excited the penis, normally flaccid, becomes erect due to the corpus spongiosum and corpora cavernosa becoming distended with blood under pressure and the condition maintained by closure of the veins. This is the result of a reflex action. In this condition the penis is inserted into the vagina of the female and eventually by a series of muscular contractions accompanied by considerable physical and mental excitement by both sexes culminating in the orgasm, ejaculation takes place and semen passes into the vagina thereby completing the sexual act. Millions of spermatozoa are contained in this discharge (it has been estimated at 250,000,000!) but only one is necessary to bring fertilisation. Many swim into the uterus and some reach the Fallopian tubes while others remain in the vagina and ultimately degenerate. Those which reach the uterus and Fallopian tubes can fertilise an ovum up to a few days after coition has taken place. Though several spermatozoa may enter an ovum, the head of only one fuses with the nucleus. Thus a zygote is formed and once this has occurred pregnancy begins.

Artificial insemination in which semen is injected into the female vagina is widely practised with cattle and other domestic animals. The stored semen is frozen and in this condition the spermatozoa remain alive for several years. When required for use the semen is, of course, thawed before injection. This method is sometimes used with human subjects when a husband is unfertile but only under strict medical supervision. Naturally in the case of human beings such a practice is beset with innumerable psychological and other difficulties and objections and certain conditions must be fulfilled such, for example, as freedom from congenital disease and the correct Rh factor.

In February, 1969 it was announced that Dr. R. G. Edwards of the Cambridge University Physiological Laboratory, working in collaboration with Dr. P. C. Steptoe of Oldham General Hospital and others, had succeeded in fertilising human ova *in vitro, i.e.* outside the living body, in test-tubes. Both ova and spermatozoa, provided by voluntary donors, were cultured in fluid from Graafian follicles.

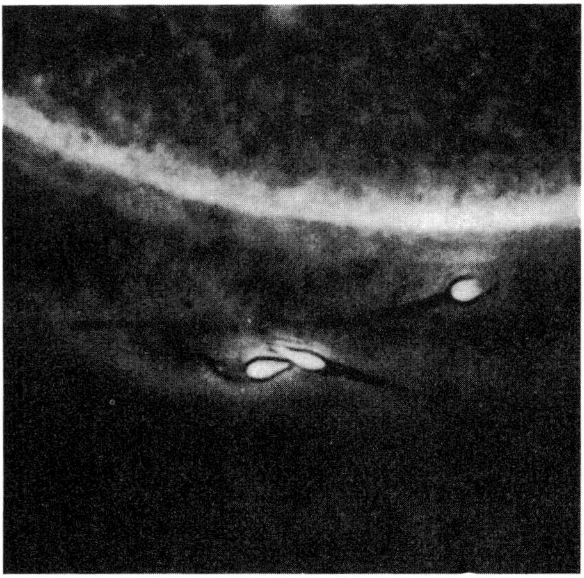

Fig. 17.14. **Micrograph of the fertilisation of a living human ovum in vitro** *i.e.* **outside the human body. A spermatozoon is swimming through the zona pellucida. Two others remain outside the membrane.**

(By courtesy of Dr. R. G. Edwards of the University of Cambridge Physiological Laboratory and Central Press Photos Ltd.)

It may be possible to culture the fertilised ova *in vitro,* at least to the blastocyst stage (see below) as has already been done with mice—or even further, once suitable techniques have been discovered. This would provide a means of making a detailed study of the early development of the human embryo which, in turn, could lead to the discovery of the reasons for abnormal development and the failure of these embryos to reach maturity. Furthermore sterility, due to blockage of the Fallopian tubes, for example, may be overcome if the ova fertilised *in vitro* could be successfully implanted in the uterus and if these zygotes then completed their development to maturity. But there are innumerable technical difficulties to be overcome and success is not anticipated in the foreseeable future. Dr. Edwards has reported that he has fertilised more than seventy preovulatory oocytes eight of which were transferred to a woman's uterus but none of them proved successful. No successful cases are known.

Contraception—It will be obvious from the above account of fertilisation that the chances of its occurring as the result of coition are extremely high. Contraception or the prevention of fertilisation is possible by various artificial means. One method, the use of an oral contraceptive, popularly known as "the pill' is stated to be 100 per cent. effective but it has been reported as having ill effects in some cases, three deaths a year from thrombo-embolism resulting from the taking of it. This appears to be related to the blood groups of the women taking the pill, for investigation into a large number of cases seems to indicate that the chance of women in groups A, B and AB suffering in this way is about three times greater than for those in group O. It must therefore be taken only under medical supervision. It has been estimated that fifty million women throughout the world are on the pill. Other methods are the wearing of a condom or sheath by the male, the fitting of a cervical cap over the neck of the uterus, the insertion of spermicidal chemicals into the vagina, the rhythm method in which intercourse is restricted to certain so-called "safe periods" between menstruation, coitus interruptus and sterilisation. Of these the safest method is the sheath. The uterine cap is also safe if correctly fitted over the cervix to prevent the entry of spermatozoa and in conjunction with spermicides. These chemicals alone and the rhythm method are not very reliable. Not only is the latter very restrictive but the safe periods *i.e.* when fertilisation is least likely between menstrual periods cannot be determined with any degree of certainty. This, incidentally, is the only method approved by the Roman Catholic Church. Coitus interruptus is very unsafe and is highly undesirable in every way. Finally, sterilisation provides permanent sterility and should be regarded as terminal. In the male it is a simple and short operation in which the vasa deferentia are severed in the scrotum and the free ends tied off under local anaesthetic (*vasectomy*). But the female operation is more complicated and involves an abdominal operation under a general anaesthetic in order to sever the fallopian tubes and tie off the free ends. Both are said to be reversible though the chances of success are not above 50 per cent. Claims were made in California in the Autumn of 1975 that a new method of rejoining the severed vasa deferentia involving micro-surgery had been performed on sixteen men who had previously undergone the vasectomy operation and that all their wives had since become pregnant. A great deal of additional evidence will be necessary, however, before the certainty of success can be guaranteed in all cases. Surprisingly, it has been stated that there is quite a high demand for sterilisation operations. Avoidance of conception in countries such as India in which food shortage, starvation and consequent disease is a serious problem is plain common sense and the size of families should be strictly limited to prevent over population. Fortunately, voluntary sterilisation on a very large scale is performed in India and other countries where it is so important. In any case it is often desirable or even essential that family size should be limited and it is only in recent years that instruction on this matter by properly qualified people has come into use in this country.

Pregnancy tests—During early pregnancy the secretion of the hormone **gonadotrophin** by the pituitary gland at the base of the brain, which controls the secretion of the hormones which play their part in pregnancy, is considerably increased. Excess of this hormone is excreted in the urine and this serves as the basis for various tests for pregnancy. In one such test a small quantity of the urine under test is injected into immature female mice. A few days later the mice are killed and dissected. If the subject under examination is pregnant, it will be found that the ovaries of the mice have matured as the result of the gonadotrophin in the urine. If she is not pregnant the ovaries remain in their immature condition. In another test the urine is injected into the female South African Clawed toad. If the woman is pregnant ovulation in the toad occurs within twenty four hours. Failure to ovulate indicates the absence of pregnancy. Nowadays a more recently discovered method giving quicker results is used. When gonadotrophin is injected into a female rabbit, it produces antibodies which counteract it. The antibodies are mixed with the urine of the woman under test. If she is pregnant the gonadotrophin is neutralised by the rabbit's antibodies which thus become inactive but if she is not pregnant, there being no gonadotrophin in her urine, the antibodies remain active. A drop of the urine/antibody mixture is then added to a drop of a colloidal solution of latex coated with gonadotrophin. If the woman is pregnant, no further visible change takes place but if she is not pregnant, the antibodies being still active, cause immediate clouding due to the gonadotrophin/latex particles coalescing. It has been stated that this test, from which results can be obtained in a matter of minutes, is 99 per cent. accurate.

Pregnancy and the Development of the Zygote

Once fertilisation has taken place, the zygote secretes a membrane around itself to prevent entry by further spermatozoa. It begins to divide and eventually forms a mulberry-like mass of cells known as a **morula**. This passes down the Fallopian tube into the uterus which it reaches a week or so after fertilisation. By this time it has become differentiated into a single outer layer of cells called the **trophoblast** and in **inner mass** of cells and contains a fluid, the whole embryo at this stage being known as the **blastocyst**. The wall of the uterus has been prepared for the reception of this embryo by becoming thickened, a process brought about by the progesterone produced by the corpus luteum. The cells of the trophoblast break down the uterine mucous membrane and the embryo becomes implanted in the uterine wall. The cells of the inner mass undergo a series of complex changes and fluid collects. This results in the formation of a fluid-filled sac called the **amniotic cavity**, the wall of which, the **amnion**, fuses with the trophoblast.

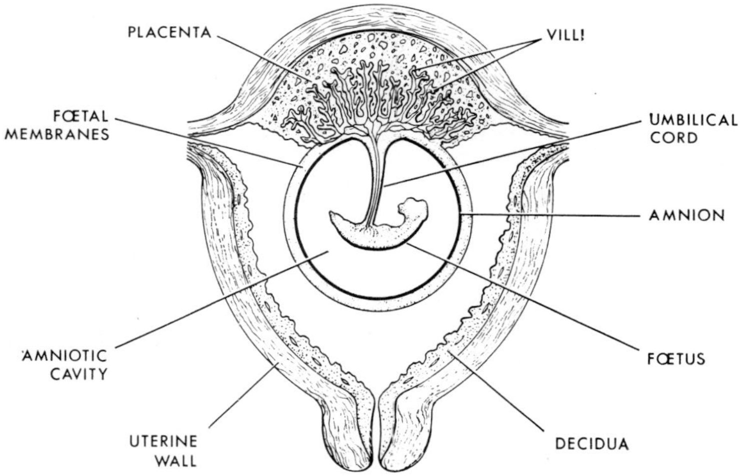

Fig. 17.15. **Early embryo in uterus.**

Inside the amniotic cavity the embryo develops protected against mechanical injury by the amniotic fluid. The embryo, now known as a **foetus**, remains attached to the trophoblast by the **umbilical cord** which contains blood vessels, and villi develop from the trophoblast and penetrate into the **decidua** as the uterine membrane is now called. These villi enlarge considerably where the umbilical cord enters the trophoblast and this and the decidua form a disc-shaped structure called the **placenta** which becomes the organ of foetal mutrition, respiration and excretion. Elsewhere the villi disappear and the trophoblast and decidua form a second foetal membrane called the **chorion**. Foetal blood vessels develop in the villi of the placenta where they come into intimate contact with the uterine blood vessels. The two sets of blood vessels never unite

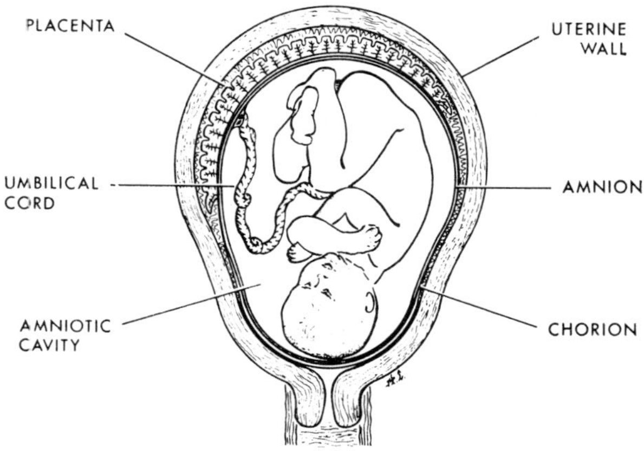

Fig. 17.16. **Mature foetus in uterus.**

and the foetal and maternal blood never mix, the nutritional substances and oxygen passing from the maternal blood to the foetal blood and the excretory substances from the foetal blood to that of the mother by diffusion.

Growth of the foetus and its differentiation into tissues, organs and systems takes place during the next few months and at the end of approximately 280 days (forty weeks) the period of pregnancy or **gestation** is at an end and the process of birth begins.

During the early stages of pregnancy, nausea and vomiting may occur, particularly in the early morning, combined with a frequent need to micturate and exhaustion is common. Careful attention must be paid to diet, regular bowel action maintained and regular exercise without fatigue must be taken. During the first three months the uterus enlarges to about three times its normal size and little discomfort is noticed but during the last three months when the foetus increases in size and in weight to about 3·3 kg. (seven and a quarter pounds) or more, its movements within the uterus can be felt and its heart beat detected. At the end of the term periodic contractions of the muscular uterine wall take place, an indication that birth is imminent.

Parturition

Parturition is the term used for the process of birth or **confinement**. The uterine contractions become more pronounced causing what are known as "labour pains", the os uteri dilates, the foetal membranes rupture and the amniotic fluid escapes through the vagina. This is followed by more severe labour pains and the muscular contractions eventually expel the foetus, normally head first, through the vagina. As the baby is still attached to the placenta by the umbilical cord, the latter has to be ligatured and cut. The scar left on the abdomen is the **umbilicus** or **navel**. Further uterine contractions follow which expel the placenta and foetal membranes: these constitute the "after birth". Once the baby is born it begins to breathe and usually cries. The whole process in a normal birth takes several hours and is usually longer with a first baby. After four or five weeks the uterus returns to its normal size but menstruation does not start again until the end of lactation.

There are cases in which the space in the pelvic girdle is too narrow to permit the passage of the baby and it is therefore necessary to make an incision in the abdominal wall in order that the baby may be born. This is known as a **Caesarian birth**, so-called because it was alleged that Caesar was born in this way. The number of possible births by this method is limited, though four or more have been recorded. In 1968 sextuplets were born to a Birmingham woman who had been taking a fertility drug and of the six three survived. In a similar case in 1976, none survived beyond a week.

Lactation

As already stated, during pregnancy the breasts enlarge as the result of the action of oestrin and progesterone and they are activated by a hormone secreted by the pituitary gland called **prolactin**. This initiates the secretion of milk which serves as food for the infant child for about eight months. Human milk is the ideal food during this period as it contains proteins, carbohydrates and fats together with mineral salts and water, all in their correct proportions. In some cases breast feeding is undesirable, either because the milk is inadequate in quantity or quality or for social reasons. In these instances, bottle feeding is necessary.

The composition of human milk is approximately as follows:—

Protein	1·7 per cent.	Carbohydrates	6·2 per cent.
Fats	3·4 per cent.	Salts	0·2 per cent.
	Water	88·86 per cent.	

Cow's milk contains more protein, slightly more fat, less carbohydrate and more salts. It is therefore necessary to add sugar and water to cow's milk when it is given to babies. There are several proprietary infant's foods on the market in which the proportions are carefully regulated.

Disorders and Diseases of the Reproductive Systems

Male

Undescended testis—It sometimes happens that one or both of the testes fail to descend properly into the scrotum. This does not necessarily result in their being unable to function.

Hydrocoele—a collection of fluid in the scrotum which causes it to be considerably distended. The condition can be rectified by simple surgery.

Enlargement of the prostate—In later life the prostate may become enlarged with the result that micturition becomes difficult or impossible. Surgical removal of the gland is the only effective measure.

Impotence—The cause of the inability to perform the sexual act may be due to disease or it may be purely psychological.

Female

Prolapse of the uterus—This not uncommon condition is a dropping of the uterus from its normal position to one much lower down. If not relieved by the wearing of an internal support known as a *pessary* it can be repaired by surgery.

Inflammation of the lining of the uterus—In this condition the endometrium becomes inflamed. Scraping away the abnormal mucous membrane, an operation known as *curetting*, may be effective.

Cancer of the uterus—Occurs most frequently in the cervix though sometimes in the body. If *radium therapy* proves ineffective surgical removal or *hysterectomy* is essential. It is possible to detect cancer of the cervix or even a predisposition to it by taking what is called a *cervical smear.* It is a very wise precaution for women to have this taken, particularly those who have borne children, as suitable measures can be taken if a positive result is obtained before serious conditions develop. The practice, which has been in use in some continental countries for many years, is now being performed in health clinics here.

Tumours of the ovary—The presence of tumours, which are usually of the nature of cysts, necessitates the removal of the ovary, an operation known as *ovariotomy.*

Cancer of the breast—if treated in its earlier stages can be completely cured by surgical removal of the entire breast.

Premature birth—sometimes occurs and the child survives. Loss of the foetus in early pregnancy is called a **miscarriage.** If it takes place before the eighth month of pregnancy it is technically an **abortion** and this is more likely in the earlier than in the later stages of gestation. It may be due to a variety of causes such as disease of the uterus or of the fertilised ovum or even to a sudden shock.

In accordance with the **Abortion Act,** 1967, a pregnancy can be terminated if the life or physical or mental health of the mother or the unborn child would be at stake or if any living children of the mother would be adversely affected if the pregnancy continued. Two medical practitioners must confirm the necessity for an abortion to be performed but if it is a matter of extreme urgency the opinion of one is allowed. All such abortions must be performed in hospitals approved by the Secretary of State.

Normally abortions are not performed beyond twenty weeks of pregnancy at the most and then only if absolutely essential. Up to twelve weeks is considered as safe as there is always a risk of a woman bearing a malformed child or of her becoming infertile if it is performed later. In the short term about 5 per cent. suffer some ill effects but it is difficult to assess the long term at present. The operation can be performed under a local anaesthetic and the patient is able to return home after about three hours. If it is done under a general anaesthetic it is necessary for her to stay in hospital overnight. In the eleven months following the introduction of legal abortion in this country in April, 1968 the number of abortions notified exceeded 32,000. In 1974, it was 170,00 of which 109,820 were residents in England and Wales.

Sterility—The inability of women to conceive may be due to various causes of which one is blockage of the Fallopian tubes. As already stated on p. 190 it may become possible at some future date to overcome this by implantation in the uterus of ova fertilised *in vitro*. Other causes of sterility are abnormality or disease of the female sexual organs and a vaginal secretion lethal to spermatozoa. On the other hand lack of fertilisation may be caused by the male, his spermatozoa being malformed or is insufficient in numbers to ensure it.

Common to Both Sexes

Venereal Diseases

These diseases—there are several, **syphilis** (*pox*) and **gonorrhoea** (*clap*) being the most commonly known—constitute one of the greatest scourges of civilisation and though very occasionally contracted by other means they are usually the result of sexual intercourse with an infected person. The effects of these diseases on the individual, and indeed on the health of the nation as a whole, are enormous and far reaching and it is essential that everyone should have a full and proper knowledge of them right from the time of puberty or earlier. The continuance of venereal disease and its increase in recent years (see p. 200) can only be contributed to promiscuous sexual intercourse. Hence the preference of the medical profession of the name **sexually transmitted disease (S.T.D.)**. Doubtless use of "the pill" has helped to spread infection. The risks are enormous and prevention is the only certain way of eradicating this awful scourge which has reached epidemic proportions. This can only be brought about by knowledge of its causes and ghastly effects. Ignorance and moral laxity play no small part as causitive factors and, for the former at least, there is no excuse in these enlightened days. Now that we have discarded the ridiculous prudishness of our forefathers by whom sex was regarded as something indecent and a subject which must never be mentioned in polite society, there is absolutely no reason whatever why anyone should be ignorant of the facts. Free discussion and advice on television and radio on matters of sex, the posting of warning notices and places of treatment in all appropriate public places, instruction in schools and the publication of down-to-earth literature on the subject written by experts with accurate medical knowledge will go a long way in providing a proper education of the public in this matter.

Free treatment under condition of absolute secrecy is available to those who have contracted the disease but unfortunately V.D. is not one of the diseases which has to be notified by law. Unlike the Scandinavian countries there is no law in this country which compels a sufferer from the disease to seek treatment. In the former countries the number of cases has dropped astronomically since compulsion by law was first introduced, and anyone in this country who falls a victim to any of the venereal diseases should not only seek *immediate* treatment but continue with it until he or she is *completely* cured. This is not only for their own sakes but it is a duty to the people as a whole.

Syphilis—caused by a micro-organism, *Treponema pallidum* (*Spirochaeta pallida*) (see Fig. 25.5, p. 330) and may be contracted not only by sexual union but by kissing anyone with a syphilitic sore on their lips or even by contact with a cup or glass used by them. It can be transmitted by a blood transfusion from a syphilitic. The importance of blood tests on donors is therefore clearly imperative. A pregnant woman suffering from syphilis can infect her foetus which will either be born with congenital syphilis or be born dead. The disease is slow in development but may last for thirty or forty years and can cause death after the most ghastly suffering. Every system of the body may be affected as the result of it. The disease is divided into three stages. In the *primary stage* a chancre or sore appears at the site of infection and the lymph nodes become swollen. About two months later the *secondary stage*, which is an extremely infectious one, develops.

Fever, rash, headache and mental deterioration occur in this stage. The *tertiary stage*, which may appear after several months or may be delayed for several years may lead to disease of the muscles, bones, arteries and nervous system and general paralysis of the insane (G.P.I.) may result. Syphilis may be transmitted to their children by either parent suffering from it; this is known as *congenital syphilis*. Certain identification of the disease can be confirmed by a test on the blood of the suspected victim which is known as the *Wasserman reaction*. If treated with penicillin in the early stages there is a 100 per cent. chance of a complete cure being effected.

Gonorrhoea—caused by a bacterium, *Gonococcus*, creating inflammation of the urethra in the male and of the vagina in the female in the first instance and the infection may spread to the associated sexual organs and to the bladder. This is the commonest and most contagious of the sexually transmitted diseases. An average of 1.128 new cases occur every week compared with 66 of syphilis. Women often show no symptoms and are unaware of their infection until complications set in but they can infect men. The disease can cause sterility in both sexes, women developing blockage of the Fallopian tubes. Homosexuals can also contract the disease. Treatment is with penicillin. A recent survey in the U.K. showed that 1,639 people had become infected from one single original source as the result of promiscuous intercourse.

Chancroid—or *soft sore* is caused by the bacterium *Haemophilus ducreyi* and is a localised disease in which a painful ulcer develops on the external genitalia, the infection spreading to the lymph nodes in the groin. With this disease there is always the possibility of syphilis having been contracted at the same time. It usually responds to treatment with sulphonamides.

All venereal diseases can be cured by the use of appropriate antibiotics, a method which has superseded the earlier use of sulphonamides, though these are still used in the case of chancroid. The Wassermann Test must be performed in the case of syphilis for some years after apparent cure has been effected to ensure that it is complete.

Non-specific urethritis—caused by various micro-organisms, a fairly new S.T.D. which shows no symptoms in women though they are usually the carriers. It is the fastest spreading of the diseases, 68,574 cases of men suffering from it having been recorded recently. Arthritic and eye complications are but two of its possible effects.

Trichomonas vaginalis affects women and is transmitted by men. It has been stated that over a million women have the germ which is a protozoon.

The S.T.D. are no respecters of persons, 346,812 men and women of all classes visited V.D. clinics in Great Britain for the first time in 1973 and of these 50 per cent. of the men and 30 per cent. of the women were infected either with syphilis or gonorrhoea or both and these figures are probably at least 25 per cent. below the real number which has more than doubled in the last ten years. It is interesting that although the number of cases of S.T.D. are on the increase, deaths decreased from 150 in 1972 to 133 in 1973 and 121 in 1974.

Homosexuality

All human beings are to some extent a mixture of homosexual and heterosexual make up, determined in part by their genetic patterns. There is probably no such thing as a completely perfect male or female. In fact something of the female in the male is doubtless a useful ingredient, perhaps making his nature more sympathetic, and it is doubtless this female factor which manifests itself, for example, in creative artists and musicians. Equally, some maleness in the female may make her more self-reliant and better able to face life's difficulties. But excess of the opposite sex's character in either is certainly not desirable, lessening the natural attraction for the opposite sex.

The practice of homosexuality, however, now legalised in this country between consenting adult males but never illegal between females (lesbianism), must be regarded as abnormal, though homosexuals in the main would doubtless take the opposite view. The cause is probably partly glandular, partly psychological and partly environmental. Cure is difficult and is only possible when the desire for a cure exists in the individuals concerned.

It seems likely that it may now be possible to bring under control abnormal sexual tendencies including homosexuality in men by the administration of a synthetic female hormone, oestradial. This has the effect of neutralising the effects of the male hormones without causing impotence. A number of men in one of our prisons who had been convicted of serious abnormal sexual offences were treated over a period of two years with such marked success that the treatment was extended to other prisons.

PRACTICAL WORK

1. Examine **dissections of the reproductive systems of a male and female rabbit** and identify the organs described in the text. (*n.b.* the male rabbit has two separate scrotal sacs and lacks seminal vesicles. It also has a structure known as the uterus masculinus on the dorsal side of the urethra where it is joined by the vas

deferens. In the female there are two uteri and the urethra joins the vagina to form a urino-genital canal called the vestibule instead of remaining separate from it as in the human subject).

2. Examine a slide of a **T.S. of the testis**. Note the seminiferous tubules and developing spermatozoa. Look for the intersititial cells in the connective tissue between the tubules.

3. Examine a **T.S. of the penis** and identify the corpora cavernosa and the corpus spongiosum containing the urethra.

4. Examine a slide of **human spermatozoa** and note the head, middlepiece (or body) and tail.

5. Examine a **T.S. of the ovary.** Identify the germinal epithelium and the stroma containing Graafian follicles in various stages of development. Study a mature follicle and examine its internal structure and that of the developing ovum within.

6. Examine a slide showing **meiosis** or the **Bioset "Human Meiosis".** Identify the various stages.

7. Examine a **section through a lactating mammary gland.** Note the lobules held together by connective tissue and lined by epithelium. Some of these lobules will contain milk droplets.

8. Examine slides of **Treponema pallidum** and **gonococcus** in infected tissues.

9. The **Bioset** entitled **"Venereal Disease"** has a series of good micrographs of S.T.D. bacteria.

Chapter XVIII

INHERITANCE, VARIATION AND RACE

HEREDITY

It is an accepted fact which is taken for granted that though a child shows marked differences from its parents it always bears some bodily resemblances to them and often children show some similarity to their grandparents. This is not necessarily restricted to body characters such as facial features, build, height and so on but extends to intellectual and psychological make-up. This is well illustrated in the Darwin family right through from Charles Darwin's grandfather to his own grandsons, all of whom were (or are) men of considerable scientific ability.

The means by which this comes about is a matter of considerable interest and importance and we have to thank an Austrian monk, **Gregor Mendel** (1822–84) for setting the ball rolling in the latter part of last century when he made several important discoveries as the result of experiments he performed with plants in the garden of the monastery of which he was abbot. The results were not sufficiently widely published and this was at a time (1866) when the heated controversy on Darwin's *Origin of Species* was in progress and so their importance was unfortunately overlooked until the beginning of this century when a Cambridge biologist, Bateson, and others realised the significance of Mendel's discoveries. Since that time an enormous amount of research work has been done in plant and animal **genetics**, as the study of inheritance is called, and in the modern field of molecular biology some quite startling discoveries have been made. The inheritance of body characters is under the control of what Mendel called *factors*. We now know that these are protoplasmic units in the chromosomes and we call them **genes**. These are situated in definite places on the chromosomes called **loci** and are transmitted from parent to offspring in the gametes. The gene was first isolated from cells of the African Clawed Toad at the Institute of Genetics in Edinburgh and later in Harvard from a bacterium, *Bacillus coli.* While this opens up great possibilities for the human race, for example in curing hereditary diseases if techniques are discovered by which human genes could be injected, the risks are enormous and, in fact, frightening if used by an unscrupulous person or body. But it may well be decades before research has discovered how to change the genes of lower animals, let alone man's, so that hereditary health can be controlled or personality altered. As we already know the chromosomes are in pairs in the nuclei of the body cells, or **soma**, one member being derived from one parent and one from the other and the members of a pair are known as **homologous chromosomes**. There will be two genes for a particular character occurring on identical loci on the separate members of the pair. If the two genes are similar they are said to be **homozygous** and the individual a **homozygote** for that particular character but **heterozygous** if they are different and the individual is then a **heterozygote**. These genes are referred to as **allelomorphs** or **alleles**. For example, there may be a gene for black hair and an alternative one for red hair. These are alternative allelomorphs. Now we have already seen that the gametes contain the haploid number of chromosomes and, to continue with our illustration, the gene for hair colour will either be for black hair or for red hair; it cannot be for both. At fertilisation, therefore, the zygote may have two genes for black hair, two genes for red hair or one of each. A child possessing two genes for black hair or two genes for red hair is a homozygote but one having one gene for black hair and one for red is a heterozygote. What coloured hair will this child have? This is the kind of question we have to solve in genetics.

Two laws were formulated by Mendel and his *first law* is known as the **Law of Genetic Segregation**. It states, in modern terms, *the genes for two allelomorphs are separated in the formation of the gametes so that only one appears in each gamete.* The *second law,* known as the **Law of Independent Assortment**, concerns two or more pairs of genes and states that the *behaviour of the genes for any one pair of allelomorphic characters is independent of the distribution of the members of other pairs on different chromosomes and they therefore undergo independent assortment.* It should be added at this point that there are exceptions to this due to what are known as crossing over and linkage. (These terms are fully explained on pp. 207 and 208.

Let us now examine some specific examples and first with one pair of allelomorphic characters. This is known as **monohybrid inheritance**. A **hybrid** is the offspring from parents carrying different allelomorphic characters.

We will suppose that a right-handed man with no family history of left-handedness marries a left-handed woman both of whose parents are left-handed. Both husband and wife are homozygotes. The husband's gametes will contain the gene for right-handedness (R) and the wife's will contain that for left-handedness (r). It is customary to use the same letter for allelomorphic genes, a capital letter being used for the dominant gene and a small letter for the recessive one (see below).

It will be seen that the children's genes are Rr and an identical result would be obtained if the father was rr and the mother RR: it has no connection with sex. Their children who constitute what is called the **First Filial (F1) Generation** are all right-handed. The

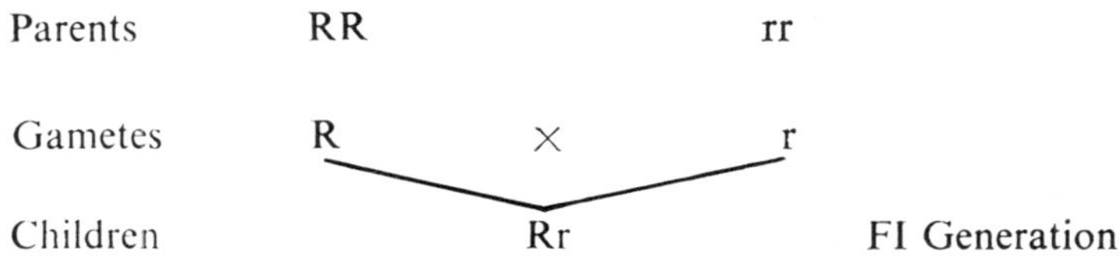

gene R is therefore said to be **dominant** to the gene r which is **recessive**. A character is said to be dominant when it appears in a heterozygote to the exclusion of its alternative allelomorph and a character is recessive when it fails to appear in a heterozygote in the presence of its alternative allelomorph.

The father (RR) and the children (Rr) are identical with regard to this particular character and are therefore said to be **phenotypically** the same but as their gene set-up is different they are said to be **genotypically** different. The terms **phenotype** and **genotype** are respectively used for these individuals depending as to whether the mere character or the genetic constitution is being considered.

Now let us consider the result of a marriage between such heterozygous types (Rr), obviously from different parents or they would be brother and sister.

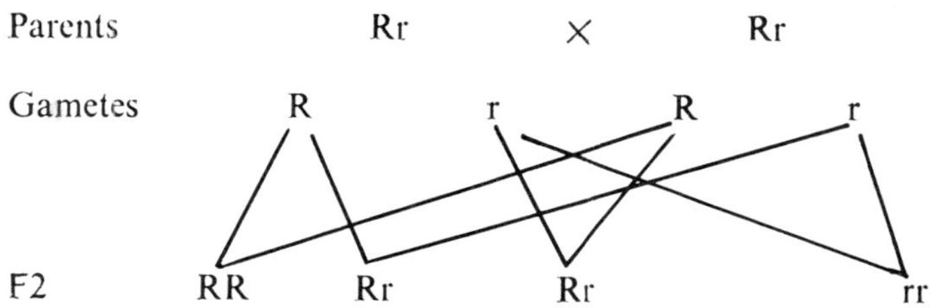

This is the **Second Filial (F2) Generation** and it will be seen that one child will be left-handed, being a homozygote and three will be right-handed. Of these three one will be a homozygote (RR) and the other two will be heterozygotes (Rr). In turn all three could pass on the right-handed gene to their children but two of them could also pass on the left-handed gene. It should be pointed out at this stage that it does not follow that this would be the result obtained in one family of four children. If, however, the progeny of a thousand or more such families were investigated, the result would correspond at least very closely to that shown above.

We must now consider the results of marriage between the heterozygous Rr and the individuals of the parental types. Such a mating is called a **back cross**. First we will examine such a back cross with a homozygote of the dominant type.

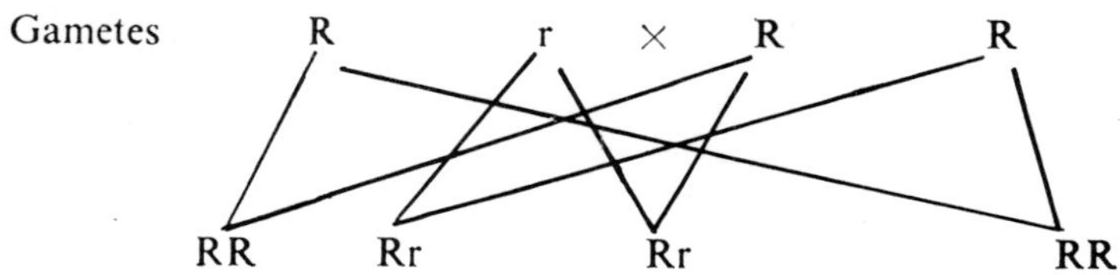

All the children will be right-handed and 50 per cent. of them will be homozygotes and 50 per cent. will be heterozygotes.

Finally consider a back cross with a homozygote of the recessive type.

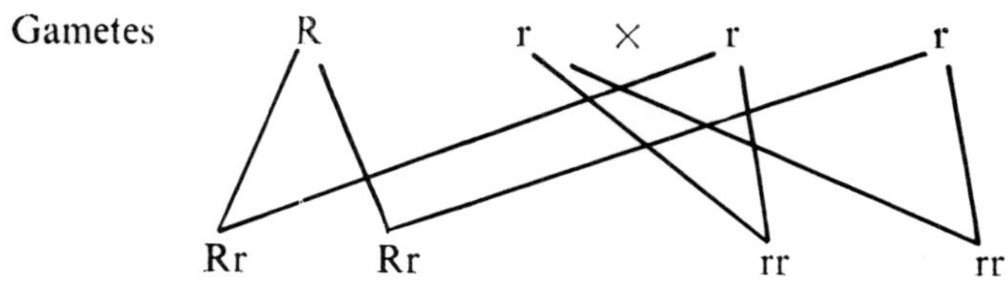

Half the children will be right-handed and half will be left-handed.

Again these ratios will be obtained only when a large number of families are examined.

In illustration of Mendel's Second Law we must consider the inheritance of two pairs of allelomorphic characters, and this is known as **dihybrid inheritance**. Let us take the case of black hair (H) which is dominant to red hair (h) together with brown eyes (E) which are dominant to blue eyes (e). The genetic constitution of an individual with black hair and brown eyes will be HHEE and one with red hair and blue eyes will be hhee if both are homozygotes. The children of the F1 generation will carry all four genes and will all have black hair and brown eyes.

Gametes	HE	×	he
F_1		HEhe	

According to Mendel's Second Law, assuming that there is no linkage and no crossing over takes place, nothing will interfere with the random distribution of these characters and they will not be retained in their original combinations in the subsequent (F2) generation. But there are only four possibilities *viz.* Black hair and brown eyes (HE), black hair and blue eyes (He), red hair and brown eyes (hE) and red hair and blue eyes (he). To work out the F2 generation we must use what is known as the "chess board" method.

F2	HE	He	hE	he
HE	HE HE	He HE	hE HE	hE HE
He	HE He	He He	hE He	he He
hE	HE hE	He hE	hE hE	he hE
he	HE he	He he	hE he	he he

HHEE	(= black hair and brown eyes)	9
HHee	(= black hair and blue eyes)	3
hhEE	(= red hair and brown eyes)	3
hhee	(= red hair and blue eyes)	1

Note that this shows 9 with both dominant characters, 3 with one dominant and one recessive, 3 with the other dominant and the other recessive and 1 with both recessives. This is a combination of the 3:1 ratios. If the **trihybrid ratio** is considered, in which three pairs of allelomorphic characters come into play, we shall still get a combination of this ratio namely 27:9:9:9:3:3:3:1.

Clearly in accordance with Mendel's Second Law there is an obvious means whereby genes which have appeared separately can be brought together in new combinations, a fact which is of considerable advantage in allowing beneficial characters to be brought together in an individual. Equally, it may be argued, harmful characters can be distributed, though to a lesser degree as they are usually recessive. Furthermore dominance is not always complete and genes are in fact influenced by neighbouring ones as a **gene complex**. In fact, the final result of the development of an individual is the product of the action of all the genes acting in concert. Recessive genes may remain dormant for several generations and then reappear. Such is the case with the disease haemophilia.

It does sometimes happen that there are more than two allelomorphic phases of a gene which produce **multiple allelomorphs**, though only two can appear in the same individual, one on each corresponding locus of a pair of homologous chromosomes.

Mendel experimented with different varieties of peas and studied such alternative allelomorphs as tallness and dwarfness of the plants, yellow and green colour of the seeds and the round and wrinkled nature of the seed coats. He discovered that tallness of the plants, the yellow colour of the seeds and round seed coats were the dominant characters.

If you work on lines similar to those used above for human characters, you will arrive at similar results, which were those obtained by Mendel. Some examples of dominance and recessiveness in human beings are shown in the following table:—

Dominant	Recessive
Brown eyes	Blue eyes
Dark hair	Fair hair
Wavy hair	Straight hair
Right handedness	Left handedness
Normal lip	Hare lip
Normal blood	Haemophilia*
Normal pigmentation	Albinism
Colour perception	Colour blindness*
Congenital cataract	Normal eye

* Sex linked characters

Crossing Over

Crossing over occurs when interchange between corresponding

parts of two chromosomes takes place during meiosis, thus producing a different combination of genes in the gametes and therefore of characters in the offspring. We will see what happens in the process. It will be recalled that at one stage in meiosis homologous chromosomes, already split longitudinally into two chromatids, are closely coiled round one another. A break probably occurs in corresponding

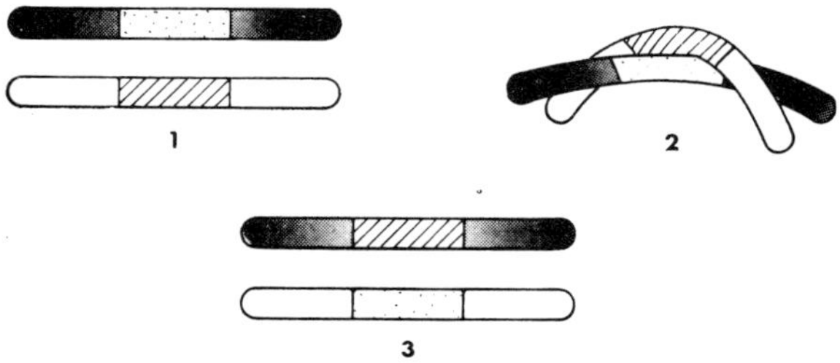

Fig. 18.1. **Crossing over.**

positions on the adjacent chromosomes and reciprocal interchange of genetical material takes place. This results in certain characters previously occurring together failing to do so in the completed gametes, the alternative allelomorphs replacing one or more of these characters. The significance of the behaviour of the chromosomes during meiosis is now evident. At fertilisation gametes with different gene complexes are united and consequently the zygote gives rise to an individual with characters, some of which are different from those of the parents, but some are similar to them.

Linkage

Linkage takes place when two genes on the same chromosome are always inherited together as if they were one and never separate in any offspring. Let us take a hypothetical case. Suppose that red haired people always had blue eyes and black haired people always had brown eyes and that red haired people with brown eyes and black haired people with blue eyes were unknown. Clearly the two characters, hair colour and eye colour would be inherited as if they were a single character *i.e.* as monohybrid and not dihybrid inheritance. We find that certain characters are common to one sex only and this is known as **sex linkage**, for sex is an inheritable character as much as any other. The inheritance of haemophilia and colour blindness are examples of this and will be discussed later.

Inheritance of Sex

It has already been stated that man has twenty-three pairs of chromosomes. Those which are responsible for the inheritance of body characters are in homologous pairs and are known as **autosomes**. But one particular pair differs from the others and these are the **sex chromosomes** which determine whether the offspring shall be male or female. In the female they are identical to one another and are referred to as the **X** chromosomes but in the male one is identical to the female's **X** chromosome but the other is smaller and is called the **Y** chromosome. So that a woman is **XX** and a man **XY**. In gametogenesis therefore as the result of meiosis all the ova will contain one X chromosome but half the spermatozoa will contain an X chromosome and half will contain Y. If, at fertilisation an X-bearing spermatozoon fuses with an ovum the zygote will be XX and will be female but if a Y-bearing spermatozoon fertilises an ovum the zygote will be XY and will be male. Sex is thus determined at the moment of fertilisation and clearly it is the male of the human species who determines the sex of the children.

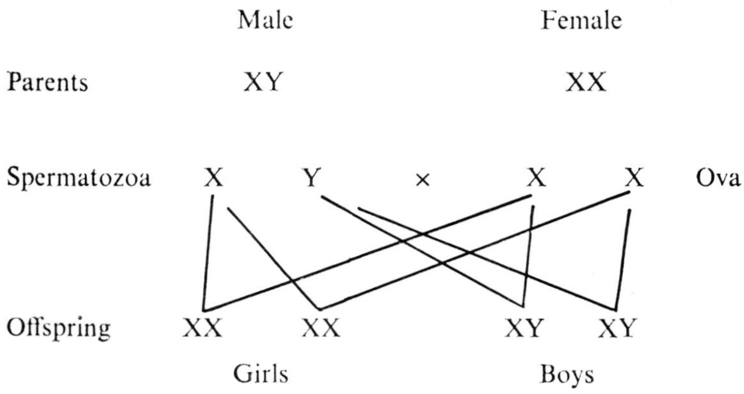

Obviously the chances of a baby being a boy or a girl are equal. Preponderence of one sex or the other in the population is due to outside factors and influences such as disease and war. Cases of semihermaphroditism, referred to as *inter-sex* are known, but these are the result of abnormal development of an embryo due to other causes and whose sex has already been determined at fertilisation.

The development of the secondary sexual characters is under the control of hormones secreted by specialised cells in the testes and ovaries and the development of the gonads is initially controlled by the sex chromosomes, there being only female-producing X chromosome in the female but male-producing Y chromosomes in the male. Thus the fertilised ovum, capable of becoming either a male or a female, will be directed by the X and Y chromosomes to develop ovaries and not testes or testes and not ovaries as the case may be. Having developed the appropriate gonads, the hormone producing cells within them control the development of the secondary sexual characters of that sex at the time of puberty.

Abnormalities in chromosome number have been observed in a small percentage of mentally ill criminals and in a murder case in Australia a few years ago a man was acquitted on the ground of insanity, his mental condition being attributed to the fact that he had an extra Y chromosome which was responsible for brain damage. In fact most violent crimes are committed by people with abnormal chromosome numbers. This does not necessarily mean that a male with XYY chromosomes will be a violent criminal but it may exert an aggressive influence on his behaviour. Extra X chromosomes may also occur in the male (XXY). This is not apparent before puberty when it results in the imperfect development of the testes and may cause the formation of small breasts. Such individuals are often mentally retarded. An extra X chromosome can occur in the female causing masculinity and infertility. These individuals are also mentally subnormal. Other abnormalities in the sex chromosomes can cause **hermaphroditism**. Mongolism is caused by an extra chromosome (see p. 283), and abnormality in the autosomes can be responsible for metabolic disorders.

SEX LINKAGE

Two inheritable diseases which appear in man but rarely, if ever, in women are red/green colour-blindness and haemophilia. These are clear cases of sex linkage, the genes for the diseases being located in the sex chromosomes and they are usually recessive. It is interesting to note that in spite of the fact that only men inherit the disease, it is the women who are responsible for this inheritance though they themselves do not suffer from it.

Referring to the sex chromosomes, if XX and XY represent normal females and normal males respectively and X′ represents the chromosome carrying the colour-blindness gene, then the result of marriage between a normal male and a colour-blind female carrier (but not a sufferer) will be as follows:—

Parents Carrier Female XX' XY Normal Male

Gametes X X' × X Y

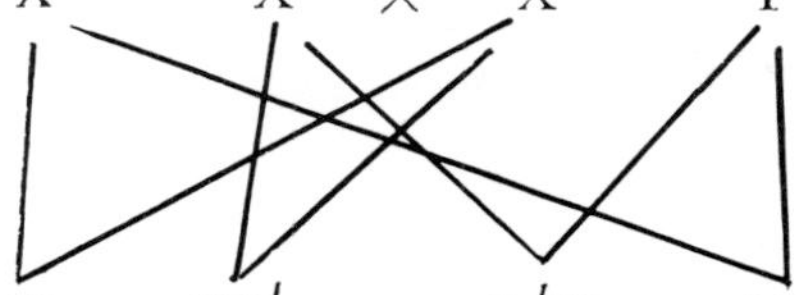

XX XX' $X'Y$ XY

Children Normal Female, Carrier Female, Colour-blind Male, Normal Male

This, it will be seen is 1 normal female, 1 normal male, 1 carrier female and 1 colour-blind male.

Now suppose that a colour-blind man marries a female carrier.

Parents Carrier Female XX' $X'Y$ Colour-Blind Male

Gametes X X' × X' Y

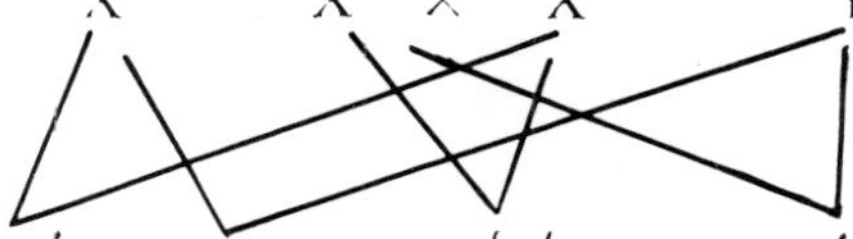

Children XX' XY $X'X'$ $X'Y$

Carrier Female, Normal Male, See below, Colour-Blind Male

1 carrier female, 1 normal male, 1 colour-blind male and, it would seem, 1 colour-blind female but colour-blindness in females is extremely rare and it is highly probable that the combination $X'X'$ is fatal to the developing embryo.

If you now consider the inheritance of haemophilia on exactly similar lines, identical results will be obtained and again, haemophiliac women ($X'X'$) are unknown.

The Inheritance of Blood groups

You will recall that human blood can be classified into four Groups A, B, AB and O based on the presence in the red corpuscles of agglutinogens (antigens) and of agglutinins (antibodies) in the plasma and that reaction occurs between some of these resulting in agglutination. These blood groups are under the control of multiple allelomorphs and A and B are both dominant to O. The genetical constitution of Group A may be AA or AO, that of B may be BB or BO while AB is AB and O is OO. If the blood groups of the parents are known it is clearly possible to predict the possible groups of the children as will seen from the following table:—

Parents' Groups	Children can only be
A & A	A or O
A & B	A, B, AB or O
B & B	B or O
AB & AB	A, B or AB
A & AB	A, B or AB
B & AB	A, B or AB
AB & O	A or B
A & O	A or O
B & O	B or O
O & O	O

The children cannot belong to any group other than those shown above. This can be used in medical jurisprudence when the paternity of a child is in dispute, but only negative evidence is, of course, admissible. If, for example, the husband and wife are both group A and the child is group B, clearly the husband is not the father of the child. But if the child is also group A, he could be its father though the parent could equally be another man whose blood is group A.

Rh positive blood is dominant to Rh negative but complications can arise during pregnancy if the mother of an Rh positive child is Rh negative as this may result in the death of the foetus.

VARIATION

Continuous Variations

If the heights of a large number of men in a particular race are measured it will be found that they vary between two extremes with few at the two extremes and with the greatest number at a particular height between them. These records can be plotted as a graph with the heights on the x axis and the numbers at any particular height on the y axis (Fig. 18.2). It will be seen from the graph that the majority of the men are 5 ft. 8 ins. in height and this is known as the **norm**. It is, of course, the average height. These variations in height are inherited as the result of

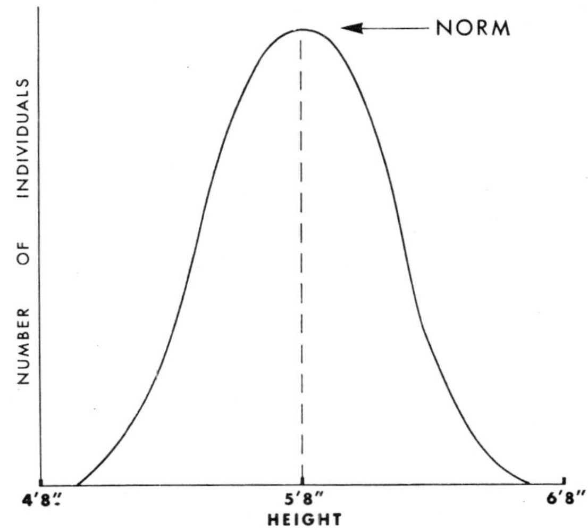

Fig. 18.2. **Continuous variation heights of men.**

the action of genes in the chromosomes. This is an example of what is called **continuous variation**. Marriages between tall and short parents often tend to give rise to children who approach the norm and it does not follow, for example, that two tall parents will necessarily produce equally tall children.

A natural question which will occur to you, no doubt, is are characters acquired during the life time of an individual inherited? All the evidence goes to show that they are not although it was formerly thought to be the case. Even Charles Darwin accepted the theory first put forward by Lamarck in the 18th century that acquired characters were passed on to subsequent generations. Let us take a simple example. A white man who spends his adult life working in the tropics will develop pigment in his skin to protect him against the harmful rays of the sun and natives are born with dark skins. Children of the white man are born with white skins though they will, of course develop the pigment if they are brought up in the same climate, but not if they return to their native land. Does a man or woman who learns how to paint or play a musical instrument and who becomes a brilliant painter or musician beget brilliant painters or musicians? Not necessarily so and if they do so become it is not because their parents developed these characters during their lifetimes. No such character can develop unless it is inherently or potentially there in the genes of the parents. Environment can certainly play a very important part in producing variation of character *during the lifetime* of an individual but only in that individual. These characters are not passed on to subsequent generations though if the children are brought up in a similar environment they will, in time, develop similar characters. The importance of a good family environment and of good education cannot be over estimated. No character will be transmitted to the next or subsequent generations unless the genes are affected.

In conclusion let us examine the cases of **twins**. Only one ovum is normally produced during each menstrual cycle but if two are formed and each is fertilised the result will be twins. They will show similarities to one another and to their parents and they may be of the same or of different sexes. Alternatively, twins may arise from a single fertilised ovum from which the cells at the first division have separated and developed into two individuals. These are known as **identical twins**. Having the same chromosomes and genetic set-up they develop along similar lines and are always of the same sex. There is always a close psychological link between such twins. The so-called "Siamese twins" are produced in a similar manner but the separation is incomplete. They can in some cases, be successfully separated, depending on the way in which they are united. The original Siamese twins were so-called because they were born in Siam though the mother was half Chinese and half Malaysian, the father being fully Chinese. The twins, some five feet in height, were both male and were connected by a band of tissue little over 12 cm. in length between an umbilicus common to both and the lower end of the thorax. Strange as it may seem, they married two sisters and between them produced twenty-two children, all of whom with two exceptions were normal! The twins died within two hours of each other at the age of sixty-three.

Mutations

Periodically individuals appear bearing characters markedly different from those of their parents. Often these are of the nature of deformities. For example, children are sometimes born with extra fingers or toes (*polydactyly*) or a child may be a haemophiliac when there appears to have been no history of it in the family. These characters arise spontaneously and clearly indicate the inception of a sudden genetical change. They are known as **mutations** or **dis-continuous variations**. They are usually disadvantageous. We had an instance of this in our own Royal family. Daughters of Queen Victoria passed on the condition of haemophilia by marriage to members of the former Spanish Royal family yet there was no evidence of haemophilia in Queen Victoria's pedigree. Evidently it arose as a mutation either in Queen Victoria herself or in one or other of her parents. She did not pass on the disease to any of her own sons. The explanation as to how mutations come about is obscure beyond the fact that it is a sudden change and that it is inheritable. It could be that during nuclear division in gametogenesis a portion of a chromosome breaks off and joins up in a different place or it could be caused by a change in the chemical nature of a gene. In 1927 H. J. Müller discovered that mutations were induced in the fruit fly *Drosophila* by ionic radiation from X-rays and that these mutations were similar to those which occurred spontaneously. So ionising radiations which occur naturally from cosmic rays in outer space through the earth's atmosphere, from rocks containing radio-active substances and in man's own tissues may well have been at least partial causes of mutations in man's past history for he has always been exposed to them and his genes could certainly have been affected by them. Cases have been recorded in which mutations are dominant as is the case with polydactyly but they are usually recessive. With the possible exception of haemophilia they are very infrequent but once they have arisen they are always liable to recur. *Sickle-cell anaemia,* so-called because the erythrocytes become sickle-shaped due to abnormal haemoglobin, is a dominant mutation. It is almost entirely confined to Negroes. Another dominant mutation is *thalassemia* which occurs amongst Mediterranean people and in which foetal haemoglobin continues to be produced after birth, usually causing death in childhood.

Mutations occur at very rare intervals and those beneficial to a species give it better chances of survival. Such beneficial mutations in different individuals become united in their offspring and, fortunately, most harmful mutations are recessive. One thing is absolutely certain and that is that no new character can be inherited unless the genes are affected to produce such change. Mutations have been artificially induced in some animals not only by X-rays but also by formaldehyde and mustard gas (used in the 1914–18 war but fortunately not in the last Great War). Amongst survivors of the atomic bombing of Hiroshima at the end of the 1939–45 war there has been a marked increase in the number of cases of cancer, particularly leukaemia, and there is some evidence which indicates that undesirable mutations in the offspring of survivors of the bombing may have arisen from the effects of radiation on the genes of the parents who were pregnant at the time, some of the children being mentally retarded and some with smaller heads than normal. But there is no evidence so far of any effect on their genes and therefore of future generations. It has been shown that the cells in the red marrow of the bones are very sensitive to radiation and are easily destroyed by the excessive development of leucocytes which results from this. Furthermore, sub-lethal doses may damage the genes and so cause mutations in offspring. Radio-active fall-out due to the release of radio-active substances into the atmosphere from nuclear explosions emits radiations which may enter the body direct or by food contaminated by it and do untold harm. The most dangerous appears to be Strontium 90 which becomes fixed in bones and may remain there indefinitely and this leads to bone cancer. Plutonium, produced in nuclear reactions, is also very suspect and serious research is still continuing on its harmful effects on the body. Others are comparatively transient. The danger of this radio-active material lies in the fact that damaged or broken chromosomes may result (this has been observed) and union may take place between the broken parts and other chromosomes. The effect of this on the genes must be obvious.

It is important that the use of X-rays should be carefully controlled, as there is considerable risk associated with it due to radiation and the protection of testes and ovaries is particularly important as the germ cells could be adversely affected by the rays with a risk to future generations. Equally important is the use of X-rays during pregnancy as their use in the early stages can cause an abortion or deformity in the child. A national survey was set up in 1975 to investigate the adverse effects, if any, of X-rays on genes. Conclusions are not expected for at least two years.

RACES

All human beings belong to the Genus and species *Homo sapiens* and yet there is an enormous amount of variation between the members of the species. This has given rise to the production of **races** consisting of people with common physical characteristics. What are the criteria which determine to which of these races a particular individual belongs? Perhaps the most obvious is the colour of the skin but this in itself is not a deciding factor. The depth of colour depends on the amount of the pigment *melanin* present in the Malpighian layer of the epidermis. When it is present in a very small amount we have a white skin and if the amount is large a black skin whereas in yellow and olive skinned peoples it is more moderate in quantity. Next, perhaps, is the character of the hair which varies from the wavy, fair hair of Europeans to the coarse straight hair of Mongols with the curly hair of an intermediate texture in Negroes. The hair of Europeans is oval in cross section while that of Mongols is round. The colour of the hair, like that of the skin, is dependent on the amount of pigment present. In Europeans it is fairly small giving brown hair or very small giving blonde hair but in Negroes and Mongols whose hair is black it is present in abundance. Facial features also serve to distinguish races. In Europeans the face is long, noses are narrow and slightly convex and the lips are thin whereas in Negroes the face and nose are broader and shorter and the lips thick. In Mongols the face is broad and flat and the cheek bones are characteristically prominent. The shape and colour of the eyes are also characteristic and it is interesting to find that blue eyes are peculiar to Northern Europeans. The ratio of the length to the breadth of the skull, technically known as the *cephalic index*, the volume of the cranium, or *cranial capacity*, and the extent to which the forehead slopes and the lower jaw protrudes, *prognathism*, all help to determine the shape of the head. There are also some noticeable differences in height as will be appreciated if Pygmies are compared with Europeans. Finally the predominance of blood groups varies in different races. For example, the majority of Europeans and Negroes are in Group O and the least in Group AB while the percentage of Group B is three times as great in Negroes as it is in Europeans.

All these factors play their part in the classification of races but there seems to be no universal agreement as to which and to what extent each should form the basis for such classification. It is perhaps difficult to define a pure Race for it must be borne in mind that intermarriage between different races often occurs and this produces children showing characteristics of both races, for a great deal of emigration has taken place over the ages so that one race tends to merge into another. So we must not pay too much attention to geographical distribution.

Race Classification showing Characteristics and Distribution

Nordic—Fair, wavy hair. Fair skin. Narrow face and nose. Long head and prominent chin. *Northern* and *Western Europe.*

Alpine—Darker wavy hair. Fair to brown skin. Round, broad head and narrow nose. *Central Europe.*

Mediterranean—Dark, wavy hair. Fair to brown skin. Long head and narrow nose. *Mediterranean countries, N. Africa, Near and Middle East, India, S. Asia.*

Australoid—Dark, wavy hair. Sallow or brown skin. Long head and narrow nose. Prominent brow ridges. *Australia, S. Asia.*

Negroid—Black, curly hair. Black skin. Round head and broad flat nose. Thick lips. Protruding lower jaw. *Africa.*

Mongoloid—Dark, straight hair. Yellow to reddish skin. Round head with small flat nose. Slanting eyes. Prominent cheek bones. *E. Asia, China* and *Japan, S. America.*

Chapter XIX

CELL STRUCTURE AND THE CONCEPTS OF MOLECULAR BIOLOGY

Modern research using the electron microscope has enormously enlarged our knowledge of cell structure and recent discoveries of biochemists in the field of molecular biology our knowledge of the functions of the cell contents and it is hoped that the following very much simplified account will give the student an insight into this fascinating and far-reaching subject.

A very simple definition of a cell is a unit of protoplasmic matter containing a nucleus though, as we shall see in Chapter XXV, this does not visibly apply to the bacterial cell, but there are many cell inclusions in the cytoplasm which are of considerable importance and what of the chemistry of the molecules of which the protoplasm is composed? The all-important modern study which inter-relates biochemistry and cell structure with the concept of self-replicating molecules which guide the cell to synthesise proteins in a definite orderly manner is known as **molecular biology**. These are the matters which we shall consider in this chapter.

The **electron microscope** is an instrument in which beams of electrons are focussed just as light is focussed in the optical microscope. The electrons are produced by a heated filament and adjustment is by an electro-magnetic field which can be varied as required. The much shorter wave-length makes considerably higher magnifications possible. A maximum direct magnification of 500,000 times can be increased to 4,500,000 if the specially fitted binoculars are used. Sharp images of extremely minute objects down to 3 to 5 $m\mu$* can be obtained on a fluorescent screen or on a photographic plate and this has made possible a very detailed study of the cell inclusions.

More recently a remarkable scanning instrument known as the **Stereoscan electron microscope** has been developed. The image, which is produced in a different way from that in the conventional type can be seen directly and in three dimensions. The magnification, though less than that possible with the ordinary kind of electron microscope, being about 100,000 times as a maximum, is adequate for most purposes. These cover a very wide field which includes medical research. It is hoped that many hitherto baffling problems may prove soluble with the use of this instrument, which, incidentally, is an entirely British invention.

* $m\mu$ = millimicron = 0·000001 mm.

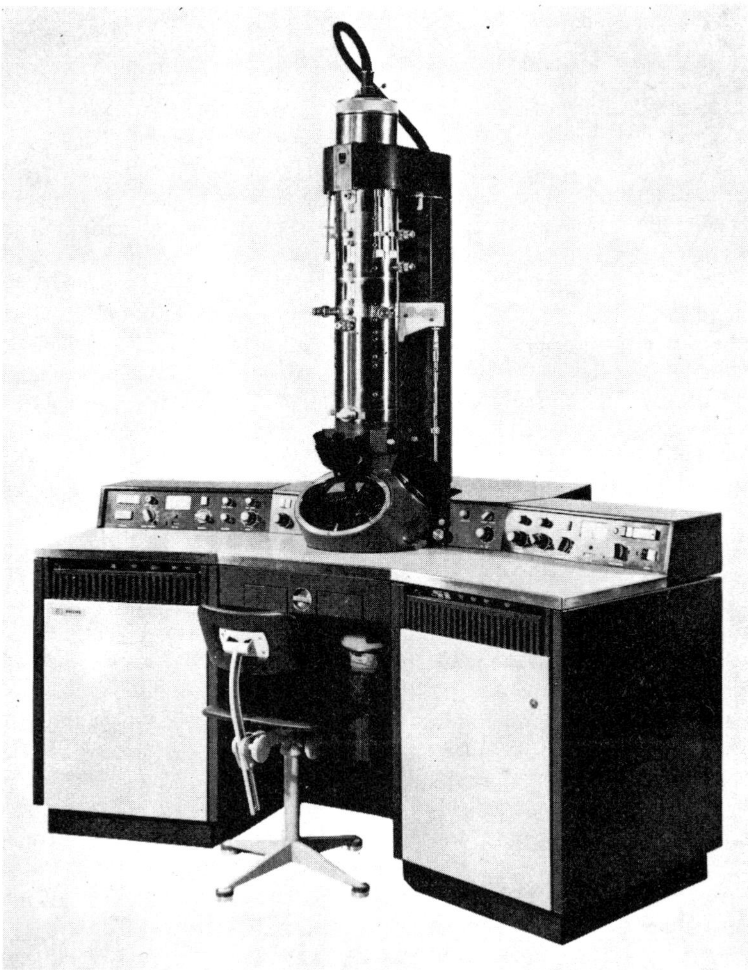

FIG. 19.1. **An electron microscope.**
(*By Courtesy of the M. E. L. Equipment Company.*)

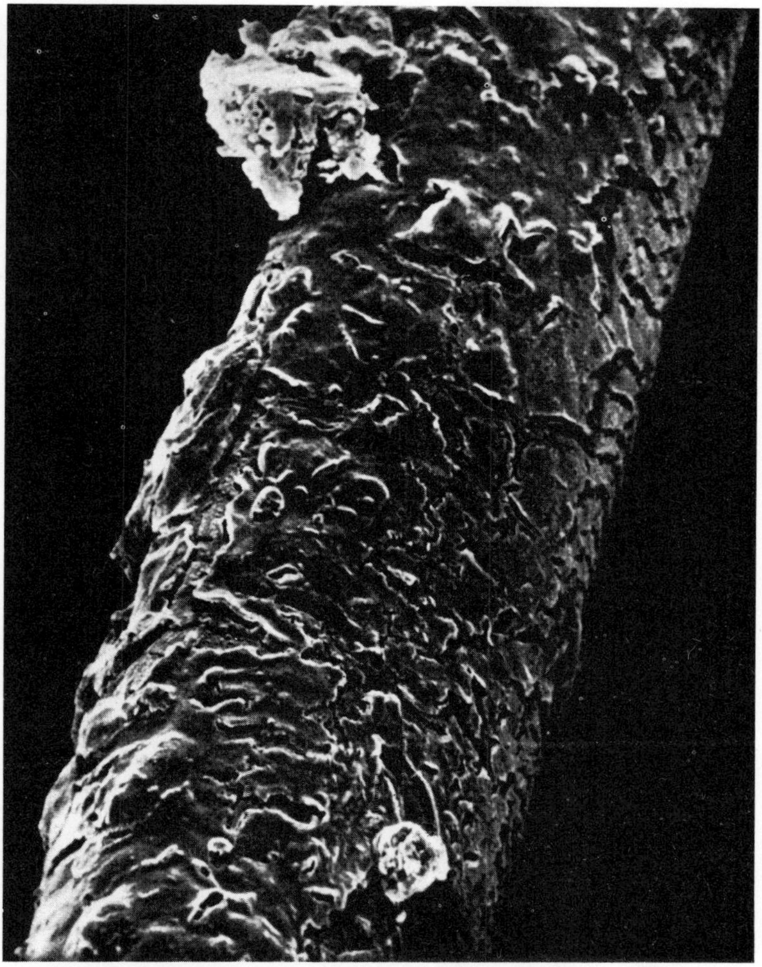

Fig. 19.2. **Electronmicrograph of part of a human eyelash as seen with the stereoscan electron microscope. Original magnification × 940.**

(By courtesy of Cambridge Scientific Instruments Limited.)

Cell Structure

The animal cell is enclosed in a **plasma membrane** through which oxygen, water and dissolved substances move into the cell and carbon dioxide and excretory substances out of it. This plasma membrane is considered to be composed of lipid material lined on both sides with protein and appears to be around 70 Å* in thickness. Within the plasma membrane is the **cytoplasm** in which lies the **nucleus** and various **cell inclusions**. The cytoplasm is composed of a matrix known as **hyaloplasm** containing an **endoplasmic reticulum**. The latter consists of small sacs or vesicles enclosed in membranes and on the surface of these lie granules known as **ribosomes**. These contain **R.N.A.** (*ribonucleic acid*) and proteins and are the sites of protein synthesis from amino-aicds. In addition to ribosomes are very minute **microsomes** which are mainly com-

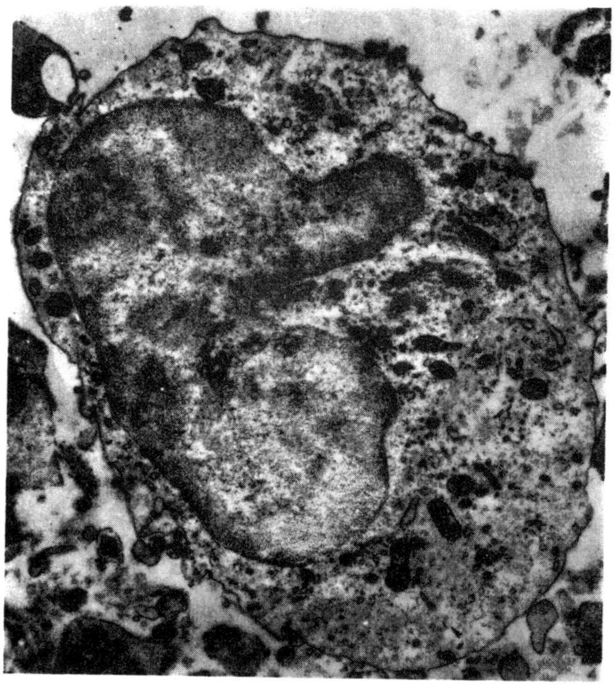

Fig. 19.3. **Electronmicrograph of an animal cell × 12,000.**

(By courtesy of Dr. D. Kay and Dr. J. C. F. Poole.)

* Å = Ångström = 10^{-7} mm.

posed of R.N.A. of which they may form a store. Visible under the optical microscope, somewhat rod-shaped or rounded inclusions occur very freely; these are the **mitochondria**. The electron microscope shows them to contain a series of membranes which divide them up into transverse sacs. They are self-replicating and the sacs contain the enzymes necessary in the various stages of respiration in the cell. The mitochondria are responsible for the transfer of the released energy. The **Golgi body (Golgi apparatus)**, composed of a series of elongated sacs and vacuoles is a more compact

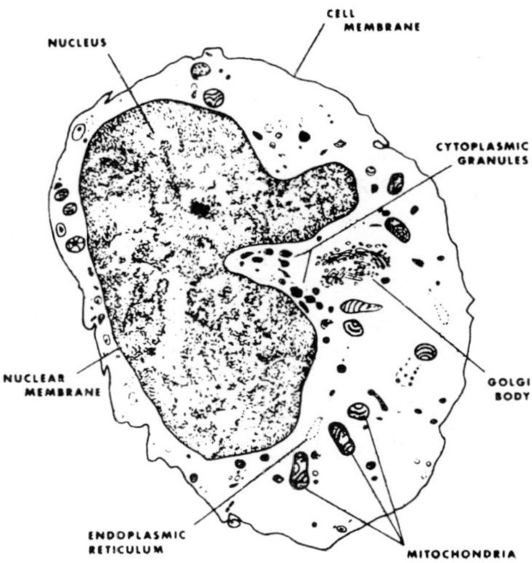

Fig. 19.4. **Drawing of electron-micrograph on opposite page to show cell inclusions.**

cell-inclusion and though the function is uncertain it has been suggested that it is concerned with secretion as it is well developed in secretory cells but this is not by any means certain.

The most prominent structure lying in the cytoplasm is, of course, the **nucleus** enclosed in a **nuclear membrane** in which openings have been observed with the electron microscope. The nucleus is often spherical but, as has already been seen, it can assume other shapes as in the leucocytes of the blood. Within the **nucleoplasm** or **nuclear sap**, as the protoplasm in the nucleus is called, are one or more spherical bodies called **nucleoli**. They contain R.N.A.

while the nucleoplasm itself contains **D.N.A.** (*deoxyribonucleic acid*) and protein. Also dispersed in the nuclear sap is the substance called **chromatin** which may be present in the form of a network in the resting nucleus but at nuclear divisions a definite number of **chromosomes** is visible and it may be that, though invisible, they exist as such in the resting state. As already seen in our study of genetics, the number of chromosomes is constant for any one species of plant or animal and in man this is twenty-three pairs. The chromosomes contain D.N.A. and proteins such as *histone.* It is possible that the nucleoli are concerned in the transference of nucleic acid between chromosomes and the cytoplasm.

The Molecular Biology of the Cell

It is well-known and obvious fact that when living organisms reproduce they give rise to new organisms like themselves and this is due to the **replication** (self-duplication) of genetic substances which are handed on intact. We have seen that the **chromosomes** in the nucleus contain **genes** which are responsible for the inheritance of specific characters. These genes contain D.N.A. and more and more evidence is accumulating which shows that this D.N.A. stores the necessary information which is replicated and which directs the synthesis of the materials necessary for the development of the new organism in a particular form. **Messenger R.N.A.,** it seems, is the carrier of the genetic information from the D.N.A. to the ribosomes (which contain **ribosomal R.N.A.)** and directs protein synthesis. A third kind of R.N.A. called **transfer R.N.A.** (or 3-R.N.A.) passes the amino-acids on to the surface of the ribosomes. R.N.A. contains the sugar *ribose* whereas D.N.A. contains *deoxyribose.* Both contain *phosphoric acid* and organic bases and whereas D.N.A. contains the bases *adenine, cytosine, guanine* and *thymine.* R.N.A. has *uracil* in place of thymine. The D.N.A. molecule consists of sugar molecules alternating with phosphate groups, each sugar molecule being attached to one of the four organic basic groups already mentioned. Furthermore the bases are joined to one another in pairs, the two different bases in each pair being in equal amounts. The molecule is composed of two long threads each composed of a chain of the units just described and each thread is exactly complementary to the other. The threads are wound round each other in spiral fashion forming a helix. As a result of this arrangement the fibres are able to replicate the D.N.A. though the method by which this is done is not fully understood. It appears that the order of the bases in the threads decides the particular genetic messages in the chromosomes.

bacteria and bacteriophages has added a great deal to our knowledge of the nucleic acids. Viruses, it will be remembered, are extremely minute filter-passing bodies, probably on the border-line between the living world and the non-living, occurring in crystalline form and capable of replication provided they are within living matter. Diseases such as smallpox, poliomyelitis, measles, mumps and the common cold and many plant diseases are caused by them. Research has shown that they have a protein structure combined with R.N.A.—some contain D.N.A.—and that it is the R.N.A. which is primarily responsible for infection. Some interesting experiments have been performed with bacteria which have made an important contribution to our knowledge of protein synthesis and inheritance. These minute cells are devoid of normal nuclei and show little structure in their protoplasm beyond the presence of D.N.A. and ribosomes containing R.N.A. There is an enclosing cell-wall lined by a plasma membrane and the D.N.A. is in the form of loosely entwined threads in the centre of the cell. Bacteria have always been regarded as reproducing by fission—hence the name Schizómycetes—but, under the electron microscope, conjugation has now been observed between them by the formation of a conjugation tube. Through this tube D.N.A. passes from one bacterium to the other and exchange of genes takes place which ultimately results in the production of new types. It is thought that in this way penicillin resistant forms of bacteria have developed. D.N.A. has been extracted from some bacteria and added to cultures of different types which has resulted in the aquisition of characters similar to those of the type from which the D.N.A. has been extracted.

As to bacteriophages, which cause bacterial colonies to disintegrate, these have been shown to be virus infections of the bacteria. The phage attacks the wall of the bacterium and its D.N.A. passes into the host. This breaks down the D.N.A. in the bacterium and uses the available material to produce more phage, the phage D.N.A. making it possible for the bacterium to produce enzymes which synthesise a base which previously occurred only in the phage itself. Ultimately all the bacterial material is used up by the phage and the bacterial cell dies. Furthermore different strains of phage can produce hybrids when growing together and phages contain genes which control their infecting capabilities.

We must now turn our attention to the **chemistry of respiration**. This is the process by which energy is liberated in cells by the breaking down of glucose to carbon dioxide and water*. But this

* "A" Level candidates study the process in greater detail in the General Section of their syllabus.

is not a simple straightforward process as indicated by the equation $C_6H_{12}O_6 + 6O_2 = 6CO_2 + 6H_2O + 674$ Calories though this summarises what takes place. The process is a complex one and takes place in several stages. First *glycogen*, stored in muscles and in the liver, is hydrolysed to *glucose* $(C_6H_{10}O_5)_n + n\,H_2O = n\,C_6H_{12}O_6$. The glucose combines with phosphate radicals forming *glucose diphosphate* and this is broken down into two molecules of *phospho-glyceraldehyde* which is oxidised to *pyruvic acid* $CH_3{\cdot}CO{\cdot}COOH$. The energy is utilised to form a compound of *adenosine* (a compound of the base *adenine*, present, as already seen, in D.N.A.) first with two phosphate radicals and known as *adenosine diphosphate* (**ADP**) and then with three forming *adenosine triphosphate* (**ATP**) in which two of the phosphate groups are bound together with considerable energy. The energy which would otherwise be set free as heat is stored in the ATP and becomes readily available to the cell. Clearly, therefore, the formation of ATP is essential. These processes do not require oxygen, the oxidation consisting of the removal of H atoms by enzymes known as *dehydrogenases* and these combine with oxygen atoms to form water. The enzymes are contained in the sacs of the mitochondria which thus generate high-energy-charged molecules. When the oxygen supply is deficient, owing to a high rate of energy consumption, the pyruvic acid is decomposed into lactic acid and this liberates more energy. If there is plenty of oxygen, however, the pyruvic acid is oxidised to carbon dioxide. The energy from ATP can be utilised immediately for protein synthesis and other vital processes. The whole series of reactions described above constitute collectively the first stage in respiration which is known as **glycosis.** Energy-rich molecules of ATP are also formed in a second stage known as **Krebs Cycle** (or the **citric acid cycle**) in which carbon dioxide and water are formed by aerobic oxidation of pyruvic and lactic acid.

To sum up, the D.N.A. in the nucleus enables it to control the general metabolism of the cell as well as the pattern of its reproduction. Energy is provided by the ATP as the result of enzyme action in the mitochondria for the synthesis of proteins by the R.N.A. The R.N.A. carries genetic information from the D.N.A. to the ribosomes where the proteins are synthesised to a particular pattern.

PRACTICAL WORK

1. Examine any available electronmicrographs which are available and try to identify the structures mentioned in the text.

2. The **Bioset "Chromosomes and Genes in Action"** consists of a series of eight stained micrographs and electronmicrographs of chromosomes in varying stages of activity, ribosomes and a D.N.A. molecule. If available these should certainly be studied as they are most instructive.

Chapter XX

HORMONES AND THE ENDOCRINE SYSTEM

There are certain glands in the body which are devoid of ducts and often called **ductless glands** but they are also known as **endocrine glands**. In some cases they are not compact bodies but consist of groups of cells in another organ. We have already come across some of these, the islets of Langerhans in the pancreas, the interstitial cells in the testes, and others. The compact glands are the **thyroid**, the **parathyroids**, the **thymus**, the **adrenals** and the **pituitary**. Collectively they form the **endocrine system**. All are provided with a good blood vascular supply which carries their secretions to other parts of the body. These *internal secretions*, as they are sometimes called, are responsible for the co-ordination and control of the behaviour of cells which may be distant from the glands. They may therefore be regarded as "chemical messengers" and they supplement the control which is exerted by the nervous system. They are usually known as **hormones**, a name derived from a Greek word meaning to excite. This suggests that they have a stimulating effect and this is so in the majority of cases but some are inhibitory in their action. The modern term **autacoid**, universally understood but not yet universally adopted in place of hormones, is therefore preferable, restricting hormones to those which stimulate and using the word **chalones** for those which inhibit.

As will be seen from the following account, the action of hormones—we shall continue to use this term for all of them—is very varied. Some of them, for instance the interstitial cells of the testes, act only in certain places while others, such as those which influence growth, affect the entire body. Excessive secretion, **hypersecretion**, and deficient secretion, **hyposecretion**, can have far reaching adverse effects on the body.

THE THYROID

This gland is situated in the neck. It consists of two lobes lying one on each side of the trachea and connected together by an isthmus. The lobes are composed of closed spherical vesicles lined by cubical epithelium which secretes the hormone **thyroxin** which is stored in the form of **thyroglobulin**. Thyroxin contains iodine

All the photographs in this chapter were kindly provided by Mr. E. C. Butler, F.R.C.S. and Mr. P. G. Nunn of the London Hospital Medical Museum from the collection of Dr. Donald Hunter.

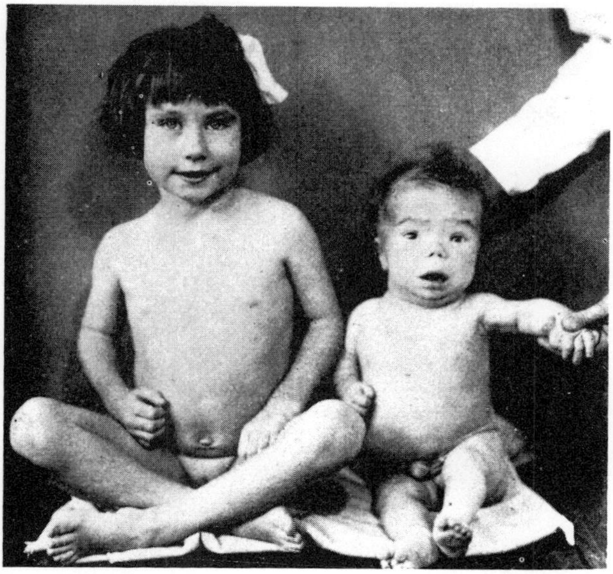

FIG. 20.1.

Cretinism. Girl aged four years with normal control on the left.

Note the umbilical hernia in the cretin.

(By Courtesy of Mr. E. C. Butler, F.R.C.S. and Mr. P. G. Nunn.)

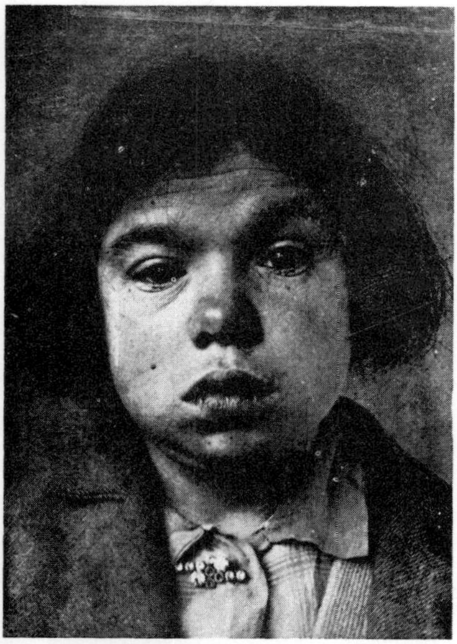

Woman aged 24 years. Treated with thyroid between ages of 5 and 16 when treatment was stopped.

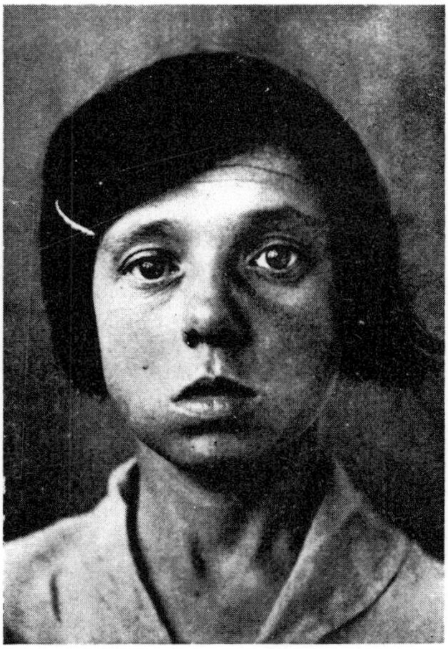

Same woman after 4 months further treatment with thyroid.

FIG. 20.2. **Cretinism.**

(By Courtesy of Mr. E. C. Butler, F.R.C.S. and Mr. P. G. Nunn.)

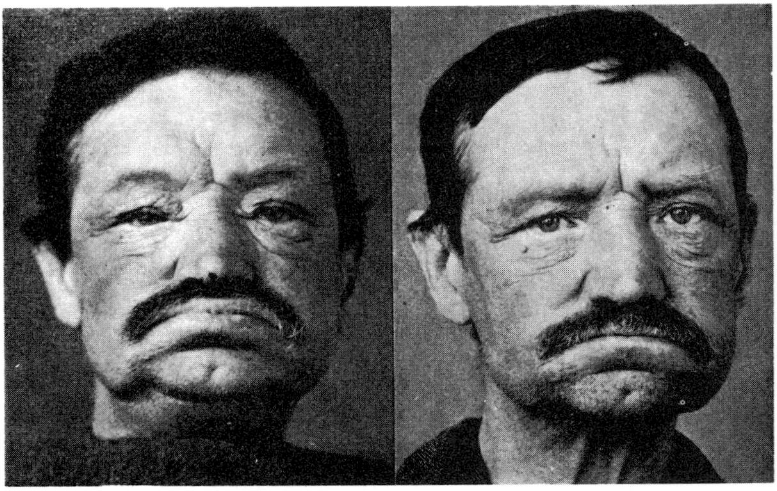

Before treatment. After treatment with thyroid.

Fig. 20.3. **Myxoedema.**

(By Courtesy of Mr. E. C. Butler, F.R.C.S. and Mr. P. G. Nunn.)

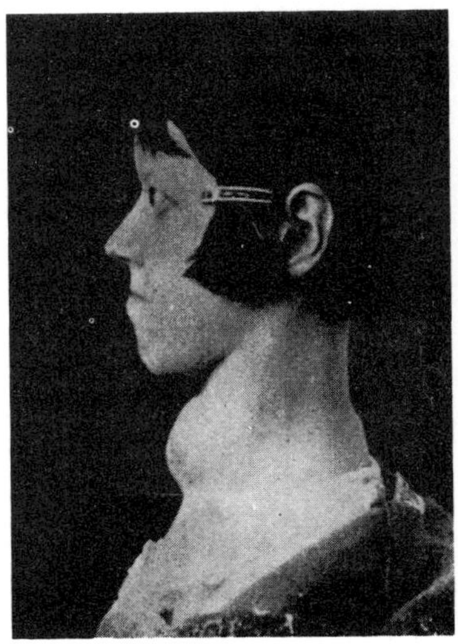

Fig. 20.4.

Exophthalmic Goitre.

(By Courtesy of Mr. E. C. Butler, F.R.C.S. and Mr. P. G. Nunn.)

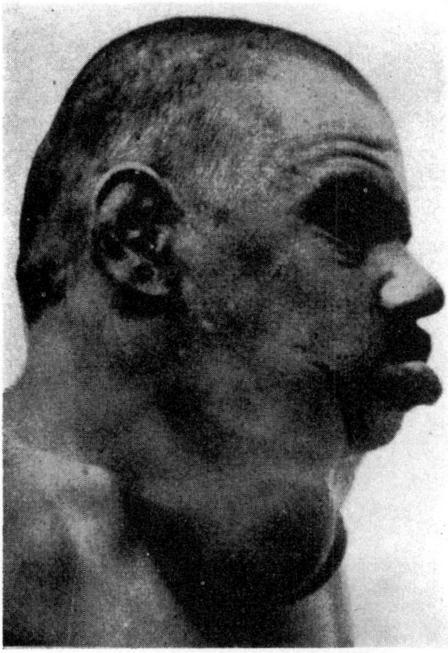

Fig. 20.5. **Simple (endemic) goitre. Swiss cretin aged 39 years.**
(By courtesy of Mr. E. C. Butler, F.R.C.S. and Mr. P. G. Nunn.)

and its function is to regulate the general rate of metabolism in the body. *Hyposecretion in infancy* leads to the condition known as **cretinism** in which growth and mental development are retarded. This can be rectified if treatment with thyroid is given sufficiently early. *In adult life hyposecretion* causes **myxoedema** when the sufferer becomes obese and bald and is slow in action and thought. Again, treatment with thyroxin produces beneficial results. *Hyper-secretion* increases the rate of metabolism producing a restless, excitable individual with increased rates of heart beat and respiration and who consequently becomes easily exhausted. The cause of this over-secretion is the enlargement of the gland indicated by a swelling in the neck and, accompanying this, is a protrusion of the eyeballs. This is **exophthalmic goitre** (*Grave's disease*) and it can be treated by surgical removal of part of the enlarged gland or by the administration of a drug called *thiouracil.* Enlargement of the gland may also be due to a deficiency of iodine in the diet or water supply. The gland enlarges in an endeavour to produce more thyroxin in order to restore it to normal level. This is known as **simple**

goitre. It is also called **endemic goitre** because it occurs in certain localities, namely in regions where there is an iodine deficiency in the water. Examples of such places are mountain villages in Switzerland, in the Tyrol and, at one time, in Derbyshire where it became known as "Derbyshire neck". There should be no reason for this condition nowadays as iodine deficiency can be made up by using iodised table salt.

THE PARATHYROIDS

These are four small glands lying practically buried in the thyroid, two on each side. The secretion, **parathormone**, regulates the calcium content of the blood. *Hyposecretion* causes a decrease in blood calcium resulting in muscular spasm or *tetany*. Administration of parathormone effectively treats the condition. *Hypersecretion* increases the blood calcium which is taken from the bones and which therefore become weakened by decalcification.

THE THYMUS

Unlike the other endoctrine glands the thymus is normally found only up to the age of puberty when it begins to atrophy. If the gland persists after puberty, a sudden shock may cause instant death. The gland is situated between the thyroid and the heart and lies in front of the great vessels. It contains a large number of leucocytes. The action of its secretion is uncertain but it is thought to delay the maturing of the sexual organs until puberty and if this is the case the secretion is a chalone. Sometimes the gland becomes enlarged and this is accompanied by enlargement of other lymphatic tissue in the body. This condition is called *status lymphaticus.*

THE ADRENALS

Also called the **supra-renals**, these glands lie immediately above the kidneys and are shaped rather like cocked hats. Each has a yellowish outer part, the cortex, enclosing a reddish-brown medulla. The **cortex** produces hormones concerned with the development of the testes and ovaries and of the secondary sexual characters. Abnormalities in the adrenal cortex of the embryo can lead to intersexuality. Other hormones influence the balance of salts in the body while further secretions, **cortisone** and **hydrocortisone** control carbohydrate metabolism. The injection of cortisone has proved of considerable benefit in cases of rheumatoid arthritis and in Addison's disease, but in certain conditions its effects are only temporary and therefore continued use may prove necessary. Unfortunately it sometimes has bad side-effects in some cases.

Hyposecretion, due to disease of the gland, interferes with the proper balance of salts in the body and with carbohydrate metabolism. Muscular weakness, reduced blood pressure, digestive disturbances and loss of weight result and the skin becomes pigmented, sometimes to a bronze colour. This is *Addison's disease. Hypersecretion* causes precocious sexual maturity before the age of puberty while in adults it may result in the development of the secondary sexual characters of the opposite sex.

The **medulla** secretes **adrenalin** which has a similar effect to stimulation of the sympathetic nervous system. This secretion is produced under stress of emotion, fear, fright, anger and so on. It causes constriction of some blood vessels such as those in the skin but dilates others namely those in the voluntary muscles thus giving them a greater supply of blood. The force of the heart beat is increased and glycogen stored in the liver is converted into glucose and circulated to the muscles. Thus, provided with these conditions, the body is put into a state of "fighting fettle" ready for dealing with the situation in hand.

Hypersecretion, which may take place during emotional stress, produces these effects in a more marked degree and the heart may "thump", the blood pressure rise temporarily and the skin becomes pallid. *Hyposecretion* results in inadequate toning-up for the emotional state and it is doubtful if adrenalin is secreted except under stress of some kind or other. This hormone is often injected with local anaesthetics to reduce or prevent bleeding as in operations on the eye and nose and in the extraction of teeth.

THE PITUITARY

The pituitary gland or *Hypophysis cerebri* at the base of the forebrain consists of an anterior lobe and a posterior lobe each producing different hormones.

The **anterior lobe** secretes a large number of hormones, some of which control the secretions of the other endocrine glands. For this reason it is often referred to as the "master gland". These secretions are called **thyrotrophic hormones, adrenotrophic hormones, gonadotrophic hormones** and so on. A **somatotrophic hormone** controls growth. *Hyposecretion* of this hormone retards bone growth and produces **dwarfism**. *Hypersecretion* during the growth period results in **gigantism** and heights of seven feet and more have been reached as the result of this. If the excess secretion occurs when full growth has been attained, the bones of the hands, feet, face and jaw are affected, producing a condition known as **acromegaly**. A **diabetogenic hormone** controls carbohydrate metabolism and being

Fig. 20.6.

Pituitary gigantism and dwarfism.

Both are adults who took advantage of their abnormalities by performing on the stage.

(By Courtesy of Mr. E. C. Butler, F.R.C.S. and Mr. P. G. Nunn.)

Fig. 20.7.

Acromegaly. Man aged 47 years.

(By Courtesy of Mr. E. C. Butler F.R.C.S. and Mr. P. G. Nunn.)

antagonistic to insulin, its action is to increase the blood sugar but its interaction with insulin maintains this at a constant level. A further hormone, **prolactin,** is responsible for the initiation and maintenance of the secretion of milk by the female breasts during lactation.

The **posterior lobe** secretes hormones under the general name of **pituitrin**. One, **vasopressin**, stimulates the contraction of the involuntary muscles in the intestines and bladder. Blood pressure is temporarily increased by this hormone which causes constriction of blood vessels and so brings about this increased pressure. During parturition **oxytocin** effects contraction of the uterine muscle. There is also an **antidiuretic hormone** which causes a reduction in the amount of water in the urine by increasing the amount of reabsorption in the renal tubules. *Hyposecretion* causes the disease *diabetes insipidus* (not to be confused with *diabetes mellitus*) in which great quantities of urine are passed and considerable thirst has to be endured. Injection of pituitrin relieves the condition. It therefore appears that this hormone controls the water balance of the body.

THE ISLETS OF LANGERHANS

These groups of cells in the pancreas secrete **insulin** which controls the metabolism of sugar and glycogen and therefore the sugar content of the blood. It makes possible the oxidation of glucose to liberate energy in tissue respiration and the conversion of glucose to glycogen in the liver and muscles for storage. At the same time it prevents the formation of sugar from amino-acids. Thus it regulates these conversions in the tissues and so maintains the corrrect balance of blood sugar.

Hyposecretion causes an increase in the blood sugar. This is called **hyperglycaemia** and is a symptom of the disease *diabetes mellitus.* In this disease the tissues are unable to use sugar or to store it as glycogen and it appears in the urine and is thus excreted from the body. Thirst is considerable and there is considerable wasting because amino-acids are then used to form more carbohydrate. It can be treated successfully, though not cured, by appropriate injections of insulin, regulated according to the severity of the disease. A form of insulin which can be taken orally in some cases has now come into use.

Hypersecretion results in a decrease of blood sugar because the excess of insulin causes an increase in the utilisation of sugar. This condition is called **hypoglycaemia** and it can give rise to coma. The taking of sugar will rectify the ill-balance that has occurred.

HORMONES OF THE TESTES AND OVARIES

The interstitial cells of the testes secrete the male hormones or **androgens** of which **testosterone** is most important. This stimulates the growth of the accessory sexual organs and the development of the secondary sexual characters. The ovary secretes the female hormones or **oestrogens** such as **oestrin.** They are responsible for the maturing of the primary sexual organs and of the secondary sexual characters. **Progesterone** controls the menstrual cycle. The corpus luteum secretes this hormone and this, as we have seen, effects the satisfactory implantation of the zygote in the wall of the uterus and the development of the placenta and prevents ovulation during pregnancy.

HORMONES OF THE ALIMENTARY CANAL

Cells in the pyloric region of the stomach secrete **gastrin** which activates the gastric glands and **secretin**, the first hormone to be discovered and produced in the duodenum, stimulates the secretion of pancreatic juice.

Disorders of the Endocrine System

These have already been described in the foregoing chapter when explaining the effects of the hyposecretion and hypersecretion of the various hormones.

Chapter XXI

THE SENSES

Sensations

We are made aware of our environment by the reception of stimuli arising from it which produce sensations in our bodies and these sensations give rise to such appropriate responses as need be. The sensations are of different kinds and intensities, for example sight and hearing are quite different from pain and pleasure. We have special sense organs which receive these stimuli and nerves which carry messages or impulses from them to the brain which is able to interpret them. This is consciousness. The organs of special sense are the eye, the ear, the sensory epithelium of the nose, the taste buds in the tongue and the sense receptors in the skin. It is difficult to say which of these is most important but perhaps sight is our greatest gift with hearing a close second.

THE ORGANS OF SPECIAL SENSE

I. THE EYE AND VISION

The eye is an organ forming an optical system which focuses rays of light by means of a lens on to a sensitive screen called the retina. It is almost spherical in shape and is kept in place in the orbit, which protects it, by six muscles. A pad of fat behind the eyeball also cushions it against mechanical shock. The muscles enable the eyeball to be moved within the limits of action of these muscles, in any direction. The **superior rectus** turns it upwards, the **inferior rectus** downwards, the **internal rectus** towards the nose and the **external rectus** outwards. The **superior oblique** and the **inferior oblique** provide for movement in a cross-wise direction. When the tone in all these muscles is equal the eye looks forwards and the two eyes are co-ordinated in this respect. Contraction of

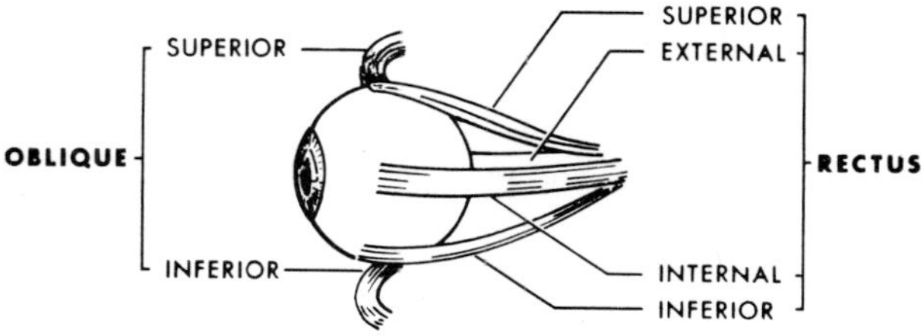

Fig. 21.1. **Muscles of the eye.**

the muscles brings about movement of both eyes. Protection is also afforded to the eyes by movable eyelids which are lined by a thin mucous membrane called the **conjunctiva** which continues over the front of the eyeball. Glands along the edges of the eyelids secrete a lubricating fluid which prevents them from sticking together when they are closed. Eyelashes along the edges of the eyelids help to prevent the entry of dust particles and minute insects and the eyebrows tend to exclude sweat from the forehead.

In the inner corner of each eye is a small swelling which is a vestigal nictitating membrane or third eyelid, a structure well seen in some lower animals in which it passes over the eyeball

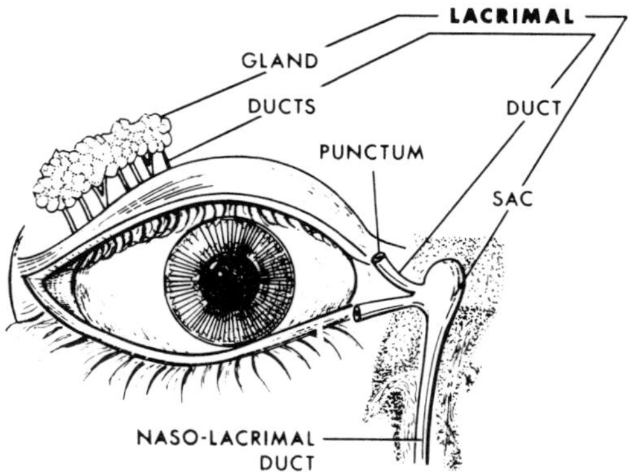

Fig. 21.2. **Lacrimal gland and ducts.**

to keep it cleansed. In man this cleansing is made possible by the **lacrimal*** **fluid** coupled with blinking which keeps it in motion. This fluid is secreted by the **lacrimal gland (tear gland)** which lies above the eye in the upper outer corner of the orbit. Several ducts allow the secretion to flow on to the upper surface of the eyeball and it is carried away by two **lacrimal ducts** which it enters by minute openings called **puncta**. These ducts lead into a **lacrimal sac** lying in the orbit and from this a **naso-lacrimal** duct carries the secretion into the nasal cavity. Irritant vapours such as those emitted by onions and strong emotions may cause the lacrimal fluid to be secreted at a greater rate than that at which the ducts can carry it away with the result that tears flow down the cheeks.

* Or **lachrymal.**

The eye is composed of three coats, the outermost being the white, opaque **sclera** composed of dense fibrous tissue. In front, the central part of this coat is transparent to permit the entry of light and this region is called the **cornea**. The middle coat is the **choroid**. This is highly vascular and contains a dark brown pigment which makes the interior of the eye a dark chamber and prevents reflection around it within. Towards the front of the eye it joins the ciliary body and the pigmented circular **iris**. The latter contains circular and radial fibres and is therefore capable of contraction. The **pupil** is a hole in the centre of the iris which appears black because the interior of the eye is dark. The contraction of the circular fibres in the iris diminishes the size of the pupil and

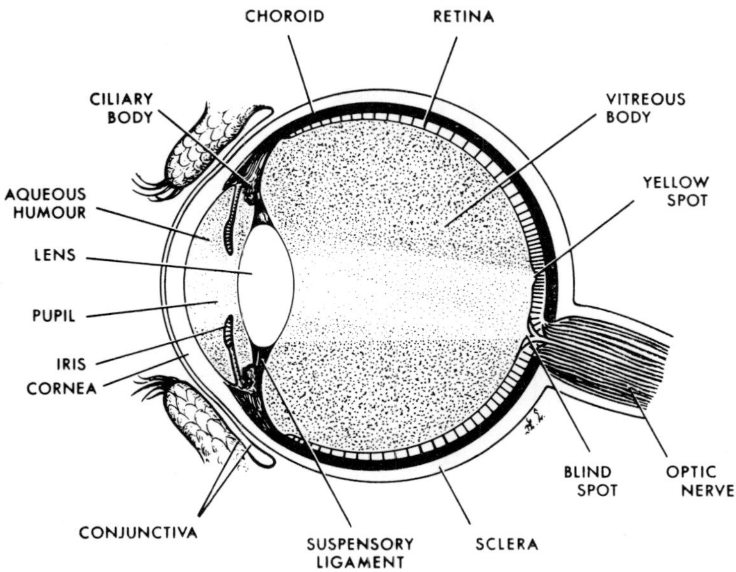

Fig. 21.3. **Section through the eye.**

this occurs in bright and intense light, serving to protect the delicate lining of the eye. When the radial fibres in the iris contract, the pupil enlarges, as in twilight and darkness, in order to allow more light to enter. Contraction of the iris also takes place when the eye is examining near objects and enlargement occurs when distant objects are viewed.

The **ciliary body** contains **ciliary muscles** and is connected to a biconvex **lens** situated behind the iris by an elastic **suspensory ligament** which stretches round its edge. The lens is a transparent, fibrous laminated structure enclosed in a capsule and is more convex on its posterior surface than it is in front. It divides the eye into two chambers. The region in front is filled with a watery fluid called **aqueous humour** and the region behind is occupied by the gelatinous **vitreous body** or **vitreous humour**. Both serve to maintain the shape of the eyeball.

The innermost coat of the eye is the transparent **retina** on which the images produced by the lens are formed. It consists of several layers but the actual **photoreceptors** are in the outermost layer

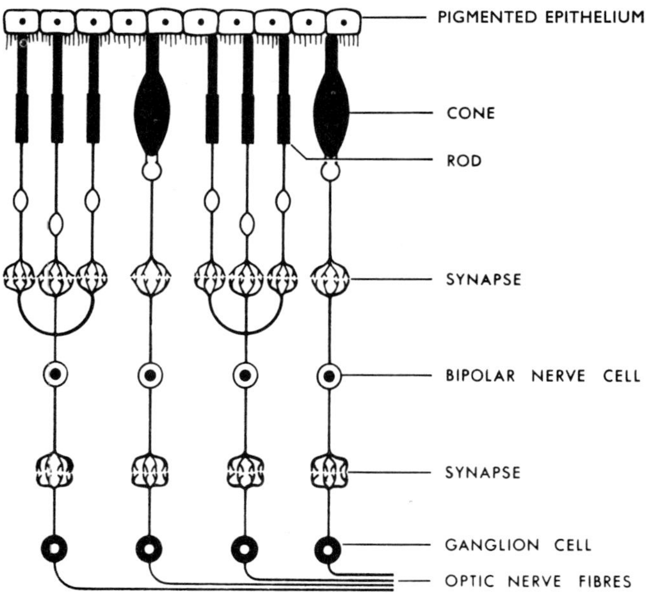

Fig. 21.4. **T.S. of the retina.**

which is covered by a layer of *pigmented epithelium* to prevent halation. Fibres arise from the photoreceptors which from synapses with *bipolar nerve cells* in the next layer. These in turn form synapses with *ganglion cells* in the next layer, the axons of which form the *fibres of the optic nerve.* Though the retina stretches as far forward as the iris, the sensory constituents reach only as far as the ciliary body.

The photoreceptors are of two kinds known as **rods** and **cones**. The rods which predominate near the periphery are sensitive to low light intensity and function particularly in twilight and darkness whereas the cones are thought to serve in daylight vision and to be sensitive to colour. The rods contain a substance called **rhodopsin** or **visual purple** which gives the retina as a whole a purplish colour. This substance is bleached by light but develops in darkness and it has the effect of improving night vision by stimulating the activity of the rods. You must have experienced the inability to see at all clearly on first entering a cinema in daytime and noticed that vision is shortly improved. This is due to the development of visual purple. On leaving the cinema and entering bright daylight, you are at first dazzled by the brightness of the light but very soon normal vision is restored as the visual purple is bleached.

Exactly opposite the optical centre of the lens is a thinner area of the retina known as the **fovea centralis** or **yellow spot**. This is the region of acute vision and it is devoid of rods. The **optic nerve** leaves the eyeball just on the median side of the fovea and an artery also enters at this place which is called the **optic disc** and, as it contains no photoreceptors, it is known as the **blind spot**.

The Physiology of The Eye

The lens of the eye focuses diminished, inverted images of near or distant objects on the retina and the impulses thus set up are conveyed by the optic nerve to the brain which enables us to understand the objects at which we are looking and thus to obtain a picture of them. Each eye forms a separate and slightly different image as the objects are examined from a slightly different angle and the two images overlap. This **binocular vision**, as it is called, gives clarity and depth to the image.

Focussing of near and distant objects is made possible by changing the convexity of the lens because its distance from the retina cannot be altered. When viewing distant objects it is at its minimum convexity and converging power is then at a minimum but when viewing near objects the converging power is increased by increasing its convexity. These changes are brought about by changes in tension in the suspensory ligament and the natural elasticity of the lens. The eyes are at rest when viewing distant objects, the circular muscles in the ciliary body relaxed and the suspensory ligament taut thus making the lens less convex. When viewing a near object on the other hand, the ciliary muscles contract and the suspensory ligament is slack. The effect of this is to increase the convexity of the lens. At the same time the circular muscles in the iris contract and diminish the size of the pupil which reduces the width of the cone of light which enters. If this did not take place objects close at hand would not all be in focus at the same time. This ability to focus near and distant objects is called **accommodation**.

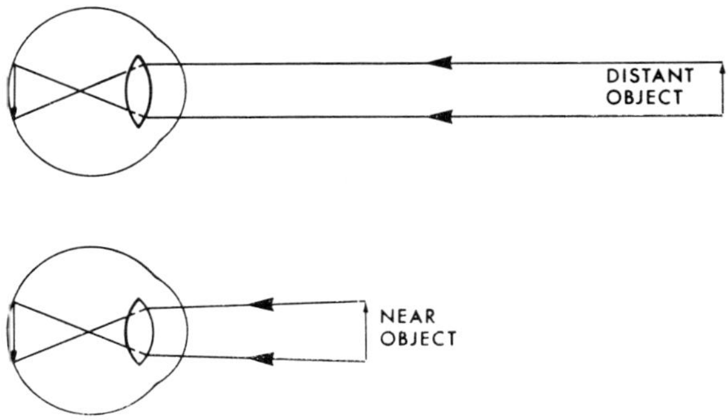

Fig. 21.5. **Accommodation.**

Clear vision is attained by the image being focussed on the yellow spot and when looking at a particular object, say a picture on a wall, this is focussed on the yellow spot and seen clearly. At the same time objects around the picture are focussed on the surrounding part of the retina and are but vaguely seen.

When the eyes are closed after looking at a bright light, the bright image remains for a short time. But if, instead of closing the eyes, one looks at a light surface, the image appears to be dark, and will be of the colour complementary to that of the object. These are known as after-images and the first kind is said to be a **positive after-image** and the other a **negative after-image**.

The farthest distance from the eyes at which an object can be seen clearly is infinity but depends on various factors such as illumination, curvature of the earth, obstacles in the way and so on. In practice it is anything from about 6 m. (almost 20 ft.) upwards. This is known as the **far point**. The shortest distance from the eye at which an object can be clearly focussed is called the **near point** and is about 25 cm. (10 in.). Images focussed on the retina persist for about 1/8 second. This explains why spokes of a rapidly rotating wheel are not visible as individual structures and in fast falling rain the separate drops may not be visible. In both cases the next image has reached the retina before the last one has left it.

It will be seen that the eye may be compared with a camera for in both a lens focuses the image of an object on a sensitive screen, in the eye the retina and in the camera the film, but whereas the eye can enable colours to be seen, in the camera a film is either for black and white only or for colour. Both eye and camera are dark inside to prevent reflection, in the former its being effected by the choroid and in the latter by matt black paint. Both have a means of regulating the entry of light, the iris in the eye and the iris diaphragm in the camera but whereas the former is a reflex action the latter requires manual adjustment. Similarly with focussing, accommodation in the eye takes place automatically by alteration of the convexity of the lens but in the camera it is necessary to adjust the distance between the lens and the film in order to obtain a sharp focus. However, it will be obvious that both work on similar principles.

Diseases of the Eye and Disorders of Vision

Myopia or **short sight**—an inability to focus distant objects. An inherited tendency to produce an eyeball which is too long causes the image to be focussed in front of the retina in the vitreous body. The near point is also brought closer. This condition can be precipitated by intensive close work. It can be rectified by the

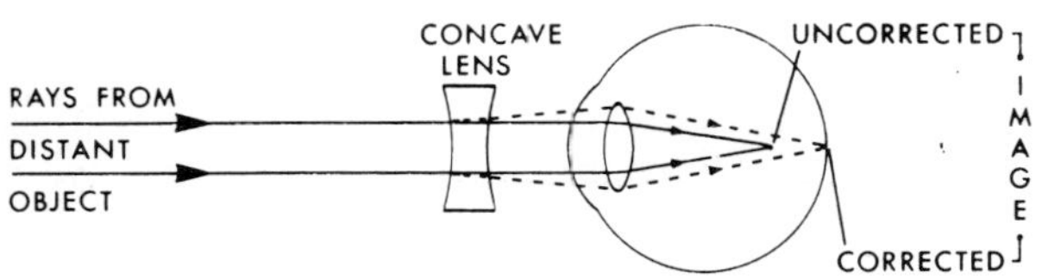

Fig. 21.6. **Myopia (short sight) and its correction.**

wearing of spectacles of concave lenses of suitable focal length which, being concave, increase the divergence of the rays and focus the image on the retina.

Hypermetropia or **long sight**—the inability to focus near objects and in this case the image is produced behind the retina as the eyeball is too short. The near point is further away and a book has to be held at a greater distance for it to be possible to read. Distant objects can be focussed as the ciliary body adjusts the lens. The wearing of convex lenses of suitable focal length which converge the rays of light on to the retina corrects this condition.

Long sight is normal in babies because the lens becomes fully developed before the rest of the eyeball but sight usually becomes normal as growth proceeds.

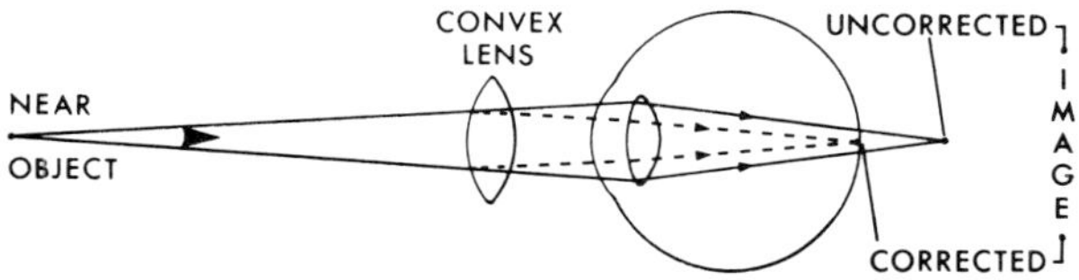

Fig. 21.7. **Hypermetropia (long sight) and its correction.**

Astigmatism—due to the cornea becoming deformed so that its curvature ceases to be uniform, results in images in different planes being focussed to different extents. It often accompanies myopia and hypermetropia and is rectified by the use of cylindrical lenses *i.e.* lenses which are cylindrical on one side though concave or convex on the opposite side if long or short sight is also present.

Presbyopia or **old sight**—a lack of power of accommodation and this is brought about by a hardening of the lens which consequently loses its elasticity. Near-by objects cannot be clearly focussed and the condition is rectified by the use of convex lenses. As the name implies old sight appears and progresses with the advance of age.

Squinting—in which the two eyes look in different directions at the same time is caused by unequal tone in the eye muscles. People who are "cross-eyed" suffer from double vision.

Colour blindness—so-called. This frequently used name for this disease is really a misnomer as most affected people have some colour vision. A better term is **colour defective vision**. The commonest form of the disease, which occurs more frequently in men and rarely in women is red/green colour blindness. It is, as we have already seen, inheritable and the character is dominant. Red and green appear as the same colour but the defect appears in different degrees and other colours are usually confused with these two colours.

Night blindness—due to non-development of visual purple, can be caused by a deficiency of vitamin A which promotes the development of the pigment.

Cataract—the term used for the condition in which the lens of the eye becomes opaque caused by the deposition of calcium salts in the lens. Ultimately blindness in the affected eye results from this. It can be inheritable and the gene for cataract is dominant. It usually occurs only in old age but may appear at birth in which case it is known as congenital cataract.

Blindness—may be partial or total and can be brought about by a degeneration of the yellow spot though it may be caused by cataract, to the cornea becoming opaque or to **glaucoma** in which the nerve fibres are destroyed by excessive fluid pressure in the eye. Haemorrhage from various causes may cause blindness.

Conjunctivitis—inflammation of the conjunctiva which may be brought about by irritation caused by the entry of foreign agents or micro-organisms. The latter is extremely infectious and is known as "pink eye".

Epiphora or **watering of the eye**—caused by hypersecretion of the lacrimal gland or by blockage of the lacrimal duct, for example by a detached eyelash.

Detached retina—a serious condition which can lead to blindness but it can sometimes be cured by a very delicate eye operation using an electronic method.

Stye—an inflamed eye-lash follicle in which pus collects.

II. THE EAR

HEARING AND BALANCE

The Structure of the Ear

The **outer ear** is composed of an external **auricle**, which in some mammals is adapted by its structure for the reception of sound waves though it plays little part in it in man, and an **external auditory meatus** or auditory canal at the base of which is the ear drum. The auricle is composed of skin supported by elastic cartilage except in the lowest portion, the lobe, which contains fibrous tissue and fat. The external auditory meatus is nearly 3 cm. in length and is slightly curved. Glands in its wall secrete wax or **cerumen** which entraps any particles of dust or dirt which may enter the canal and at the entrance are small hairs which serve a similar purpose.

The **middle ear** or tympanic cavity lies in the temporal bone of the skull and is air-filled. The **ear-drum** or **tympanic membrane** closes the entrance at the base of the external auditory meatus. It is composed of fibrous tissue lined externally by modified skin and internally by a mucous membrane. The middle ear contains three small bones known as **ossicles**. Lying in contact with the inside of the ear-drum is the handle of the **hammer bone** or **malleus**, the first of the three ossicles, the head of which is in contact with the **anvil bone** or **incus**. This, in turn, articulates with the **stirrup bone** or **stapes**, the base of which covers the **oval window** or **fenestra ovalis** which is the entrance to the inner ear. Leading from the tympanic cavity to the naso-pharynx is the **Eustachian tube**. Its purpose is to equalise the pressure on the two sides of the ear-drum.

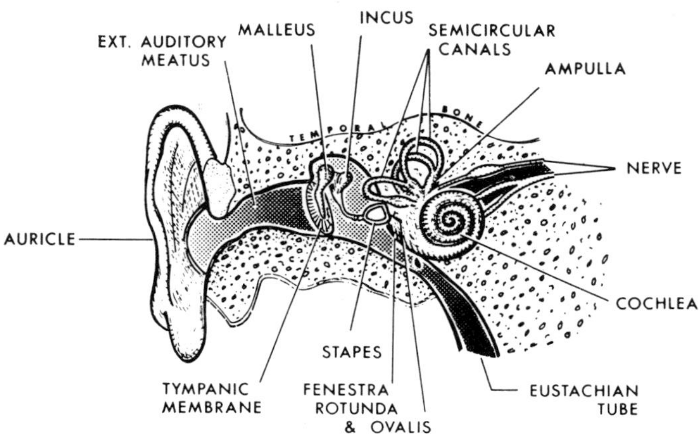

Fig. 21.8. **Structure of the ear.**

The **inner ear** also lies in the temporal bone and consists of what is called the **membranous labyrinth**. This lies surrounded by a fluid known as **perilymph** in cavities in the bone which constitute the **bony** or **osseous labyrinth**. This consists of three parts, the **vestibule** in the centre, the **cochlea** in front and the **semicircular canals** above the vestibule, all of which are in communication with each other.

The vestibule contains two membranous sacs, the larger one, the **utricle**, communicating with the semi-circular canals and the smaller **saccule** communicating with the cochlea. The **membranous semi-circular canals** lie inside the bony semi-circular canals and at the beginning of each is a swelling called an **ampulla**. They contain a fluid called **endolymph**. Two of the canals are vertical at right angles to each other and one is horizontal.

The cochlea is so-called because it is shaped like a snail's shell and consists of a spiral canal of two and a half turns wound round a central pillar called the **modiolus.** A lamina of bone round the modiolus partially divides the cochlea into two and a further division by two membranes from this bone divides it into three channels. The upper membrane is known as **Reissner's membrane**

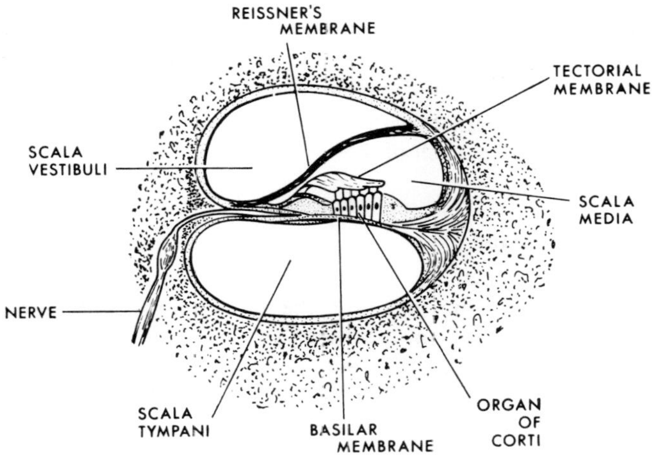

Fig. 21.9. **T.S. through part of the cochlea.**

Fig. 21.10. **L.S. cochlea uncoiled and shortened. Diagrammatic.**

and the lower one is called the **basilar membrane**. The upper channel is called the **scala vestibuli**, the central one the **scala media** and the lower one the **scala tympani**. The fenestra ovalis, the membrane of which is covered by the base of the stapes in the middle ear, is the entrance to the scala vestibuli and at the end of the scala tympani is the **round window** or **fenestra rotunda**, also covered by a membrane. The scala media is the only channel to contain endolymph, the other two being filled with **perilymph**. Lying on the upper surface of the basilar membrane is the organ of hearing known as the **organ of Corti**. This is composed of a series of arched rods, the **rods of Corti**, forming a kind of tunnel lined by sensory epithelial cells bearing stiff hairs which are covered by what is called the **tectorial membrane**.

The Physiology of the Ear

The ear is concerned with two functions, *hearing* and *balance* and we shall consider each of these separately.

Hearing

The Nature of Sound

Sound is produced by vibration and the greater the frequency of these vibrations the higher is the **pitch** of the note. It travels at the rate of 332·4 m. per sec. (about 1,100 ft.) in air, but the velocity is affected by wind, humidity and temperature. The **loudness** of a sound depends on the energy of the sound waves which control the *amplitude* or maximum displacement of the waves. If this is too great a sense of pain may develop. The **quality** is controlled by the number and strength of the overtones which sound with the note produced or *fundamental* as it is called. A musical note has a definite pitch but a noise has none and is the result of a succession of irregular vibrations. An echo is simply a reflection of sound.

The units in which loudness of sound is measured are known as **decibels***. Zero is the threshold of hearing *i.e.* almost complete silence. Ordinary conversation produces 60–70 decibels and this is reduced to around 15 when whispering. Pneumatic drills used in road work create something like 120 decibels and over 90 to 95 can be damaging to the human ear if continuous. Excessive noise may have adverse psychological effects and can cause fatigue. Pain is produced by 120 decibels. The whole question of noise, its effects and control, is considered under Air Pollution on p. 401 seq.

The Mechanism of Hearing

Sound waves which are collected by the outer ear cause the

* The definition of a decibel is somewhat too involved to enter into here. Briefly it is a measure of loudness and this is a physiological sensation. It increases with the intensity of sound (though not directly) which is a measurable physical quantity. Unit difference of loudness on a fixed scale derived from these facts is called a bel and a decibel is one-tenth of this.

tympanic membrane to vibrate and these vibrations are transmitted through the ossicles to the fenestra ovalis which sets up vibrations with a similar frequency in the perilymph in the inner ear. As the membrane covering the fenestra ovalis is pushed inwards or outwards so that covering the fenestra rotunda is forced outwards or inwards respectively. This is necessary to allow for the displacement of the perilymph. As we have seen, the organ of Corti is situated on the basilar membrane and so the hairs in this organ are stimulated. Nerve fibres originating from these cells form the **auditory nerve** and carry impulses to the brain. The hairs at the end of the organ of Corti nearest to the fenestra ovalis are sensitive to notes of high pitch and those at the opposite end to low notes with sounds of intermediate frequency between. Thus we have the sensation of hearing and are able to distinguish notes of different pitch and to hear several at the same time. The human ear is able to detect vibrations between 20 and 20,000 per sec. (about 10 or 11 octaves). Outside these limits sounds are inaudible to us. Having an ear on each side of the head provides us with **binaural hearing** which enables us to sense the direction from which a sound has come.

Balance

The membranous semi-circular canals are lined by epithelium provided with long, stiff hairs which lie in a layer of mucus. Inside the saccule and utricle are similar hair cells and amongst the hairs are calcareous bodies called **otoliths** which move as the head is moved and the hairs are bent owing to the inertia of the endolymph in the canals. Impulses are thus sent through the nerve fibres which arise from the hair cells and these fibres form a branch of the auditory nerve which passes them to the brain and we become conscious of changes in movement of the head.

Disorders and Diseases of the Ear

Deafness—may be due to excessive wax in the ear which has become hardened or to some other obstruction. It can also be caused by damage to the ear-drum by excessive noise. The extreme limit of noise tolerance is 140 decibels. Difference of pressure on the two sides of the ear-drum can cause deafness and this is experienced at changes of altitude as for example when driving over a mountain pass and during ascent and descent in a 'plane. It can also be produced by infection passing up the Eustachian tube from an infected throat.

Earache is usually the result of inflammation of some part of the middle ear and may result from this. Thickening of the ear-drum will also interfere with hearing. Again the trouble may lie in the inner ear itself and this is one of the commonest causes of deafness, or it may be in the auditory nerve. One form of deafness is hereditary.

There are various forms of hearing aids, the commonest kind being one in which a small microphone converts sound into electrical energy which is reconverted into sound in an amplified state in a receiver fitted into the external ear. Periodic replacement of the small batteries used to provide the electrical energy is necessary.

Mastoid disease—arises in the mastoid antrum, a cavity lying in the mastoid bone which is part of the auditory capsule. It is a very dangerous condition which demands immediate treatment with antibiotics or surgery lest the infection passes into the cranium.

Vertigo or **giddiness**—If one turns round and round rapidly in one direction the fluid in the semicircular canals is set in rapid motion in one direction. If one then stops suddenly, different impressions are received by the brain from the ears and the eyes and giddiness results. Disease of the semicircular canals is not common but if it occurs, dizziness, inability to maintain the body's equilibrium and a reeling gait result. In *Menière's disease* the giddiness comes on suddenly and is severe and this is caused by the semicircular canals ceasing to function.

III. THE NOSE AND SMELL

The two nasal cavities are lined by mucous membrane and with ciliated epithelium in the lower region and hairs at the entrance to remove dirt and infective agents from the air as it passes up the cavities. The **olfactory cells** forming the sensory epithelium are restricted to the upper region of the cavities. These cells terminate in hair-like processes and give rise to nerve fibres which pass through pores in the cribriform plate of the ethmoid bone and form the **olfactory nerves**.

The sense of smell is poorly developed in man but he is able to distinguish pleasant or aromatic odours from unpleasant ones, such as those possessed by certain gases and by putrefying material, and whether a smell is feeble or pungent. These vapours or gases dissolve in the mucus in the nasal cavity and this stimulates the hair-like processes of the olfactory cells and sets up impulses which are carried to the brain. This is the reason why, after exposure to a strong odour, the smell appears to continue for a short time when away from its source. The ability to identify a particular powerful smell is soon lost after exposure to it though other odours may be recognised. The stimulus in the case of smell is obviously a chemical one. The proximity of the nose to the mouth enables us to decide on the suitability of anything taken as food.

Disorders and Diseases of the Nose

Loss of smell—often the result of prolonged exposure to a powerful odour. It also occurs when one is suffering from a cold.

Inflammation—often follows infection by common cold viruses. The mucous membrane becomes swollen and its secretion is considerably increased. This is the cause of loss of smell under these conditions. Sneezing is a sudden expulsion of air in order to remove irritants from the air passages.

Polypus—a soft swelling borne on a stalk which grows in the nasal cavities. They are easily removed.

Nose bleeding—not a serious condition unless it continues for a prolonged period.

IV. THE TONGUE AND TASTE

The tongue is composed of muscular tissue and is covered by a mucous membrane which bears a number of papillae on which the organs of taste, known as **taste-buds**, are situated. These papillae are situated on the **dorsum** or upper surface of the tongue and are of three kinds. The mushroom shaped **fungiform papillae**

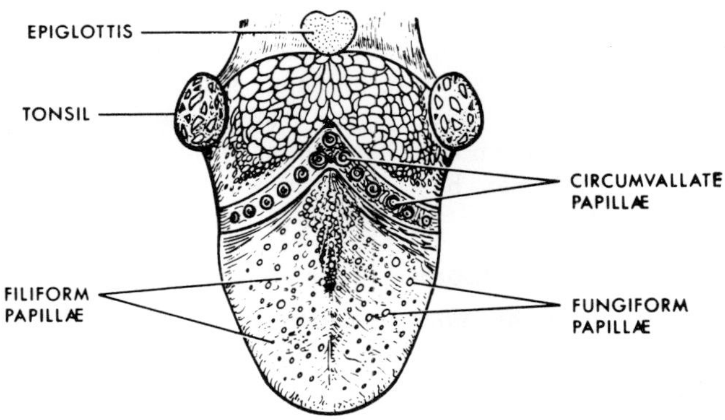

Fig. 21.11. **The tongue.**

occur on the sides and tips of the tongue, the simple **filiform papillae** all over the surface and the circular **circumvallate papillae**, which are the largest, are arranged like an upturned V lie near the root of the tongue. Each is surrounded by a moat. Whereas the filiform papillae are mostly concerned with touch, the others deal with the sense of of taste.

The **taste buds** are ovoid structures and consist of cells of two kinds, long spindle shaped **gustatory cells** in the centre, each with a taste hairlet on its free end and all opening into a narrow **gustatory pore** and the flattened **sustentacular cells** around the outside which

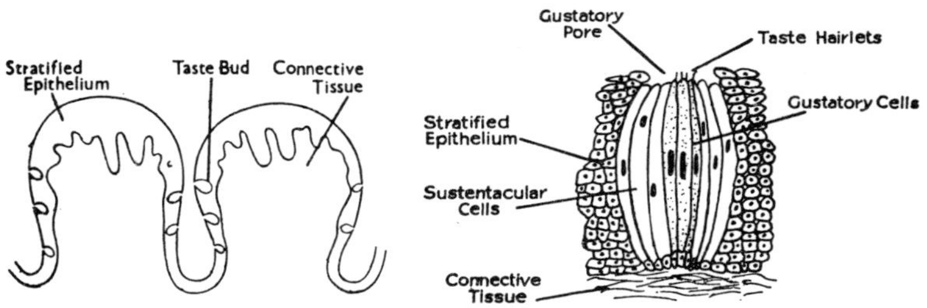

Fig. 21.12. **T.S. tongue, diagrammatic, and tastebud.**
(From Wallis—Practical Biology.)

give support to them. All substances must be in liquid form before they can be tasted and these substances dissolve in the saliva. The hairlets receive the stimulus, a chemical one, and nerve fibres carry the impulses which are thus set up, to the brain.

Taste is of four kinds, *sweet, sour, bitter* and *salt.* Sweet taste is perceived at the tip of the tongue, sour at the sides, bitter at the back and salt all over it. Other tastes, for example that which we describe as savoury, are really odours and they arise from stimulation in the nose. They cannot, therefore, be appreciated, as sense of smell may be lost, when we have a cold or when the nose is pinched.

Disorders of the Tongue

Inflammation of the tongue—causes the tongue to become red and swollen.

Fur on the tongue—often occurs in fevers that may be caused by gastritis, constipation or throat infection.

V. THE SKIN AS A SENSE ORGAN

Situated on the ends of nerve fibres in the dermis of the skin are sense receptors of *pressure, touch, heat* and *cold* and *pain* and it is by means of them that we become aware of changes in our external environment. These receptors are of four different kinds. The simplest are the **pain receptors** which are just nerve endings in the form of plexuses. **Pressure** is perceived by **Pacinian corpuscles** which are bulbous in shape with the nerve ending enclosed in a capsule composed of several concentric layers. **Touch** is dealt with by **Meissner's corpuscles**. These are situated in the dermal papillae immediately beneath the Malpighian layer of the epidermis and the nerve endings are enclosed in a capsule composed of connective tissue. These receptors are very close together in the tips of the fingers but are further apart in other parts of the body. Heat and cold

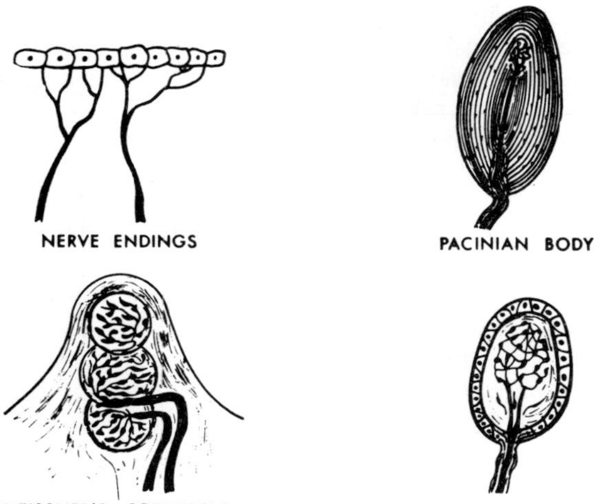

Fig. 21.13. **Sense organs of the skin.**

are thought to be distinguished by two different kind of receptors known as **end-bulbs**. In each the nerve endings are enclosed in a simple capsule and they are particularly abundant where the skin is thin, as for example on the back of the hands.

It should be noted that some of the receptors and nerve endings also occur in structures other than skin such as connective tissues and muscle and it is by means of these that the extent of muscular contraction and limb position is perceived.

PRACTICAL WORK

The Eye and Vision

Structure of the Eye

1. Examine a **bullock's eye** with the muscles attached and identify the individual muscles, the sclera, cornea, iris, pupil and optic nerve.

2. Cut a longitudinal section of the eye by cutting through it carefully, using small dissecting scissors. Identify the internal

structures and then remove the retina and note the choroid underneath.

3. Examine an **anatomical model of the human eye.**

4. Examine a microscopial slide of a **section through the retina** and distinguish between the rods and the cones.

Vision

1. **Image produced by the lens of the eye.** Cut a small portion out of the back of a bullock's eye to serve as a window and cover the gap with a piece of translucent paper. Hold a lighted splint or candle in front of the eye and note the inverted image produced by the lens on the paper. The image can be brought into focus by moving the source of light backwards and forwards until the sharpest image is obtained. The best results will be obtained if this is performed in a darkened room.

2. **To show that the image produced is inverted.** Make a hole with a pin in a piece of card and hold the card close to the eye so that you can see through the hole. You should adjust the card so that you look into the sky so that there are no objects to interfere with your view. Look through the hole with one eye, keeping the other one closed. Now bring a mounted needle into your line of vision between your eye and the card, so that the handle of the needle touches your eyebrow, holding the needle vertically. Slowly move it up and down, taking care not to look at the needle but maintaining your view through the hole. An inverted image of the needle will be seen through the pinhole.

3. **To demonstrate the blind spot.** Draw a cross and a large dot about 7 to 8 cm. apart on a piece of card as shown in Fig. 21.14 or use this figure. Hold the card 40 to 50 cm. away and with the left eye closed keep your gaze fixed on the cross. The dot will also be seen.

Fig. 21.14. **To demonstrate the blind spot.**

Now draw the card slowly towards you, still keeping your gaze on the cross. At one point the dot will disappear and the card will appear white where it was formerly seen. At this point the image falls on the blind spot and is therefore invisible. The dot will reappear on bringing the card closer to the eye.

4. **The movements of the iris.** Examine the pupil of a fellow student's eye. Then shine a torch in his eye for a few seconds and note that the pupil becomes smaller. Now get him to cover his eye with his hand for a few seconds and examine the size of the pupil on removal of his hand. It will be seen to have enlarged.

5. **After images** Look at a bright light for a few seconds and then close the eyes. Note the *positive after image.* Again look at a bright light for a few seconds and then immediately transfer your vision to a piece of white paper. You will see a *negative after image.*

The Ear and Hearing

Structure of the Ear

1. Examine an **anatomical model of the human ear** and identify the parts mentioned in the text.

2. Examine a **T.S. of the cochlea** under the microscope under the low power and identify the scala vestibuli, scala media, scala tympani and the basilar membrane. Look for the cells of the organ of Corti with their sense hairlets.

Hearing

1. Hold a watch close to another student's ear and then move it away slowly until he can no longer hear the tick. Record the maximum distance of hearing. Repeat with the other ear then with those of other students and compare results.

2. Plug the ears with cotton wool and then hold a watch in contact with the forehead. The tick of the watch can be heard as the sound is transmitted through the bones of the head. Repeat the experiment holding the watch between the teeth.

The Tongue and Taste

Taste buds

Examine a microscopical slide of **a section of the tongue showing taste buds** and identify the gustatory cells with their sense hairlets and the supporting sustentacular cells.

Location of Different Tastes

Wash out your mouth with water. Then get a fellow student to touch your tongue in different places with a rod dipped in (i) glucose (ii) lemon juice (iii) quinine and (iv) salt and thus ascertain where the four different kinds of taste are identified. Repeat the experiment yourself on another student.

The Skin as a Sense Organ

1. Examine microscopical slides showing **sense organs in the skin**, if available.

2. Heat a knitting needle in hot water and then touch different places on the back and front of the hand, the arm and the face of a fellow student. Ask him to tell you when he feels the heat least and most of all.

3. Immerse the end of a knitting needle in ice for a short time and repeat the above experiment to ascertain where and to what extent cold is felt.

4. Blindfold another student and then touch his skin lightly in various places with a mounted needle. He should tell you when he feels pain from the prick. You will then be able to discover the regions most sensitive to pain.

5. Apply a pair of dividers to various places on a blindfolded student's skin such as the palm and back of the hand, tips of the fingers and face. Vary the width between the points and ask your student whether he feels one or two. Thus you will be able to find the relative sensitivity of the skin in different places.

Chapter XXII

THE NERVOUS SYSTEM

If the body is to function at its maximum efficiency, the various systems of which it is composed must also work at their highest possible efficiency and they must do so in complete harmony with one another. Further the body must be able to respond to changes in its external and internal environment so that this efficiency will not be impaired. We have already seen how the sense organs inform the body of such changes and that the endocrine glands are responsible for chemical co-ordination in the body. We must now investigate the working of the nervous system, the role of which is to control and co-ordinate the working of the other systems and effect appropriate responses to changes in environment.

The nervous system consists of the **brain** and **spinal cord** which constitute the **central nervous system** and the **cranial** and **spinal nerves** which connect them with all parts of the body and form the **peripheral nervous system**. Stimuli which are received by sense organs set up impulses which the **sensory** (or **afferent**) **nerves** carry to the central nervous system. Here they are sorted out and appropriate action is taken by the body as the result of impulses sent out by the central nervous system along **motor** (or **efferent**) **nerves** to the various muscles or organs. In many cases a nerve contains both sensory and motor fibres and these are known as **mixed nerves**. Impulses in some cases are initiated in the brain as the result of the working of man's mind and intelligence. Internal control of involuntary actions such as the rate of heart beat and peristalsis come under the control of nerves and ganglia which constitute the **autonomic nervous system** which is connected with the spinal system by communicating nerves.

Nerve cells are found in the brain and spinal cord and there are concentrations of them elsewhere in what are known as **ganglia**. Such are found on sensory nerves. The nerve cells in the brain and spinal cord are supported by connective tissue containing cells known as **neuroglia**. It might be as well if we recall from our study of histology in Chapter II that the nerve unit, a **nerve cell** with its processes known as **dendrons**, **dendrites** and an **axon**, is called a **neuron** and that these may be *unipolar, bipolar* or *multipolar*. Most unipolar nerve cells are sensory and multipolar motor, the bipolar type occurring but rarely in the human subject *e.g.* in the retina. It will also be remembered that communication between neurons occurs at a **synapse** where the dendrites interlock without actual contact, being separated from one another by protoplasm which allows impulses to pass in one direction only. A **connector neuron** occurs between a sensory neuron and a motor neuron. The axon from a nerve cell continues as the **axis cylinder** (or **axon**) of a **nerve fibre**. Dendrons carry sensory impulses to the cell body and axons carry motor impulses away from it. There are three kinds of sensory cells (i) **exteroceptors** which respond to changes in the external environment and are found in the skin, eye, ear and nose (ii) **enteroceptors** in internal epithelia and concerned with the viscera and (iii) **proprioceptors** such as the nerve endings in stretch-receptors in muscles and tendons and stimulated by mechanical movements of the body.

In the cranial and spinal nerves the fibres are enclosed in a fatty sheath and these are **myelinated fibres** whereas in the **non-myelinated fibres** of the autonomic system this sheath is lacking.

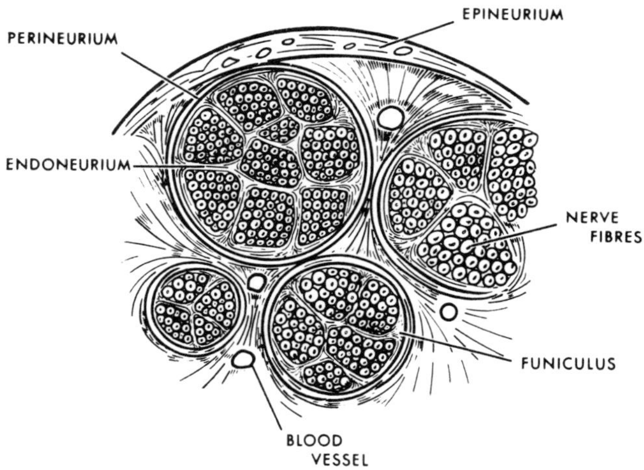

Fig. 22.1. **T.S. nerve.**

The nerve fibre is enclosed in a thin **neurilemma**. A **nerve** is made up of bundles of fibres called **fasiculi**, bounded by connective tissue called **perineurium**, rather like an electric cable, and between the fibres within the bundle is connective tissue known as **endoneurium**. The external coat of the nerve which encloses all the bundles is called the **epineurium**. We have already seen the endings of sensory neurons in the skin and sense organs. The motor neurons terminate in arborisations and in voluntary muscles these are enclosed in protoplasmic **end-plates** amongst the muscle fibres. Two kinds of nervous tissue are found in the central nervous system, **grey matter** composed of neurons and **white matter** consisting of nerve fibres.

THE NATURE OF THE NERVE IMPULSE

What is a nerve impulse and how is it transmitted along a nerve fibre? If two electrodes are placed apart on the surface of a nerve through which no impulse is passing, there is no difference in electrical potential between the electrodes but when excitation takes place and an impulse passes along the fibre a difference in potential can be demonstrated as a minute electric current travels along it. Furthermore this tiny current travels at the same rate as the impulse. It would not be unreasonable, therefore, to wonder if a nerve impulse is a minute electric current. The obvious question which now arises is what causes this difference of potential? Electric currents pass as the result of the movement of ions, an ion being an electrically charged atom or radical. Now on the outer side of the axon membrane of a resting nerve fibre is a concentration of positive ions and on the inside is a concentration of negative ions. There is an excess of sodium ions (Na^+) on the outside but inside potassium ions (K^+) predominate. Consequently there is a

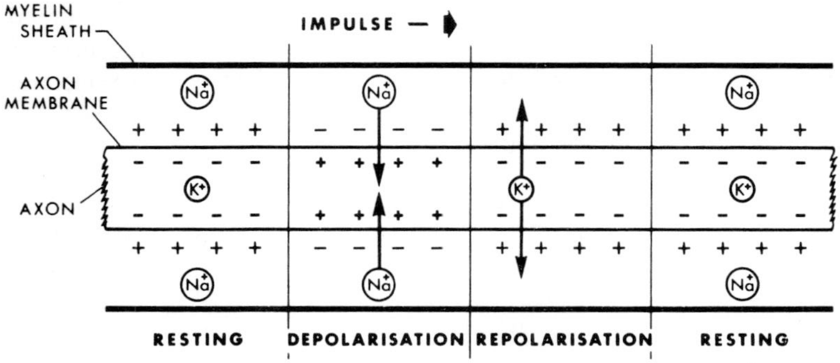

Fig. 22.2. **Movement of ions and electrical charges in a nerve fibre during the passage of an impulse.**

potential difference of about 80 millivolts between the two sides of the membrane, the outside being positive and the inside negative and the membrane is said to be polarised.

When excitation takes place by application of a stimulus, the potential difference is lowered and the permeability of the membrane is altered causing an inflow of Na^+ ions through it. This continues until the inside is slightly positive. The impulse is then immediately started and this produces similar changes in the membrane ahead of it. This occurs along the entire fibre and a wave of depolarisation passes along it. This wave of depolarisation is the nerve impulse and it travels at a speed of up to 100 m. a second. The speed varies according to the thickness of the fibres, travelling more quickly in the larger ones. Once this wave has started at any one point, the permeability of the membrane to Na^+ ions decreases again and K^+ ions flow outwards more slowly and lasting longer. Thus the negative charge on the inside is restored. When the permeability of the membrane has been restored to its original state and the balance of the ions on each side of it is back to the condition of the resting fibre with excess of Na^+ ions on the outside, the membrane is once again depolarised at that point.

The passage of a nerve impulse is not, however, quite as simple as this for there are chemical changes taking place as well. A substance known as **acetylcholine** is said to be produced when an impulse passes along a nerve and it has been claimed that this carries it across the synapses, for the impulse does not follow an uninterrupted path through the nerve in the same way as an electric current passes along a wire. Acetylcholine is released when a nerve impulse reaches the end plate in a muscle and this causes the muscle to contract. An enzyme, **cholinesterase**, produced by the end plates is responsible for the almost instantaneous removal of the acetylcholine to avoid its accumulation which would prevent further contraction. It should now be realised that the minute electric current passing along a nerve fibre is not the impulse itself but is really a manifestation of it.

THE BRAIN

The brain is situated in the cranial cavity and weighs about 1,400 gm. in the average adult. It is protected by the cranium and by three membranes known as the **meninges** in which it is enclosed. The outer one, the **dura mater**, is thick, fibrous and tough and lines the inside of the cranium. The innermost one, the **pia mater**, is thin and highly vascular and closely invests the brain. Between them is the thin **arachnoid** which secretes a fluid between itself and the pia mater which is known as **cerebo-spinal fluid**. It serves to protect the brain by acting as a cushion and also serves as a means of providing the brain tissue with nutritive matter.

The bulk of the brain is composed of the cerebrum and beneath it is the much narrower **brain stem** continuous with which, at the lower, hinder end is the spinal cord. The part of the brain stem continuous with the spinal cord is the **medulla oblongata** and, following this, is the **pons** which is broad in the centre and narrower

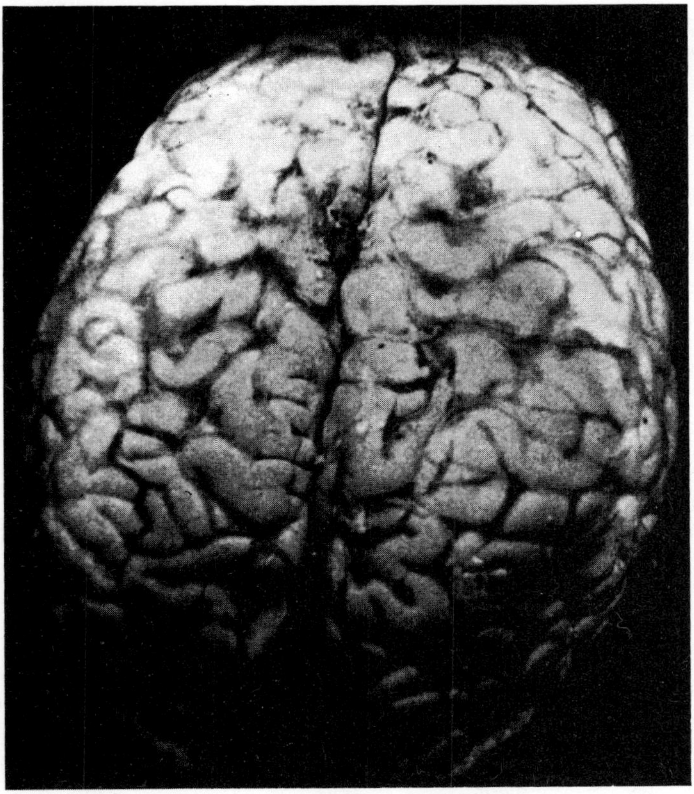

FIG. 22.3. **Human brain—from above.** *Specimen in the author's laboratory*

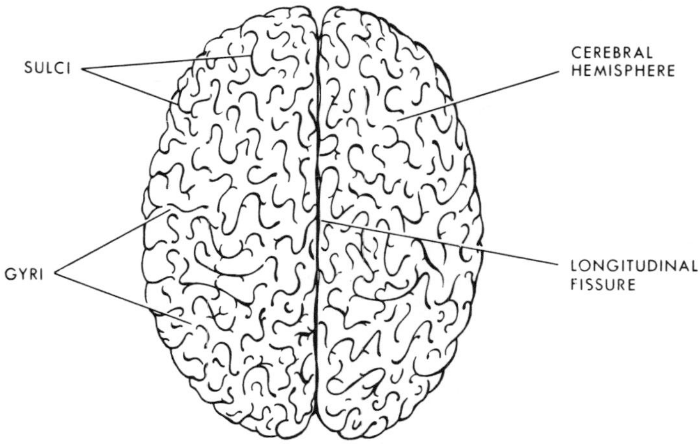

FIG. 22.4. **The brain. Upper surface.**

at each end where its fibres enter the **cerebellum** lying behind it. The medulla and the cerebellum constitute the **hind-brain.** Following this is the **mid-brain**, composed of four small bodies known as the **corpora quadrigemina** together with the brain stem. This part of the brain is joined to the cerebellum by two bridges of nervous tissue called the **superior peduncles**. Two more join the cerebellum to the brain stem and these are the **inferior peduncles**, the pons constituting the **middle peduncle**. In front of the mid-brain the brain stem continues as the **thalami**, above which are two very large lobes, the **cerebral hemispheres** or **cerebrum**. This with the thalami constitute the **fore-brain**.

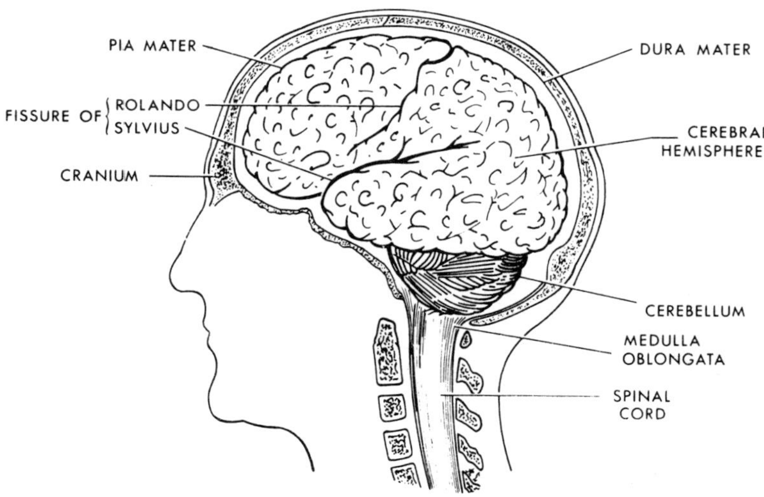

Fig. 22.5. **The brain. Side view.**

The Fore-Brain

The two cerebral hemispheres of the cerebrum are separated by a deep groove called the **longitudinal fissure**. At its base is a transversely running band of white fibres, the **corpus callosum**, which binds the two hemispheres together. The outer part of the cerebrum, the cortex, consists of grey matter and this is considerably increased in quantity by a large number of fissures called **convolutions** or **sulci** which throw it into folds or **gyri**. The cortex contains millions of neurons which are devoid of axons and which are in several layers. It is fortunate that there are so many as they cannot be regenerated if damaged.

Two prominent fissures are visible on the surface of the cerebrum. The **fissure of Rolando** (or **central sulcus**) runs obliquely downwards and divides the cerebrum into **frontal** and **parietal lobes**. The **fissure of Sylvius** (or **lateral sulcus**) follows an upward course from the base of the cerebrum and runs obliquely forwards, separating the parietal lobe from the **temporal lobe**. On the under side of each hemisphere two **olfactory tracts** run forwards and terminate in **olfactory bulbs**.

The inside of the hemispheres consists of white matter and a very large number of fibres run to and from the cortex and the brain stem and between the two hemispheres. Each hemisphere contains a cavity known as a **lateral ventricle** in which is cerebo-spinal fluid as is the case in all the ventricles of the brain. Inside the

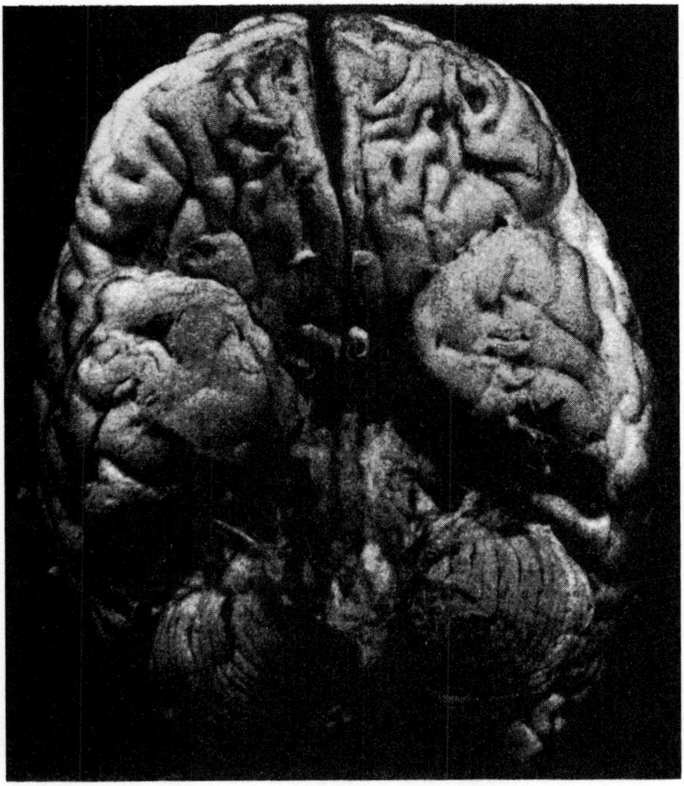

Fig. 22.6. **Human brain—from below.** *Specimen in the author's laboratory.*

thalamus is the **third ventricle** with which each lateral ventricle communicates by a **foramen of Monro**. Immediately behind its posterior wall and situated in the centre is the **pineal body**, the function of which is unknown. It has been suggested that it is an endocrine organ but no hormones have as yet been extracted from it. It is probably a vestigial third eye as is evident in some of the lower animals. The floor of the thalamus, known as the **hypothalamus**, bears a stalk externally, the **infundibulum** to which the **pituitary gland**

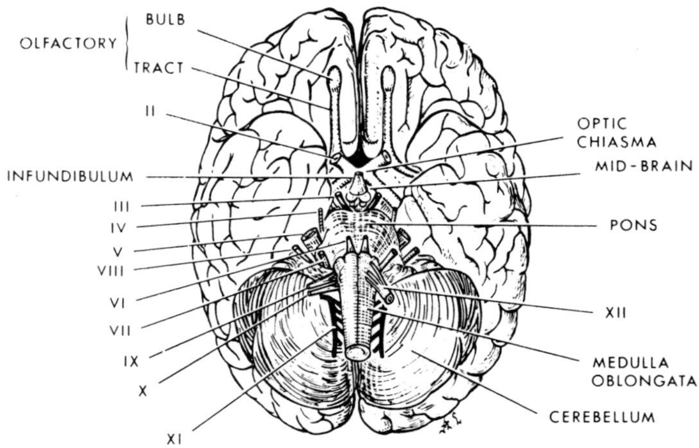

Fig. 22.7. **The brain. Under surface.**

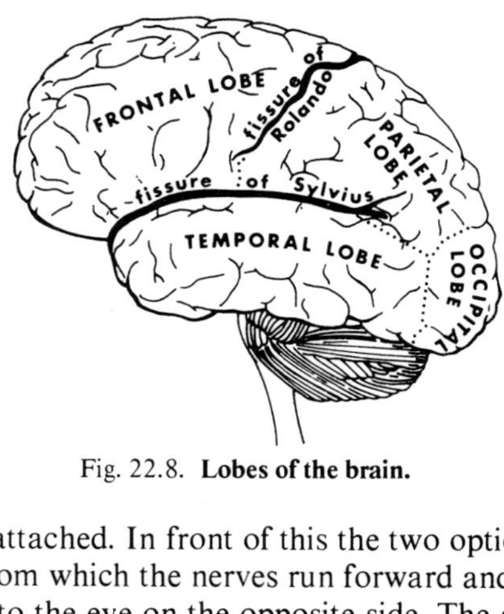

Fig. 22.8. **Lobes of the brain.**

(hypophysis) is attached. In front of this the two optic nerves emerge as **optic tracts** from which the nerves run forward and cross over one another to pass to the eye on the opposite side. The crossing-over of these nerves is known as the **optic chiasma**.

The Mid-Brain

This is composed of the brain stem and four small swellings known as the **corpora quadrigemina**. A narrow canal called the **aqueduct of Sylvius** passes through the mid-brain from the third ventricle.

The Hind-Brain

The **medulla** is that part of the brain stem which is continuous with the spinal cord and the cerebellum lies behind it. Inside the medulla is the **fourth ventricle** into which the aqueduct of Sylvius enters. Three minute pores in the roof of the fourth ventricle enable cerebro-spinal fluid to pass into it and so into the cavities of the brain.

The **cerebellum** is composed of two **hemispheres** which are connected to one another by a small central **vermis**. It has an external

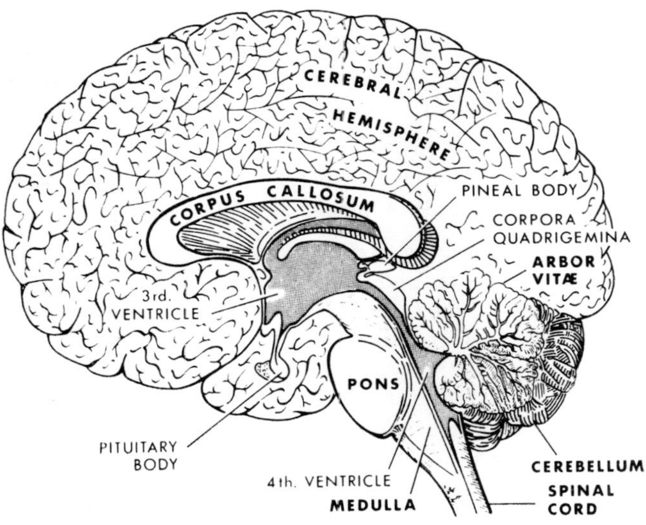

Fig. 22.9. **Vertical section of the brain through the longitudinal fissure.**

cortex of grey matter, the surface of which is covered by numerous narrow ridges, and this grey matter contains millions of neurons. Inside the grey matter is white matter and in the interior this is arranged in strands like the branches of a tree bearing leaves and it is therefore known as the **arbor vitae**. The fibres on leaving the cerebellum form up into bundles. Those connecting it to the mid-brain are the superior peduncles, those joining it to the medulla are the inferior peduncles while those connecting up with the two hemispheres constitute the middle peduncle or pons, as we have already seen.

THE CRANIAL NERVES

Twelve pairs of cranial nerves arise from the brain of which the 1st, 2nd and 8th are sensory, the 3rd, 4th, 6th, 11th and 12th motor and the remainder, the 5th, 7th, 9th and 10th, are mixed.

I. **Olfactory**—*Sensory.* Originating in the sensory epithelium of the nose, their fibres pass through foramina in the cribriform plate and enter the olfactory bulbs. The nerve of smell.

II. **Optic**—*Sensory.* Arises in the retina. The two optic nerves cross over one another at the base of the brain to form the optic chiasma as we have seen. They then form the optic tracts which go to the corpora quadrigemina. The nerve of sight.

III. **Oculo-motor**—*Motor.* Originates in the mid-brain and passes to the superior rectus, inferior rectus, internal rectus and inferior oblique muscles of the eye. Responsible for movements of the eyeball, change in convexity of the lens and size of the pupil.

IV. **Trochlear**—*Motor.* Originates in the mid-brain and supplies the external rectus muscle of the eye.

V. **Trigeminal**—*Mixed.* Arises from the pons by two roots. The dorsal root is sensory and bears a ganglion. This gives rise to three branches:—(i) the OPHTHALMIC which supplies the eyeball, the upper region of the nasal cavity and the skin of the scalp and forehead; (ii) the MAXILLARY which supplies the upper jaw, teeth and lip and the skin of the front of the face and (iii) the MANDIBULAR which supplies the lower jaw, teeth and lip, the front of the tongue and the skin around the lips.

VI. **Abducens**—*Motor.* This innervates the external rectus muscle of the eye.

VII. **Facial**—*Mixed.* The motor fibres arise in the pons and go to the muscles of the face. It is called the "nerve of expression". The sensory fibres arise in the back of the tongue and pharynx and go to the pons.

VIII. **Auditory**—*Sensory.* It has two branches:—(i) the AUDITORY which originates in the organ of Corti in the inner ear and passes to the medulla and (ii) the VESTIGIAL which arises in the semicircular canals and goes to the cerebellum. The nerve of hearing and balance.

IX. **Glossopharyngeal**—*Mixed.* Has two branches. The *sensory branch* arises in the taste buds of the tongue and goes to the medulla and the *motor branch* arises in the medulla and runs to the muscles of the tongue, pharynx and larynx.

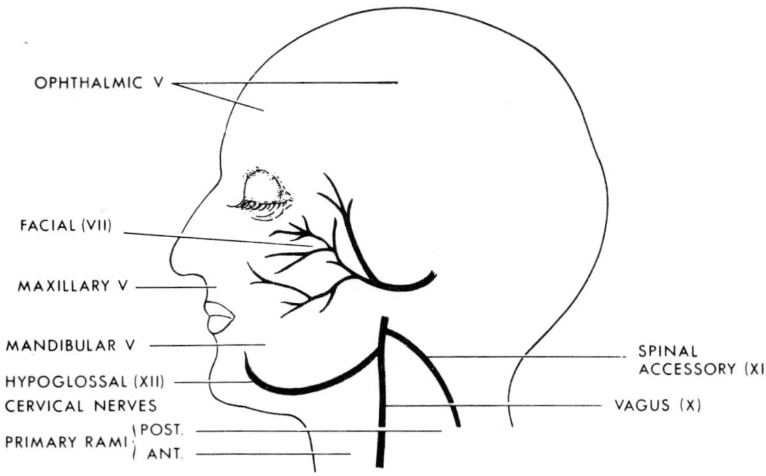

Fig. 22.10. **Chief nerves of head and face.**

X. **Vagus**—*Mixed.* Both the sensory and motor fibres end in or arise from the medulla as the case may be. There are several branches:—(i) the PHARYNGEAL—*Motor.* Supplies the muscles of the pharynx; (ii) the LARYNGEAL—*Mixed.* Supplies the larynx: (iii) OESOPHAGEAL—*Mixed.* Supplies the oesophagus: (iv) the CARDIAC—*Motor.* Regulates the rate of heart beat: (v) the PULMONARY—*Mixed. Motor branches* pass to the bronchi and sensory branches supply the lungs: (vi) VISCERAL—various branches innervate the alimentary canal and associated organs.

XI. **Accessory**—*Motor.* Arises from the medulla and supplies the neck muscles. It is therefore responsible for movements of the head.

XII. **Hypoglossal**—*Motor.* Originating in the medulla it goes to the muscles of the tongue and provides for its movement.

THE FUNCTIONS OF THE BRAIN

The Cerebrum

One of the characteristics which distinguishes man from the lower mammals is his ability to reason. The cerebral cortex, consisting of millions of nerve cells is the part of the brain concerned with consciousness and all it involves in man—reasoning, intelligence, memory, emotion, the understanding of sensations, the control of voluntary movements and, in fact, all mental activity. Definite **sensory** and **motor areas** can be mapped out in the cerebral cortex and in these areas the parts of the body covered by them as shown in Fig. 22.11. Sensory impulses travelling from the sense organs along the fibres reach the sensory areas of the cerebral cortex and this results in conscious sensation. We can distinguish between different sensations and the depth of intensity of these sensations. As the result of this we are able to decide on the appropriate action to be taken. Impulses pass to the motor areas

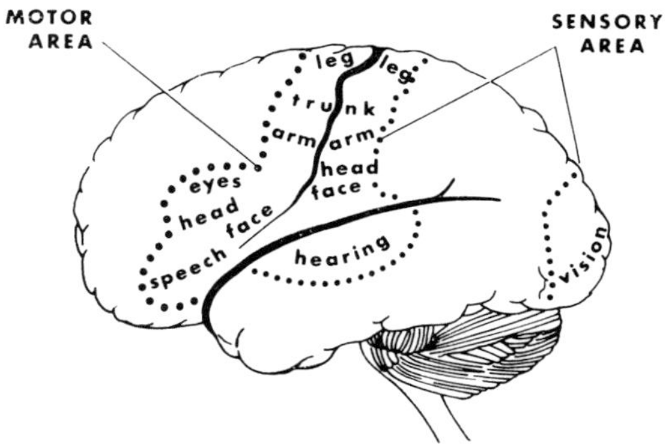

Fig. 22.11. **Areas of the cerebrum.**

and so along motor fibres to the muscles in order that these responses shall be made. Connection between the sensory and motor areas is made possible by synapses between the innumerable neurons of each area. Fibres arise in these motor areas and run down the brain stem and spinal cord. It should be noted here that the cerebellum is also concerned in voluntary movement as we shall see later on. Some of the sensory fibres in the grey matter of the spinal cord cross to the opposite side and so by means of synapses pass up in fibres of the white matter of the spinal cord and the brain stem to the thalamus and so to the cerebral cortex by means of further synapses. The consequence of this is that each side of the cortex is concerned with the opposite side of the body.

The reception of sensory impulses into consciousness produces what we refer to as our feelings, the result of which is consideration of and decision on the action to be taken. These are our *thoughts* and the decisions taken are the result of our power to reason and our intelligence, even though they may sometimes be the wrong ones.

To make a decision we must have had one or more similar experiences on previous occasions and we must therefore have a store of records of such events. This is what we call *memory*. Permanent impressions must be made for we are often able to recall incidents, sometimes of no importance or interest, which took place many years beforehand. Some sensory impulse of a similar nature is received and this brings back to mind the distant previous event. It seems that as time passes more and more new impressions are made and the older ones, at any rate those of minor importance, tend to be submerged. Even so, how difficult it is sometimes to remember things of immediate importance, as you know only too well when you are sitting for an examination! Constant repetition, which means repeated stimulation, improves the memory.

Several theories have been formulated and discarded to explain memory. Some permanent impression or record must be made, possibly as the result of molecular changes in the cells, which is capable of repeating the impulses originally produced. Molecular biology has come to our aid and experimental work lends support to a theory that R.N.A. molecules in the cells of the cerebral cortex undergo changes in their constitution when sensory impulses reach them. If a similar impulse is received by the brain at some future time, conditions are again produced in the cells which create a similar state to that initiated by the original impulse. However, we are still a very long way from solving the problem of memory but it is probably safe to say that, though much is as yet conjecture, this theory carries more than a germ of the truth.

We are still left with the question as to how the mind functions. What is consciousness and what do we mean by the mind? Consciousness, surely, is our awareness of our environment. When we are asleep or under an anaesthetic we are not aware of anything around us *i.e.* we are unconscious and the mind is not at work. The mind must therefore be a process of the brain which produces or is the result of our consciousness. But it is far more than this for it is by means of our mind that we are able to learn, to understand, to remember, to exercise intelligence, to make judgements and decisions and to feel our emotions. It is the "working" of the brain and this continues during the whole of our waking hours.

An instrument known as the **electro-encephalograph** has enabled us to show that minute electrical disturbances are continually taking place in the brain and that these vary according to the mental state. A number of electrodes are fixed at different places on the surface of the head and these are connected to the instrument which amplifies these disturbances and records them graphically, the record being known as an **electro-encephalogram (E.E.G.)**. There is a more or less normal rhythmic fluctuation in potential of 10 cycles per second. These are called **alpha waves**. At times there is absence of this alpha rhythm. Opening and closing the eyes,

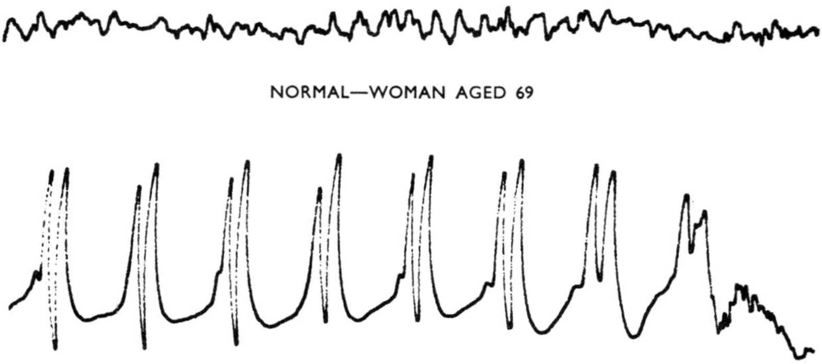

NORMAL—WOMAN AGED 69

PETIT MAL EPILEPSY—BOY AGED 7

Fig. 22.12. **Electro-encephalograms.**

deep breathing, mental effort and sudden noise cause a reduction or disappearance of the rhythm. The E.E.G. is used in the clinical diagnosis of epilepsy and cerebral tumours. A great deal of research work is going on at the present moment in the use of radioactive isotopes in the diagnosis of brain tumours and, in fact, of other organs. Considerable success has been achieved and it has been claimed that the risks to the patient are extremely small compared with the beneficial results obtained.

We often find that some mental impressions occur of which the mind is almost entirely unconscious at the time but which come fully into consciousness at some future time. These are called *sub-conscious impressions.* How often do we say "Subconsciously I was aware that . . .". It has been suggested as the result of psychological tests performed on patients who have undergone major operations on the brain that the two halves of the brain act and think separately and independently of each other but more evidence is necessary before this can be stated as factual. The view had long been held, too, that the anterior region of the cerebral cortex beyond the sensory and motor areas was concerned with consciousness but this now appears to be doubtful and that the whole of the hemispheres are concerned in this. It seems probable that the anterior region is responsible for our general "make up", our moral and mental character in general, our personal ego. The brain must be continually stimulated by incoming sensory impulses of all kinds and without these experiences there can be no memory, no reasoning, no decisions. Powerful impulses produce mental images which create our emotions.

Periodically the brain, and indeed the whole body, needs rest from its labours, consciousness is lost and we fall asleep. But the sense organs and the nerves are still active to some extent for external noises or a gentle touch may produce reflex actions without arousing us from sleep. Equally a loud noise or shaking a sleeping person will wake him up. The amount of sleep we require varies very much with the individual and with age. Babies sleep for the greater part of the day but adults require much less, eight hours being the average need until old age is reached when five or six hours may be adequate. It was said that Napoleon, even in his most active days, slept for only four hours. Sometimes our consciousness is not completely lost when we are asleep and as a result we dream. Unpleasant dreams and nightmares are caused by internal sensations. Winston Churchill is reputed to have said of dreams "children of the night of indigestion bred".

Intelligence is difficult to define but we can safely say that it refers to our capacity to understand and to learn rather than actual learning itself and to our ability to profit by experience. We inherit from our parents the potential to develop a certain degree of intelligence. This is due to genes we receive from them and will vary with different children. It may be called *genetic intelligence* and clearly this cannot be measured. In order that this degree may be developed and attained, stimulation and learning are necessary and in this environment plays no small a part. It has been shown that this development occurs to the greatest extent in the first four or five years of life. The extent to which an individual has been able to utilise his genetic intelligence may be called his actual or *observed intelligence* and this mental capacity can be measured by tests and expressed by what is known as the **Intelligence Quotient (I.Q.)**. People of average intelligence have an I.Q. of 90 to 109, any measure above this being "above average" and I.Qs. of 120–129 indicate superior intelligence. For I.Qs. below 90 see Mental Deficiency on p. 283. The value of the I.Q. as an indication of intelligence should not be overrated, however, as instances of creative thinkers and brillant scientists with I.Qs. at average level have shown. Incredible as it may seem, there are 3,000,000 people in this country who are unable to read or write.

Hypnosis is a kind of sleep in which the patient responds to suggestions put by the hypnotiser. Though not without its dangers, it has been successfully used in childbirth, in psychiatric treatment and even in dental operations. It should be performed only by specially qualified medical practitioners and is not a subject for entertainment by unqualified persons.

The Thalamus

Each of the thalami acts as a kind of junction from which impulses are relayed to other parts of the brain, creating our ability to see, hear, feel and so on.

The Mid-Brain

The two superior corpora quadrigemina deal with visual reflexes and the two inferior ones with auditory reflexes. Both reflexes result in appropriate responses, for example when you hear a sudden loud explosion you quickly and automatically put your hands over your ears. Similarly when a small fly or a particle of dust is approaching the eye you very quickly close it.

The Pons

The pons is that part of the brain stem which puts into communication the medulla, the cerebellum, the mid-brain and the cerebral hemispheres.

The Medulla

This contains centres which control respiration and heart beat and, of course, contains fibres which put the brain and spinal cord into continuous communication in both directions.

The Cerebellum

Maintenance of balance is the concern of the cerebellum. It therefore receives sensory impulses from the semicircular canals of the ears and issues motor impulses to the muscles in order that they may maintain that degree of tension or tonus which ensures equilibrium of the body. It functions as a reflex centre but impulses are able to pass to and from the cerebral hemispheres by tracts of fibres connecting the two regions together. These fibres cross one another so that each side of the cerebellum is connected with the cerebral hemisphere on the opposite side. Because of the relationship between the cerebrum and the cerebellum the former can exert such control as may be necessary over the latter when so required. When you first learn to swim or to drive a car, you think carefully about every movement which has to be made and the impulses for these movements originate in the motor areas of the cerebral cortex. With practice, however, the movements become automatic and reflex and the impulses cease to come directly from the cerebral hemispheres.

REFLEX ACTIONS

We have already come across several instances of reflex actions and a well-known example is the *knee-jerk reflex.* If you sit down and cross one leg over the other, allowing the upper one to hang freely and a sharp tap is made on the tendon just below the knee cap of the upper leg, it will jerk quickly upwards. You cannot control it and this is a normal reflex brought about in the following manner. Wrapped round the fibres of the quadriceps femoris muscle on the front of the leg are the dendrons of a sensory neuron and these constitute a capsule which is called a **stretch receptor**. The stimulus of the tap on the tendon below the knee-cap causes a stretching of this receptor and creates an impulse which passes along the sensory fibres of a spinal nerve to the dorsal root of the spinal cord. In this grey matter it crosses a synapse with a motor neuron in the ventral root and so passes along the motor fibres in the spinal nerve back to the muscle thus causing it to contract. The essential constituents of a reflex action are shown in Fig. 22.13. and these are embodied in what is known as the **reflex arc**.

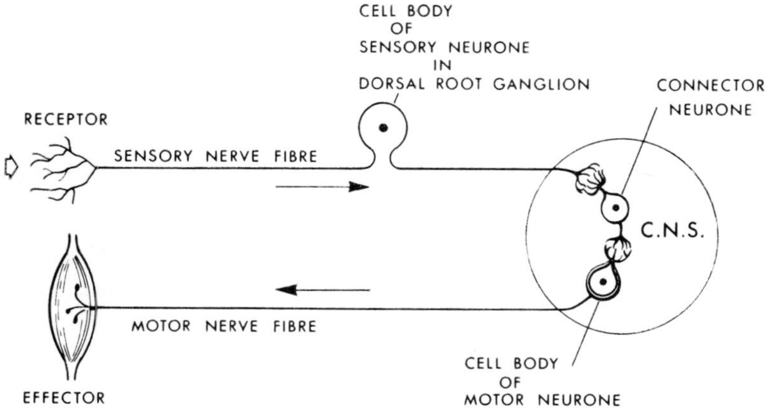

Fig. 22.13. **Diagram of the reflex arc.**

The stretch reflex is of fundamental importance in the maintenance of posture and the tonus of the muscles concerned is brought about by a series of stretch reflexes.

Another familiar example of a reflex action is the withdrawal reflex. Suppose the hand touches something very hot or very sharp, the hand is immediately withdrawn. One does not stop to think about it, the action is instantaneous. The flexor muscles of the arm contract, thus raising the arm and withdrawing the hand from the cause of pain.

All the reflexes we have so far considered are instinctive and inherited and they are known as **unconditioned reflexes**. But we develop many reflexes during our lifetime as the result of experience and repetition. Consciousness obviously participates in the formation of this kind of reflex which is called a **conditioned reflex**. These were first explained by the Russian physiologist Pavlov from experiments he performed on dogs. He rung a bell every time food was presented to the hungry animal over a period of time. Saliva is secreted whenever food is offered to a dog but Pavlov found that eventually the ringing of the bell alone, without offering food caused the saliva to flow freely. Clearly the dog had associated the sound of the bell with the appearance of food and this is where consciousness had played its part. But as the secreting of saliva on the presentation of food is a normal unconditioned reflex in the dog the secretion on the ringing of the bell must have been an acquired or conditioned reflex. This type of reflex requires repetition otherwise it ceases to be maintained and, as the result of repetition, habits are acquired. An example of this is in the driving of a car, including the many evasive actions we have to take on the road today. In fact, we can go as far as to say that a great deal of human behaviour is dependent on the acquisition of conditioned reflexes.

Further examples of unconditioned reflexes are:

The **pupil reflex** in which the iris expands when a bright light falls on the eye causing the pupil to contract.

The **swallowing reflex**, produced by food on the tongue and the posterior wall of the pharynx causing the epiglottis to close, the soft palate and the larynx to be raised and a peristaltic wave to be set up in the oesophagus.

The **blinking reflex** when rapid closing of the eyelids is brought about by minute particles of dust or other objects touching the conjunctiva.

The **coughing reflex** when, as the result of particles at the entrance to the glottis, the expiratory muscles contract, forcing the offending particles into the pharynx and mouth.

All of these, apart from the swallowing reflex, are clearly for purposes of protection. *Micturition* and *defaecation,* once started voluntarily, become reflex and in infants and in some diseases are entirely so.

THE SPINAL CORD

The spinal cord which, as we have seen, is continuous with the medulla, is enclosed in the spinal canal of the vertebral column and, like the brain, it is enclosed in the three meninges. These all serve to protect it from injury, as does the cerebro-spinal fluid secreted by the arachnoid. About 45 cm. in length and 1·9 cm. in diameter it is roughly cylindrical in shape though it becomes thinner towards its lower end and then tapers to a slender thread known as the **filium terminale**. This is surrounded by a tail-like collection of nerves called the **cauda equina**. The filium terminale begins after the first lumbar vertebra and ends at the coccyx. In the cervical and lumbar regions

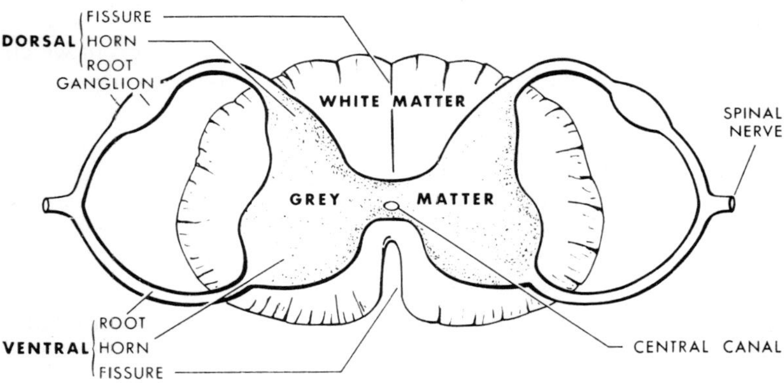

Fig. 22.14. **Spinal cord. Tranverse section.**

swellings occur and they are known as the **cervical and lumbar enlargements** respectively. On the dorsal side of the cord is a deep **dorsal** (or **posterior**) **fissure** and on the ventral side is a shallower **ventral** (or **anterior) fissure.** A transverse section of the cord shows that the white matter is external and that the grey matter is shaped somewhat like a butterfly or a rough letter H. In the centre of the cross-piece of the H is a small **central canal**. The arms of the H are referred to as **dorsal** (or **posterior**) and **ventral** (or **anterior) horns** or **cornua**. It is here that the spinal nerves have their roots. The white matter consists of sensory fibres running upwards to the brain and motor fibres passing downwards from the brain. These are arranged in definite *tracts* so that all the fibres from a particular sense organ or running to a particular muscle are grouped together. The sensory fibres of the spinal nerves pass to the dorsal horns and the motor fibres leave the ventral horns. The dorsal root of each spinal nerve bears a ganglion, the **dorsal root ganglion**, after which it is joined by the ventral root. Sensory neurons can form synapses with several connector neurons and these in turn may form synapses with motor neurons either at the same or a different level in the spinal cord.

The Spinal Nerves

There are thirty-one pairs of spinal nerves and each, as we have seen, has two roots, a dorsal root which is sensory and a ventral root which is motor. Since the two roots join it follows that the spinal nerves are all mixed nerves. These emerge between the vertebrae but those in the lumbar and sacral regions cluster together to form the cauda equina. After the union of the two roots a branch called the **ramus communicans** arises and this provides a linkage with the autonomic nervous system.

The spinal nerves are classified according to the region in which they emerge from the spinal cord and they are as follows:—

8 cervical
12 thoracic
5 lumbar
5 sacral
1 coccygeal

Each soon divides into two branches known as the **anterior** and **posterior primary branches** which contain motor and sensory fibres respectively. All the posterior primary rami supply the muscles of the back. The anterior primary rami form plexuses except in the thorax which they encircle as **intercostal nerves** innervating the intercostal muscles and skin of that region, though the last six also supply the abdominal wall. The anterior rami of the first four cervical nerves form the **cervical plexus** from which nerves arise supplying the neck muscles. The 3rd, 4th and 5th cervical nerves give rise to the **phrenic nerve** which supplies the diaphragm. The anterior rami of the 5th to 8th cervical nerves and the first thoracic form the **brachial plexus** in the region of the brachial enlargement of the spinal cord. This supplies the shoulder and arms by the **median, radial** and **ulnar nerves**. Branches supply the hands and fingers.

The anterior rami of the first four lumbar nerves form the **lumbar plexus** in the region of the lumbar enlargement and go to the legs by the large **femoral nerve** which arises from the 2nd, 3rd and 4th lumbar nerves. Likewise the **sacral plexus**, formed by the anterior rami of the 4th and 5th lumbar and first four sacral nerves gives rise to the **sciatic nerve**. This divides above the knee into a **tibial nerve (internal popliteal)** providing the calf muscles and the sole of the foot and an **external popliteal nerve** which innervate the muscles of the

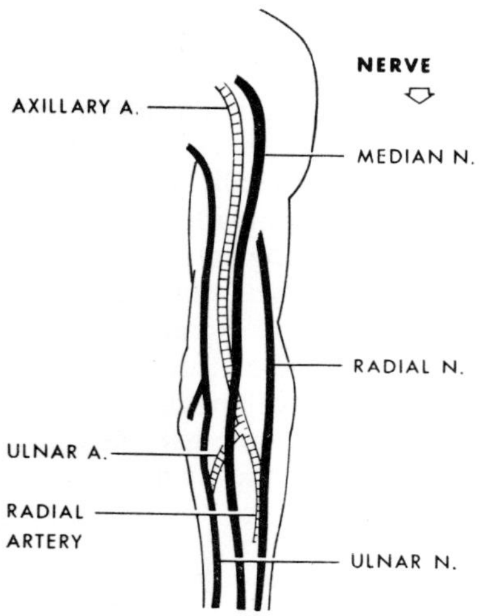

Fig. 22.15. **Chief nerves of arm.**

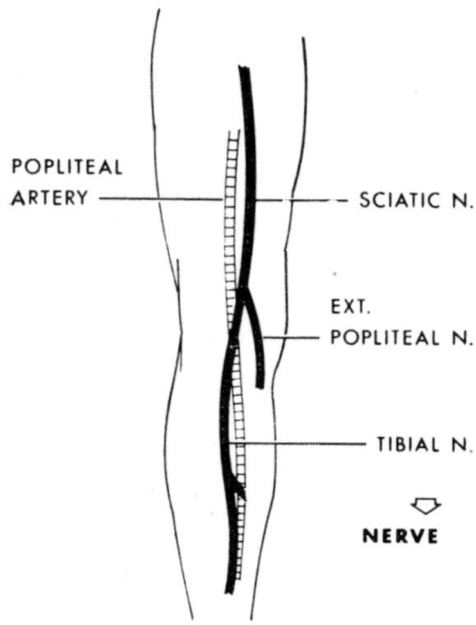

Fig. 22.16. **Chief nerves of leg.**

outside of the leg and the foot and toes. Branches of the sciatic plexus also provide nerves to the bladder, rectum, external reproductive organs and gluteal muscles.

Functions of the Spinal Cord

Sensory impulses reach the dorsal (posterior) horns of the grey matter in the spinal cord by means of fibres from the neurons in the dorsal root ganglion and pass by means of connector neurons to motor neurons in the ventral (anterior) horns and thus by motor fibres in the spinal nerves to muscles. The passage of these impulses is rapid and produces reflex actions. Thus the spinal cord is the centre for reflex actions. The brain may become aware of these because synapses make possible the passage of impulses up the sensory fibres in the white matter of the cord to the brain. The brain can control these reflexes to some extent so that they are best adapted to the particular requirements of the moment. Furthermore the brain can send impulses down the motor fibres of the spinal cord and therefore bring about voluntary movements. Generally speaking the fibres cross over in the medulla and pass down the spinal cord on the opposite side and enter the ventral horn at various levels. Sensory neurons can form synapses with several connector neurons which, in turn, form synapses with more than one motor neuron thus enabling co-ordination between antagonistic muscles. Co-ordination of movement, too, is effected by association tracts of fibres.

THE AUTONOMIC NERVOUS SYSTEM

This system which supplies the organs of the body which are outside the control of the will, such as the rate of heart beat, peristalsis and gland secretion consists of two generally antagonistic systems, the **sympathetic system** and the **parasympathetic system**. The sympathetic system is composed of fibres originating from the thoracic and lumbar regions of the spinal cord and the parasympathetic system of fibres from the brain and sacral region of the cord. The axons from the cell-bodies of these neurons in the central nervous system form synapses with other cell-bodies grouped together as **sympathetic ganglia** outside it. From these ganglia fibres arise which end in the structures and organs which they innervate.

The Sympathetic System

The first four sympathetic ganglia of the thoracic region are themselves grouped together to form a single **stellate ganglion** but those of the remaining thoracic ganglia of the thoracic region and those of the lumbar region are separate, there being one to each nerve root, and form a chain of connected ganglia on each side of the vertebral column. The motor sympathetic fibres arises in the grey matter of the spinal cord and leave it in the roots of the spinal nerves. After the dorsal and ventral roots have joined a branch, the **white ramus**, is given off which carries the sympathetic fibres to the sympathetic ganglia. Sensory sympathetic fibres arising in the viscera also run through the white ramus to the dorsal roots of the spinal nerves and so enter the dorsal root ganglion. On reaching the grey matter of the spinal cord they form synapses with connector neurons. From these connector neurons, fibres either form synapses with sympathetic motor neurons in the sympathetic ganglia at the same level or pass upwards or downwards to form such synapses at different levels, as in the cervical and sacral regions where there are no white rami, or they leave the sympathetic chain altogether forming synapses in ganglia outside the sympathetic trunk. Some of the fibres on entering the spinal cord pass up to the brain.

Some of the fibres from the sympathetic ganglia rejoin the spinal nerves by a branch known as the **grey ramus** and then follow these nerves along their normal course to the muscles of the limbs or trunk and to the arterioles, sweat glands and the arrector pili muscles of the hairs. Thus constriction of the arterioles, increased sweating and erection of hair is brought about by impulses along these sympathetic nerves.

The fibres from the sympathetic ganglia which do not rejoin spinal nerves supply the heart, bronchioles in the lungs, salivary glands, the iris of the eye and the lacrimal glands. The result of impulses passing along these sympathetic fibres is increase in the rate of heart beat, dilation of the bronchioles, decrease in salivation and of lacrimal secretion and enlargement of the pupil.

The fibres forming ganglia outside the sympathetic trunk form plexuses from which nerves pass to the viscera. The **solar plexus** situated behind the stomach provides the stomach, intestines and other abdominal viscera, the **cardiac plexus** near the base of the heart supplies the heart and lungs while the organs in the pelvic cavity—the rectum, reproductive organs and bladder—are provided for by nerves from the **hypogastric plexus** in front of the sacrum.

The Parasympathetic System

This part of the autonomic system has its centres in the mid-brain, medulla and sacral region of the spinal cord. The fibres from cell bodies in the mid-brain leave it in the oculo-motor nerve (IV) and pass to the ciliary muscles causing the contraction of these muscles and therefore an increase in the convexity of the lens of the eye for the examination of near objects, at the same time reducing the diameter of the pupil by causing the circular muscles in the iris to contract. The fibres from cell bodies in the medulla leave it in the facial (VII), glossopharyngeal (IX) and vagus (X) nerves. Those in the facial nerve go to the sublingual and submandibular glands and to the lacrimal glands, producing an increase in their secretion. The fibres in the glossopharyngeal nerve supply the parotid gland and those in the vagus innervate the organs of the thorax and abdomen, decreasing the rate of heart beat, constricting the bronchioles and stimulating peristalsis. Finally the parasympathetic fibres from cell bodies in the sacral region of the spinal cord run in the first three sacral nerves to the organs of the pelvis.

The Autonomic System as a Whole

As already stated, the two systems in general act in opposition to each other and therefore only one functions at a time. The sympathetic system is very largely concerned with self-preservation. For example it increases the rate of heart beat and causes adrenalin to be secreted into the blood and thereby puts the body into a state of active defence when conditions require it. The parasympathetic system restores the rate of heart beat to normal and causes the cessation of adrenal secretion when such conditions are no longer required. The parasympathetic system brings about expansion of the iris and contraction of the pupil when a bright light shines into the eye and the sympathetic system causes its dilation in darkness. The parasympathetic system causes the muscles of the alimentary canal to contract in peristalsis and the sympathetic system make them relax and therefore slows down this process. The muscles in the bladder wall are stimulated by the parasympathetic fibres and inhibited by the nerves of the sympathetic system.

As we end our study of the nervous system we are inevitably drawn to the conclusion, I think, that while a great deal is known of its working, there is still a vast field for investigation, particularly in regard to the working of the brain.

Disorders and Diseases of the Nervous System

Fainting or **syncope**—the result of an insufficient blood supply to the brain.

Headache—considered to be produced by pressure on the brain caused by dilation of the arteries supplying it, since the tissue of the brain itself has been proved to be insensitive to pain. Headache may be produced by sheer mental exhaustion or by more serious causes such as a brain tumour. In **migraine** the headache is very severe and is accompanied by blurred vision and sickness.

Neuralgia—pain, often severe, in regions supplied by sensory nerves. A frequent position is the side of the forehead and here it is the sensory branches of the trigeminal nerve which are responsible, though other regions are also affected.

Insomnia—Apart from such causes as pain, worry or grief, mental work or eating a large meal immediately before retiring, discomfort in bed, external noise, and, of course, illness can produce insomnia. Sleep will be induced when the mind is very active if one can make it a complete blank for a short time. Sedatives, or even a hot drink, can be taken with effect.

Neuritis—a general term for inflammation of the nerves.

Epilepsy—due to improper functioning of the brain, causes the seizures, convulsions and loss of consciousness of the characteristic epileptic fit known as *grand mal*. In another form, *petit mal*, while there is loss of consciousness for a short period, there are no convulsions. Effective control of the disease by means of drugs is now possible.

Paralysis—loss of muscular power. This has already been described under diseases of the muscular system. It may be caused by disease of the brain, spinal cord or peripheral nerves.

St. Vitus's Dance—in which peculiar unco-ordinated, jerky movements of the limbs takes place, occurs mostly in children.

Locomotor ataxis—is symptomised by an inability to exercise co-ordinated movements of the limbs and to maintain normal equilibrium. It is due to disease of the dorsal nerve roots and posterior columns in the spinal cord.

Neurasthenia—a mental condition in which the patient suffers very easily and acutely from nervous exhaustion, insomnia, depression and complete inability to make any sustained effort.

Neurosis—a term applied to those mental conditions in which a person's actions are governed by his emotions instead of by reason. Neurotics often suffer from imaginary fears.

Meningitis—inflammation of the meninges caused by bacterial infection. There are several forms of the disease, and examination of the cerebro-spinal fluid extracted by a *lumbar puncture* reveals the kind of infecting organism. One form, **cerebro-spinal meningitis** is extremely infectious and is caused by a bacterium, *Meningococcus*.

Poliomyelitis—formerly called *infantile paralysis*, is a virus infection of the ventral root of the spinal nerves and is extremely infectious chiefly due to carriers. Paralysis of the limbs and of the inter-costal muscles and diaphragm may occur and, in the latter case, the patient must be put into an *iron lung* or breathing will cease and death ensue. Immediate treatment of the disease is essential to effect a cure and save life.

Tumour of the brain—can be detected by means of the electro-encephalograph and can sometimes be removed by surgery.

Hydrocephalus—or "water on the brain" which is, in fact, the presence of excess cerebro-spinal fluid, may be brought about by a brain tumour or meningitis or it may be congenital. It often occurs in cases of spina bifida.

Concussion—a severe blow on the skull may cause bruising of the brain and cause unconsciousness. If it is not too heavy a blow, consciousness remains but a very severe headache develops. Unconsciousness may last for a few hours or for months before recovery, after which there may be loss of memory.

Hysteria—a condition in which the sufferer becomes abnormally excitable and often gives way to fits of laughter and weeping on the slightest provocation. These people are also ruled by very fixed ideas. In extreme cases spasms and even paralysis may occur. Psychoanalytical treatment is used in this disorder of the brain.

Brain fever—a term used for **sleepy sickness** or **encephalitis lethargica**, not to be confused with sleeping sickness or trypanosomiasis which is a disease occurring in Africa caused by a parasitic protozoon called trypanosoma which is carried by tsetse flies. Sleepy sickness is caused by a virus which produces severe drowsiness and even unconsciousness and paralysis.

Disseminated sclerosis—or **multiple sclerosis** is brought about by degeneration of the myelin sheaths of the nerve fibres, though the cause of this is uncertain beyond the fact that degenerative hardened patches occur in the brain and spinal cord. But in 1976 it was reported from America that a virus had been isolated which it was thought was almost certainly the cause of the disease. The symptoms are paralysis, giddiness and tremor, interference with speech and vision and abnormal movements of the limbs (ataxia). Unfortunately these symptoms usually get progressively worse and ultimately the patient becomes a permanent invalid. The discovery of the virus, however, may ultimately lead to the development of a vaccine with prophylatic and therapeutric properties.

Spina bifida—in which the neural arches of some vertebrae have not united was explained on p. 47. Cells in the spinal cord may not develop and infection may set in. Total recovery even after a very early surgical operation is seldom possible. Hydrocephalus often accompanies spina bifida.

Dyslexia—a condition occurring in children who find difficulty in reading due to visual confusion between similar letters, thus causing them to be read out of position with the result that wrong words or no proper words are read. It may be called "word blindness" and can occur in otherwise intelligent children as well as those who are mentally deficient. Many intelligent children can be taught to read quite reasonably well.

Mental disorders include **neuroses, psychoses** and **mental deficiency** of various kinds.

Psychopaths—persons devoid of conscience or any sense of responsibility who have no sense of right or wrong and are aggressive and irresponsible. They are often criminals. Psychosis is a term used for a serious disordered condition of the mind and this is manifested in various ways such as the following:—

Psychoneurosis—an inability to face up to life, sometimes resulting in suicide.

Delirium—a condition in which the conscious mind is perverted, delusions occur and speech is incoherent. It often occurs in fever. In **delirium tremens**, caused by alcoholism, the delusions are more intense and queer and trembling occurs. This may become more severe and lead to raving and violence.

Melancholia—a feeling of excessive depression unjustified by conditions.

Schizophrenia ("split mind")—dual personality.

Paranoia—persecution mania.

Kleptomania—uncontrollable compulsion to steal.

Dipsomania—an uncontrollable yearning for alcohol.

Nymphomania—an uncontrollable urge for sexual satisfaction in women.

Megalomania—a delusion of personal power or greatness.

Mental deficiency—a state in which the mental powers are retarded and below those of the average man. There are various degrees of mental deficiency from those whose I.Q. is so low that they are quite unable to look after themselves and need treatment and care in a mental hospital to those who are limited in their ability to learn and who often become misfits in society, frequently getting into trouble and unable to hold a job. I.Qs. of 79 to 70 are border-line, 69 to 50 feeble-minded and 49 to 25 are imbecile. Anyone with an I.Q. below 24 is classified as idiot.

Mongolism—a form of mental subnormality in which a child is born with slanting eyes, small facial features and head and short fingers, superficially, but only superficially, resembling the Asian Mongol races. Mongol children have an extra chromosome attached to the 21st pair and therefore have 47 chromosomes. This is, of course, a mutation. These children are usually of an affectionate nature and may live to adult or middle age. The cause of the mutation is not known for mongolism occurs in normal families, about two in every thousand births in Western Europe and America being mongols. There seems to be some connection between the condition and the age of the mother at conception, for the older she is the greater is the risk of her producing a mongol child.

PRACTICAL WORK

The Brain

Examine a model of the human brain or, better, if available, a preserved specimen. Failing either of these examine a preserved sheep's brain. Identify the various parts as described in the text and, if a vertical section is available, identify the internal structure and distinguish the grey matter and the white matter.

The Cranial Nerves

Examine the neck of a rabbit which has been dissected to display the blood vessels and nerves and identify the nerves by reference to a diagram of the dissection in a practical book.

Spinal Cord

Examine a **transverse section of the spinal cord** and note the white matter externally and the grey matter internally. Identify the details of the structure as described in the text.

Neurons

Examine a prepared **slide of neurons** showing cell body, dendrons, and dendrites. Note also any axons which are present.

Reflex Action

(i) **Knee Jerk**

Get a fellow student to sit on a chair with one leg crossed over the other with the upper one hanging freely. Tap sharply the tendon just below the knee cap on the upper leg with the side of the hand. Note that the lower part of the limb is jerked sharply upwards.

(ii) **Iris Reflex**

Refer to experiment 4 on p. 255, and repeat it if necessary.

(iii) **Reflex Action in the Spinal Cord**

Suspend a frog in which the brain has been destroyed (pithed) by its fore-limbs from the rod of a retort stand. Now stimulate one of the hind limbs by (*a*) pinching with forceps (*b*) touching it with a hot wire (*c*) touching with a rod dipped in acid and (*d*) touching it with two wires from a battery. Note that the frog quickly withdraws the leg. The brain having been destroyed, the animal is no longer conscious and the response is therefore entirely automatic or reflex.

Impulses Carried by Nerves

Dissect open the leg of the frog used in the last experiment and sever the large sciatic nerve. (Its location can be identified from a text book diagram.) Repeat experiment (iii) using one or more stimuli. There is no response which shows that the impulses are carried by the nerve.

Chapter XXIII

BASAL METABOLISM, FOOD AND DIET

We have now completed our study of man's anatomy and physiology and we must turn our attention to the means whereby his body and those of his fellow creatures can be kept in a state of normal health in the environment in which they live and how that environment can be kept as free as possible from all pollution which can in any way adversely affect man's life.

Health is defined in the Constitution of the World Health Organisation as "a state of complete physical, mental and social well-being and not merely the absence of disease or infirmity." The health of any community depends on the health of its individual members and if ever the above state of perfection is to be attained, disease is one of the foremost things which must be conquered throughout the world and this means that a constant fight must be waged against it.

Good health is the greatest of all human possessions and to quote an old Arabian proverb, *He who has health has hope and he who has hope has everything.* One of the many things essential to good health is the right kind of food in the right quantity. Diet will therefore be the first matter of our consideration in the study of health science or hygiene.

The food which provides our diet must supply us with the materials from which energy can be liberated for the working of all the systems of the body, whatever the demands on them may be. It must also provide us with fabrics for our growth as children and for the replacement of worn-out cells in adults.

BASAL METABOLISM

Before work of any kind is done, when the body is lying completely at rest, a minimum amount of energy is required to maintain the body's functions—heart beat, respiration and the maintenance of body heat—*i.e.* to keep us alive. This minimal requirement is known as **basal metabolism**. It varies according to age, sex and the surface area of the body, being higher in children and males than it is in adults and females while the rate decreases in old age in both sexes. The greater the surface area of the body, the greater the loss of heat will be and therefore the heavier the demand for energy to replace this heat loss. The surface area of the body is larger in boys and men than in girls and women; hence the higher basal metabolic rate in the male. In old age, the body functions proceed at a slower rate and therefore the basal metabolism decreases.

The energy is measured in heat units known as **Large Calories** (or **kilocalories**). A large calorie (**C**) is the amount of heat required to raise 1 kg. of water 1°C. New-born babies have a basic metabolic rate of around 30 Calories per square metre of surface area per hour. On reaching the age of puberty, when the maturing of the sex glands demands an extra supply of energy, this rises to 45 or more Calories in boys and a few less in girls. In adult age the figure falls to about 40 in men and 37 in women. The generally accepted basal metabolic rate for 24 hours is 1,700 Calories for a man and 1,450 for a woman or an all round average of 1,600 Calories. Should this demand not be met, starvation will ensue and heavy calls will be made on the body's reserves of carbohydrate and fat and when these fail on the protein in the food, to provide the energy. This will result in loss of weight and emaciation. The basic metabolic rate will be higher in a cold climate as the body has to generate more heat in order to maintain its normal temperature and the reverse is true for hot countries. Disease in which there is fever will also increase the rate, but under normal conditions of health and climate, 1,600 Calories remain as the average basic metabolism for adults.

The Measurement of Basal Metabolism

One method by which the basal metabolism can be ascertained is by means of a **human calorimeter**. This is a large, insulated, double-walled chamber fitted with an insulated air-tight door and an insulated double-glazed and air-tight window. A coiled pipe, fitted with a thermometer at each end outside the chamber, runs through the chamber and water passes through the pipe at a controlled rate. There is also an inlet pipe through which air is pumped into the chamber and to which oxygen can be added before the air re-enters the chamber to replace that absorbed while the experiment is in progress. An outlet pipe takes the air out of the chamber on the opposite side. The carbon dioxide and water are then removed from the air by passing it through soda-lime and concentrated sulphuric acid and the air with fresh oxygen added is pumped into the chamber again by the inlet pipe.

The subject undergoing the experiment remains in the chamber for several hours, lying down in bed at complete rest. The temperature of the water entering and leaving the chamber is recorded and as the water circulates at a controlled rate, the volume is known. The weight of the water, calculated from the volume, is multiplied by the rise in temperature and this gives the heat absorbed by the water and therefore generated by the subject in Calories. The time which the experiment is in progress is also recorded and the basal metabolism for 24 hours can thus be calculated.

A more modern and more convenient method of finding the energy requirements is by measuring the volume of carbon dioxide produced in a given time by collecting it in a 50-litre air-tight bag known as a **Douglas Bag**. The collected air is analysed and the Respiratory Quotient calculated.

THE CALORIFIC VALUE OF FOOD

The energy available from food is determined by igniting a known weight of it in an atmosphere of oxygen in a vessel called a **bomb calorimeter**. This is immersed in a water jacket which absorbs the heat generated by the ignition of the food. The amount of heat is obtained from the product of the weight and the rise in temperature of the water from which the amount of heat liberated from 1 gram of the food can be calculated. This is the **calorific value**. The bomb calorimeter is a strong metal vessel fitted with a heating coil internally and provided with a controlled inlet for oxygen which is under pressure. The food is weighed and introduced into the calorimeter, immersed in its water jacket containing a known weight of water in kilograms of which the temperature is taken in °C. Then the electric current is passed and the food ignites explosively. The temperature of the water is again taken after stirring, immediately it becomes stable and the number of Calories per gram can be calculated. It is difficult to avoid loss of heat from the bomb calorimeter and another method is often used in which the oxygen required to oxidise the food is measured. From this the heat produced can be calculated. Every litre of oxygen produces 4·82 C.

The average results for different kinds of carbohydrates, fats and proteins are as follows:—

Carbohydrate	4·2 Calories/gram
Fat	9·3 Calories/gram
Protein	4·1* Calories/gram

The calorific values of different kinds of carbohydrates and fats varies and these are average results. As small quantities of these foods are not absorbed in human digestion, allowance must be

* In body, allowance having been made for the urea derived from proteins, the calorimeter result is 5·6 C.

made for this in working out the diet and the calorific values used are as follows:—

Carbohydrate	4 Calories/gram
Fat	9 Calories/gram
Protein	4 Calories/gram

ENERGY REQUIREMENTS OF THE BODY

When a person is doing work he requires an extra supply of energy over and above that required for his basal metabolism and the amount of this depends on the nature of his work. These additional requirements can be ascertained by placing the subject in the human calorimeter in which he performs various kinds of work *e.g.* riding a stationary bicycle. Getting up in the morning, eating meals, walking about and other normal daily activities demand an extra supply of energy and the manual worker expends more than the sedentary worker. Brain work, tiring as it may seem, makes little demand on the energy supply but physical work can make quite heavy demands. Some idea of the body's requirements can be gained by examining the following table:—

Daily Energy Requirements of a Normal Adult doing Light Work

	Man	Woman
8 hours sleep	568 C	447 C
8 hours, normal body functions	739 C	582 C
8 hours light work	1,693 C	1,446 C
	3,000 C	2,475 C

Now compare the body's approximate requirements according to the kind of work done.

Man doing sedentary work with a little exercise	2,500 C
Man doing medium physical work with exercise	3,500 C
Man doing heavy manual labour	5,000 C

These extra energy requirements are supplied by the food and this means that a properly balanced diet is necessary.

During lactation a nursing mother must increase her calorie intake by at least one fifth.

DIET

A correctly balanced diet should consist of:—

Proteins
Carbohydrates
Fats
Mineral Salts
Vitamins
Water

Proteins

Proteins provide the ten amino-acids essential for growth and repair. *First class proteins,* of animal origin, which contain all the essential amino-acids are obtained from lean meat, poultry, fish, eggs, cheese and milk. Gelatin, though an animal protein, does not contain any of these amino-acids and is of no food value. *Second class proteins,* of vegetable origin, fail to provide the necessary amino-acids in most cases but peas and beans are good sources of protein.

There are various values obtained by biochemical and physiological methods by which the nutritive quality of proteins is expressed. The nitrogen of the protein is used as the index in these values and the most important is the *Net Protein Utilisation (N.P.U.).* This is the percentage of nitrogen intake in the food which is retained. Its value depends not only on the protein but also on the composition of the diet as a whole, the age of the recipient and other factors. *Digestibility* is the percentage of nitrogen in the food which is absorbed. *Biological Value* is the percentage of absorbed nitrogen which is retained and the product of these two values gives the N.P.U.

Carbohydrates

Although our entire energy requirements could be obtained from foods rich in carbohydrates, the amount needed would be such that it would be indigestible and we must therefore rely partly on fats for our energy supply. Vegetable foods such as potatoes, bread, rice, cereals, cakes and other foods made from flour, peas and beans all contain carbohydrates as do honey, milk and chocolate. One gram of carbohydrate yields 4·2 Calories.

Fats

Meat, fish, milk, cheese, butter and margarine are our chief sources of fat and one gram of fat yields 9·3 Calories, more than twice as much as carbohydrate, weight for weight. Though occupying less bulk than carbohydrates, fats take longer to digest. Furthermore, they cannot be completely digested unless about twice the weight of carbohydrate is also present in the food. Apart from their energy supplying property, they provide a food reserve in adipose tissue and the subcutaneous fat also acts as an insulator, reducing heat loss from the body.

Mineral Salts

These inorganic compounds, present in the body to some 5 per cent. of its weight, supply the mineral elements which are lacking in proteins, carbohydrates and fats, though they are present in some of the foods which contain them and these mineral elements are essential for growth and health. The elements necessary in the greatest amount are calcium, phosphorus, sodium, magnesium, iron, chlorine and sulphur. In addition, a few elements are needed in minute quantities only and they are therefore known as **trace elements**. They are fluorine, iodine, copper, manganese and cobalt. Although they may not actually form part of the body tissues or fluids, they are necessary for their formation.

These mineral elements are contained in many of our foods in the form of compounds. Magnesium, for example, occurs in meat, iron in green vegetables, especially in spinach, and also in meat, particularly in liver and kidneys, and in egg yolk. Potassium occurs in vegetables and iodine in fish and watercress. Sodium and chlorine commonly occur in foods in the form of sodium chloride. Let us now consider the parts played by some of these elements.

Calcium is necessary for the formation of bones and teeth and is concerned in the clotting of blood. About 66 per cent. of bone consists of inorganic salts and the bulk of this is calcium phosphate. Deficiency of calcium causes *rickets* in children in which the bones fail to harden properly and become mis-shapen. The element also has some effect on the general metabolism and an inadequate supply may cause muscular spasm.

Phosphorus is needed to form the calcium phosphate in bones. It is also necessary for the formation of nucleo-proteins which contain it.

Sodium, combined with chlorine as sodium chloride is the form in which the latter element is introduced into the body. Chlorine is necessary for the production of hydrochloric acid in gastric juice. A shortage of sodium causes muscular cramp.

Magnesium is required in bone formation as some of the inorganic matter is magnesium chloride.

Iron is needed for the formation of the haem in the haemoglobin in the red blood corpuscles. A deficiency of this element therefore causes *anaemia.*

Sulphur is a constituent of many proteins and is therefore necessary for the formation of these compounds. It must be taken into the body in sulphur-containing proteins as it cannot be absorbed from mineral salts which contain it.

Fluorine is essential for the formation of the enamel of teeth.

Iodine, as a constituent of thyroxin, is necessary for the formation of this autacoid.

Copper, Manganese and **Cobalt** play their part in tissue formation.

Vitamins

The foodstuffs we have so far considered, proteins, carbohydrates and fats, including the mineral elements and water will not of themselves maintain health. Until the end of the 18th century sailors suffered from a disease called *scurvy* and the mortality rate from it was high. Then it was discovered that the addition of citrous fruits to their diet prevented the onset of the disease. Another disease, *beri-beri* had been rife from the earliest times amongst people whose staple diet was polished rice. But in the latter part of the 19th century the discovery was made that when unpolished rice was eaten instead of the polished rice, the disease did not occur. Yet another disease causing anxiety here towards the end of the same century was *rickets.* It was observed that it occurred but rarely in Eskimos and they take a large amount of fish oils in their diet. It is also rare in countries where there is plenty of sunshine. As the result of these observations and subsequent experiments, it was shown that consumption of fish oils or exposure to sunshine prevented the disease.

Thus it appeared that there must be some substance or substances in the oranges and lemons, in the husks of rice and in fish oils over and above protein, carbohydrate, oil and mineral salt content which prevent these diseases and, in the case of rickets, which the body is able to manufacture when exposed to sunlight. It seemed obvious that the onset of the diseases was due to a deficiency of these substances.

In the early part of this century, **Sir Frederick Gowland Hopkins**, Professor of Biochemistry at Cambridge, performed experiments on rats which supported this idea. He fed one batch of rats on a balanced diet of pure proteins, carbohydrates and fats, together with the necessary mineral salts. These rats failed to thrive and became sick. Another batch were fed on a similar diet plus a few drops of milk and these animals grew normally and remained healthy. He then changed over the diets with the result that the first batch recovered and the second one became sick, developing the same symptoms as the first batch had originally shown. From these experiments Professor Gowland Hopkins concluded that additional substances in very minute quantities were essential to the maintenance of health and these he called *accessory food factors*.

Another investigator, **Casimir Funk**, who had been studying beri-beri, came to the conclusion that these substances were organic compounds belonging to the group known as *amines*, which contain the $—NH_2$ radical, and because they were amines essential to life he suggested the name *vitamines*. However, it was later discovered that this was not the case and so it was decided to drop the "e" and they have since been known as **vitamins**. A great deal of further research has been done since that time and is still going on and many new vitamins have been discovered. They have been designated by letters but as their chemical composition becomes known they are also given their appropriate chemical names and many can be prepared synthetically. Some of them, A, D, E, F, and K are **fat-soluble** and the rest, B (of which there are twelve varieties, designated B_1 to B_{12}), C, H and P are **water-soluble**.

Vitamin A or **axerophthol** occurs in the liver oils of fish and in milk, butter and eggs. It is a derivative of *carotene* ($C_{40}H_{56}$) which is found in carrots and is associated with chlorophyll in the leaves of green vegetables and in some fruits such as tomatoes and bananas. When these foods are eaten the body is able to convert this *precursor* of vitamin A, carotene, into vitamin A. It is not destroyed by cooking if it only involves limited heating but boiling destroys it. The vitamin promotes growth and prevents night-blindness as it stimulates the production of rhodopsin or visual purple. It also protects the body against infection of mucous membranes such as those in the bronchial passages and alimentary tract because it controls the secretion of mucus and these membranes are more open to infection when insufficiently moist. It thus gives protection against such infections as the common cold and gastro-enteritis. Deficiency of the vitamin causes night blindness and, if the deficiency is great, to more serious eye diseases such as xerophthalmia leading to blindness, and to an increased liability to infection of the mucous membranes. During the last war, people were encouraged to eat raw carrots to improve night vision during the black-out.

Vitamin D_2 or **calciferol** occurs with vitamin A in fish oils, butter, cheese and egg yolk. As margarine does not contain any vitamin

A, a specified amount must be added to it by law in this country. Herrings, salmon and sardines are particularly rich in it. A derivative of ergosterol ($C_{27}H_{41}OH$) in ergot and yeast, it can be synthesised from this compound in the body when it is exposed to the ultra-violet rays of the sun. Neither heat nor food preservation destroy this vitamin and cooking has no ill effect. We have already seen that lack of calcium and phosphorus in the diet causes rickets in children. Vitamin D_2 is also necessary to prevent this disease as it is concerned in the absorption of these elements. It is therefore the *anti-rachitic vitamin* and children suffering from rickets can be treated by exposure to ultra-violet light either from a lamp or directly from the sun.

Vitamin E or **tocopherol**, occurs in eggs and in some animal and vegetable oils and in the leaves of lettuce and other green vegetables and in cereals. It has been proved to be necessary to rats in order to prevent sterility in the male and abortion in the female but its necessity in the human subject is still a matter for speculation and has not yet been proved.

Vitamin K or **phylloquinone** is found in green leaves, rose hips, strawberries, cabbages, tomatoes and liver. It is essential for the clotting of the blood and is therefore the *anti-haemorrhagic vitamin.* There appears to be a deficiency in haemophiliacs and as haemophilia is an inheritable disease, there is probably an inability to absorb the vitamin.

Vitamin B Complex

Of the twelve B vitamins only five have been shown to be necessary for human beings. This complex is found in eggs, liver, milk, cheese, the germ-layer of seeds, whole-meal flour, green vegetables and yeast.

B_1 aneurin or **thiamin** is the *anti-neuritic vitamin*, preventing neuritis and the disease beri-beri which is a severe form of neuritis developing as a result of continued deficiency of B_1. The vitamin is concerned with energy release from carbohydrates. It is removed from flour by milling which takes away the husks of the wheat grains and is therefore absent from white flour but is present in whole-meal flour. It is also lacking from polished rice; hence the occurrence of beri-beri amongst people whose diet consists mainly of this.

B_2 or **riboflavin**, like B_1 is concerned in carbohydrate metabolism and is an *anti-dermatitic vitamin* as it is necessary for the healthy condition of the skin and its deficiency causes sore, cracked lips and inflammation of the tongue. It occurs in milk, cheese, eggs, liver and other foods and is not destroyed by cooking.

B_6 or **pyridoxine** is also *anti-dermatitic* since a deficency causes atrophy of the epidermis of the skin. It is found in cereals, liver and yeast.

B_7, nicotinic acid or **niacin**, is another vitamin which keeps the skin healthy and prevents the disease known as pellagra, caused by its deficiency in which the skin becomes inflamed and scaly and digestive and nervous disorders develop. It is prevalent among people whose staple diet is maize. B_7 is found in liver and other meat and yeast.

B_{12}, folic acid or **cyanocobalamin**, which contains cobalt, is necessary for the formation of blood corpuscles and is used in the treatment of anaemia. It is found in liver, cereals and yeast.

Vitamin C or **ascorbic acid** prevents scurvy, in which bleeding of the gums and other parts and loosening of the teeth occur. It is therefore the *anti-scorbutic vitamin.* Citrous fruits and black-currants are particularly rich in this vitamin which is also found in other fresh fruits and vegetables such as potatoes and cabbages. As it is destroyed by prolonged boiling, foods such as green vegetables should be cooked in a small amount of water and brought quickly to the boil.

Vitamin H or **biotin** helps to keep the skin in a healthy condition and its deficiency causes some forms of dermatitis. It is found in meat, liver and eggs.

Vitamin P or **citrin** is found in lemons and is necessary to enable the blood capillaries to withstand pressure of blood within them.

Vitamin Requirements and Deficiencies

The body requires certain minimum quantities of vitamins each day and these vary from vitamin to vitamin. Anyone on a normal diet of eggs, milk, butter, cheese, fresh fruits and vegetables will get his or her correct supply and any deficiency can be made up by taking foods which contain the vitamins concerned. It is necessary to know what the minimum requirements are, and it should be remembered that they are very small, as is the vitamin content of foods. Vitamins A and D are measured in what are known as **International Units (I.U.)**. These are different weights for the two vitamins, the minimum requirements being different.

The I.U. for vitamin A is $\frac{1}{1,000,000}$ gram and for vitamin D it is $\frac{1}{100,000}$ gram. The other vitamins are expressed in milligrams. The daily requirement of vitamin A for example, is about 4,000 I.U. and that for vitamin D around 400 I.U. Halibut oil, one of the best sources of vitamin A contains 5,000,000 I.U. per gm. whereas cod liver oil contains only 1/50 of that quantity. Vitamin tablets, each containing a specified number of I.Us. of vitamins A and D or milligrams of the other vitamins, can be purchased.

To prevent vitamin deficiency all that is necessary is to take a diet which contains the foods which provide the vitamins; it is as simple as that. At the same time it must be remembered that the food should be fresh and that some vitamins are destroyed by heat and therefore by cooking in any form. This does not apply to milk and therefore to milk puddings and the like or to eggs. Many mothers and young children may need to have their diet supplemented by such things as orange juice or halibut oil which is obtainable in the form of capsules.

Water

As protoplasm and all body fluids contain water and it forms about 70 per cent. of the body weight, it is obvious that it should be taken into the body although it cannot be regarded as a food. About a third of the amount needed is obtained in meat and fish, and particularly in vegetables and fruit. The rest must be taken in liquid form as beverages or other drinks. Water is the medium in which digestive enzymes act, it forms a large part of the blood plasma and lymph in which foods are transported in the body, it is the medium in which excretory substances are eliminated from the body in urine and sweat and, by its evaporation in sweating, it is responsible for the loss of excess heat from the body. The amount of water lost in various ways, in expiration, sweating, micturition and so on, is around 2·8 litres a day and this must be replaced. As to how much water is required per day, this depends on the body's activities. More is lost by sweating and expiration in vigorous exercise and more is lost through the skin in hot weather. Whereas one can live for a few weeks without solid food, lack of water will cause death in a matter of days.

BALANCED DIET

Not only must our daily supply of food contain the necessary foodstuffs and water but these must be in correct proportions and amounts to provide the 1,600 Calories necessary for basic metabolism plus the extra Calories required for our activities according to our age and sex and the climate in which we live. It must also provide us with the necessary vitamins in adequate amount and be free from injurous substances.

Amongst European races one seldom finds diets that are badly balanced, though, where incomes are low, starchy foods, because they are cheaper and at the same time satisfying, tend to be taken in excess at the expense of protein and fat. Esquimaux live almost entirely on whale and seal meat which provides them with energy from protein and fat and in the excessively low temperatures of the Arctic region food of high heat-giving value is, of course, a vital necessity. Rice and cereal foods form the staple diet in many eastern countries such as India, and their food therefore lacks sufficient protein and fat and has excess of carbohydrate. There is a region in the Caucasus in which it is maintained there are some two thousand people exceeding 100 years of age, it being claimed that one man who died in 1973 had reached the age of 168 but the evidence for this is not very strongly supported. The staple diet of these inhabitants is goat's milk, fruit and vegetables. In another region, living in a valley high up in the Andes of South America are people several of whom it is claimed are well over 100 years of age, a few over 110 and one of 135! These people have a largely vegetable diet but some eat small quantities of meat. The oldest known person ever recorded in Britain was a woman who was 112 years and 39 days of age at her death in 1973. Further claims of longevity up to 124 have been made! Certainly a woman died in the U.S.A. in 1973 at the recorded age of 116 years.

As many foods contain more than one food constituent, the calorific value of such foods can only be ascertained if the percentage of each foodstuff is known. This has been worked out for different foods and a few examples of commonly used foods are given in the table which follows.

It should be appreciated that there is a certain amount of waste material in many foods which has no food value at all, such as skin in meat and fish, rind in bacon and bones in animal foods. In the nutritious herring it is as much as 35 per cent. or more and in cabbage there is about 40 per cent. waste whereas in cooked ham, on the other hand, there is none. The cellulose in vegetable foods and fruit, of course, serves a useful purpose as roughage. In the table below the percentages given in the different foods are in the nutritious, edible parts of the foods. These, and the waste materials vary in the same food and those given below are average percentages.

Green plants are the ultimate sources of man's food for not only does he eat plants and products derived from them but the animals which provide some of his food depend on green plants for their own existence. Plants are the chief source of carbohydrates which they manufacture by a process called **photosynthesis**. The

Calorific Values of a Few Common Foods

Food	Calorific Value *C. per 100 gm*
Bacon	410
Beef	235
Herring	239
Potatoes	93
Bread (white)	246
Rice	364
Butter	807
Eggs	16
Cheese	428
Milk (100 ml)	65

Composition of a Few Common Foods

Food	Percentages		
	Proteins	Carbohydrates	Fats
Bacon, fried	24·5	—	53·4
Beef, roast	21·2	—	32·5
Chicken	29·7	—	7·1
Ham, boiled	14·5	—	33·2
Pork	15·5	—	40·0
Haddock	15·8	—	0·2
Herring	21·5	—	15·8
Sardines	23·7	—	22·1
Cabbage	1·6	4·5	0·4
Broad Beans	4·7	7·6	0·3
Lentils	6·8	18·3	—
Lettuce	1·5	3·4	—
Peas	5·0	7·7	—
Potatoes	2·2	18·4	0·1
Apples	0·4	14·2	0·3
Bananas	1·5	22·0	0·6
Oranges	0·8	10·6	0·2
Bread	7·8	52·7	1·2
Rice	8·0	29·6	0·3
Butter	0·4	—	85·0
Margarine	0·3	—	83·0
Eggs	14·8	—	10·5
Cheese (Cheddar)	25·0	2·4	30·0
Milk	3·3	4·5	4·0
Cream	1·5	2·0	48·2
Milk Chocolate	9·4	57·4	30·0

energy for this process is provided by sunlight. The process takes place in two main stages and in the first of these light is absorbed by the green pigment *chlorophyll* and this radiant energy is converted into chemical energy. Water, absorbed by the roots, passes up the plant to the leaves and in this first stage of photosynthesis it is split up into hydrogen and oxygen. This oxygen is set free and passes out of the leaves into the atmosphere. Thus the first stage is dependent on light. In the second stage carbon dioxide, which is absorbed by the leaves from the atmosphere and, no doubt, from the respiring cells of the plants, is reduced by the hydrogen set free in the first stage. The reactions involved in this stage are independent of light and result in the formation of complex organic compounds which finally give rise to glucose from which starch and cellulose are formed. The cellulose forms the structural material of the cell walls and the starch is stored in the leaf. This is reconverted into glucose at night and distributed to other parts of the plant to be used as its source of energy. Excess is conveyed to storage organs and is usually changed once again to starch for future use. Thus the radiant energy absorbed by the plant is stored in the plant as potential energy in the carbohydrates.

The process of photosynthesis may be *summarised* by the following equation

$$6CO_2 + 6H_2O \xrightarrow{\text{light energy}} C_6H_{12}O_6 + 6O_2$$

but this does *not* account for the fact that the oxygen evolved is derived entirely from the water, and, of course, it does not explain the light and dark reactions. Some of the intermediate compounds formed are converted into fats and some into amino-acids which are synthesised into proteins. Green plants are thus capable of synthesising proteins from inorganic matter. Food in the form of carbohydrates, proteins and fats is also stored in seeds and fruits and in some plants in underground stems like potato tubers and bulbs as in the onion or in roots as is the case with carrots and parsnips.

This process of photosynthesis is clearly of great value to man for not only does it provide him with sources of carbohydrate, protein and fat but it rids the air of the carbon dioxide which he and other animals exhale into it and replaces the oxygen which they remove from it. When we eat lamb or roast beef, for example, the meat which we eat was built up from the grass and other green plants on which the animals fed. When we eat fish we are again eating flesh which was built up from green plants, for the fish feed on smaller animals which in their turn fed on minute green plants like diatoms. The whole series of processes constitute what are known as **food chains** and all food chains originate in the first instance from green plants. Finally, as the energy derived from the sun is stored in the carbohydrates in the green plant, man and other animals are dependent on green plants as their sources of energy.

It is clear that plants and animals are completely interdependent and it has been truly stated that if there were no green plants in the world animals, including man, could not exist.

THE WORLD'S FOOD

The Food and Agriculture Organisation (F.A.O.) of the United Nations held a World Food Conference in Rome in 1974 when the whole question of the world food crisis was discussed. Measures were set in motion to deal with the immediate crisis and long term problems such as sending technical missions to developing countries to help their governments to draw up policies to cope with their problems and for the setting up of national reserves of food. Malnutrition affects some 460,000,000 people in the world.

The present world cereal crisis is due to poor harvests creating difficulties in meeting the requirement of 350,000,000 tons a year for livestock feeding. Not until 1974 did the world as a whole become aware of the extent of malnutrition and then only when it began to affect the affluent countries threatening the feeding of their livestock. Only 12·5 per cent. approximately of arable land in developing countries is under irrigation, most of it in the Far East and this, of course, results in a lowering of possible crop yield. Livestock in rich countries eat more grain than the population of India and China together.

In 1972 food output declined for the first time in twenty years while demand increased. Harvests were affected by bad weather conditions resulting in a drop of 20,000,000 tons in wheat stocks in wheat exporting countries between 1971 and 1973. In the latter year harvests were good but world cereal prices soared and rose even further in 1974. The cost of fertilisers reached a high level in 1972 and in 1973 there was a shortage. In 1974 world cereal production declined for the second time in three years and was estimated at 40 to 50,000,000 tons. World population is increasing by about 75,000,000 a year and this means that an extra 30,000,000 tons of food is required every year to keep pace with population growth.

It has been estimated that between 1970 and 1981 world food demand will increase by 2·4 per cent. annually, 2 per cent. due to population increase and 0·4 per cent. to increased purchasing power. The demand, it is stated, will be 1 per cent. in developed countries and 3·6 per cent. in developing countries. It has also been estimated that the food production rate between 1967–71 and 1985 would be 2·4 per cent. for developed countries, 2·6 per cent. for developing countries, 3·5 per cent. for Eastern Europe and 28 per cent. for the U.S.S.R. Furthermore it is predicted that the cereal deficit in developing countries will be 85,000,000 tons a year by 1985 and that if a large developed country had a crop failure in the same year, the world import demand for cereals could be as high as 170–180,000,000 tons.

The U.N. F.A.O. Conference concluded with a lengthy Declaration in which it stated that it is a fundamental responsibility of Governments to work together for high food production and for a more equitable and efficient distribution of food between and within countries. It also stated that the grave food crisis afflicting the developing countries where most of the world's hungry and ill-nourished live and where more than two-thirds of the world's population produce about one-third of the world's food, an imbalance which threatens to increase in the next ten years, is fraught with grave economic and social implications and acutely jeopardises the fundamental principle that every man, woman and child has the right to be free from hunger and malnutrition in order to develop fully and maintain their physical and mental faculties.

So the population of the world continues to rise rapidly but food production fails to keep pace with the increase and a large proportion of the people in the world suffer from hunger, malnutrition and starvation. Enormous amounts of foodstuffs are lost by devastating floods such as those which struck Bangladesh and elsewhere not so long ago and by famines due to drought as in Ethiopia.

Periodic plagues of locusts completely destroy the crops in parts of the Middle East, thereby causing starvation. The heavy rise in the cost of feeding stuffs for cattle affects the production of beef and milk and similar remarks apply to other animal foods. The high cost and world shortage of plant fertilisers add to the difficulties of adequate food production and bad harvests caused by adverse weather conditions, over which man has no control, result in world shortages of food. This state of affairs is bound to worsen as world population increases unless methods of food production are vastly improved in order to provide more and better crops and animals and unless there is a fairer and more even world distribution of food. Nor should the question of population control be ignored as far as it is humanly possible to deal with it. Indonesia started a five-year plan in 1970 in which the target was to recruit 6,000,000 new acceptors of contraception in Java and Bali, oral contraceptives and the condom being the chief methods. Male sterilisation was offered in Bali but it plays an insignificant part in the nation's programme at present. It is estimated that by the year 2,000 there will be 7,000 million people in the world. In poorer countries where diets are inadequate it is mainly the more expensive animal foods (which provide first class proteins) which are lacking in those diets. Richer countries do not suffer from this inadequacy but although some of them are able to produce sufficient for their needs, other smaller countries must perforce import from larger ones capable of producing in excess of their own requirements. Take one example in this country. We have to import some one and a half million tons of wheat a year which is about four times as much as we are able to produce ourselves. This means that we must manufacture goods for sale in those countries which are able to export the wheat to us.

In some countries prejudice, ignorance and religious scruples are responsible for inadequacy of food production. Education and family planning, of course, is the only answer in dealing with the matter there.

If the shortage of food in the world is to be overcome, the rate of production must be vastly increased, methods for producing more and better food must be devised and efficient methods of pest control and avoidance of disease in animals and crops must be put into use. Great advances have already been made as the result of research in recent years. Cattle and poultry which grow and fatten more quickly can now be bred, methods of increasing milk-production in cows, the rate of egg-laying by hens and the rate of growth of crops by the use of fertilisers have been discovered. New varieties of wheat and rice with double the normal yield of grain have been developed and in 1975 varieties of these crops and pasture grass which are independent of nitrogen fertilisers were discovered in South-Central Brazil. These plants obtain their nitrogen from nitrogen-fixing bacteria in the soil, a condition formerly considered to be restricted to leguminous plants like peas and beans. It is not yet known if these varieties can be grown in temperate climates but it should be possible in countries like India where some 20,000,000 acres are used for wheat. The shortage of fertilisers and their high cost would not then be matters of concern. Whether a similar process can be applied to rice remains to be seen and is under investigation. There is some interesting research in progress in which countries with advanced cattle industries donate semen from bulls of high genetic merit for improvement of cattle in developing countries. Crops can now be grown which have developed immunity to certain diseases and methods of pest control and of combating plant and animal diseases are now in common use.

This is a world problem of considerable urgency which must be treated as such if malnutrition, starvation and famine affecting the lives of millions of people are to be overcome and it is one of the tasks which is being vigorously tackled by Oxfam and by the World Health Organisation, the Food and Agriculture Organisation and other departments of the United Nations. Furthermore, the richer countries must provide those which are poorer with more food and this means a fairer and more even distribution throughout the world.

Land pollution is considered on pp. 404 seq.

CHARACTERISTICS OF FOODS

Meat

Beef, lamb, veal and similar animal foods are referred to as *red meat* as opposed to poultry which is called *white meat*. It is composed of muscle and consists of protein and connective tissue. The percentage of protein is high and the amount of fat depends on the age of the animal. Vitamin content is low and the water content high. In addition to muscle, liver and kidneys, pancreas (called sweetbread) and brain are used as food. The practice of beating beef-steak before cooking is to break up the muscle fibres and make it more tender to eat. Tripe is part of the stomach wall of the ox and is very easily digested when properly cooked. Liver is rich in vitamins A and D and in iron. Any toughness in meat is due to the connective tissue and this is softened by hanging as acids are produced in the meat which have this effect. Meat will remain tough if not hung for a sufficient length of time.

Fish

The amount of fat associated with the protein in the flesh of fish is very variable, being as high as 20 per cent. or thereabouts in the herring but less than 2 per cent. in haddock and other white fish and the protein content is lower than it is in meat. On the other hand the vitamin content is good and fish livers are excellent sources of vitamin D, particularly in halibut. White fish, however, having a low fat content lack the vitamins. Fish are also good sources of magnesium, phosphorus and iodine, and sardines, in which the bones have been softened and therefore edible, are sources of calcium.

Poultry

The flesh of poultry contains comparatively little fat except in the case of geese and ducks. It is more easily digested than red meat. Similar remarks apply to game.

Vegetables and Fruit

Although they contain a little protein, the chief constituent of green vegetables is carbohydrate. They are also a good source of minerals and vitamins. The fat content is low and the water content high. Green peas, broad beans and lentils are good sources of protein. Cabbage, blackcurrants and citrous fruits have a high vitamin C content and carrots vitamin A. This vitamin content applies to fresh foods as dry fruits lack it, though prunes and apricots do provide vitamin A. Spinach and watercress are rich in iron and calcium. Potatoes, though having a high water content, are good sources of protein and carbohydrate and they contain vitamin C. Cooking, however, decreases the vitamin content and this is still further lowered while they are kept hot. Vegetable foods provide good roughage owing to the cellulose they contain but they are bulky because of the high water content.

Flour, Bread and Cereals

Cereals include wheat, barley, oats, rye, maize and rice. The word *corn* has different meanings in different countries. In England it generally means wheat, in Scotland oats, in America maize and in Germany rye.

Flour and Bread

Flour, which is obtained from wheat by a process called *milling,* contains up to 74 per cent. of carbohydrate of which the greater proportion, some 72 per cent., is starch and the rest sugar. Of the 12 per cent. of protein, the bulk is *glutenin* which becomes sticky when mixed with water, the mixture being called *gluten.* The other proteins are *globulins* and *albumin.* There is also about 1 per cent. of fat and 0·5 per cent. of minerals. *Self-raising flours* are white flours containing cream of tartar (acid potassium tartrate) and sodium bicarbonate which react on the addition of water and make the flour "rise".

In the process of milling the indigestible seed coat of the wheat grain is removed and the soft remainder is then ground to flour. This is *white flour* from which *white bread* is made. *Whole-meal bread,* as the name implies, contains the entire grain and ordinary *brown bread* contains some of the seed coat. As the embryo in the grain is rich in B vitamins and this is removed by milling, whole-meal and brown bread are obviously preferable as foods though they are less digestible than white. The inner part of the grain, the endosperm, is chiefly starch, and white flour consists mainly of this, though small quantities of protein and fat are also present in the grain. Rye, barley and oats can also be used in the making of bread.

In bread-making, the flour is mixed with yeast and salt and water are added to form dough. This is well kneaded and left in a warm place for about an hour. The enzyme complex, zymase, in the yeast converts some of the starch into sugar and causes it to ferment. Alcohol and carbon dioxide are formed in this process and the gas causes the dough to rise. The gluten in the dough helps to keep the carbon dioxide within it and when, later, the dough is baked, the gluten hardens and gives support to the bread. After a second kneading, the dough is baked in a hot oven. This kills the yeast, causes the alcohol to evaporate and cooks the dough.

Wheat Products

Semolina, macaroni, spaghetti and *vermicelli* are made from wheat grains by course milling. The wheat for this purpose is grown in Southern Europe.

Barley

This is used in the making of beer. Germinating barley is called *malt* and contains maltose. Pearl barley is a product in which the husk is removed as in the milling of wheat. Barley is also an animal food.

Oats

Oats form the source of oatmeal which is obtained from the seeds by milling. They also serve as animal food.

Rye

Rye is used for making bread in some continental countries. The bread is dark in colour and has a flavour peculiar to itself.

Maize

This is also known as *Indian corn* and is used as a food in itself. *Cornflour* is obtained from the grains and has a high starch content.

Rice

Rice which grows in swampy fields in China, India and elsewhere in the East contains a high percentage of starch but the protein and fat content are low. As we have already seen, vitamins are absent from polished rice.

Tapioca

Tapioca is prepared from the tuberous roots of a plant grown in Brazil and tropical regions called Manihot utilissima. It consists chiefly of starch, mostly in a soluble form.

Sago

The pith of the Sago Palm is the source of this commodity. The tree grows in Malaysia and other Far Eastern countries.

Arrowroot

This is another starchy food, obtained from a plant, called Maranta arundinacea, grown in Brazil. Arrowroot is very easily digested.

Milk

Milk consists of the proteins *caseinogen* and *lactalbumin,* carbohydrate as *lactose* or milk sugar, fat, the mineral elements calcium, potassium, sodium, magnesium and iron in minute quantities and vitamins A, B_1, B_2, C and D. Its specific gravity varies between 1·028 and 1·034. The vitamin content changes according to the time of year and is highest in Summer when cows are out to grass. The vitamin C content is destroyed when milk is heated in air as it is easily oxidised and is therefore lacking in pasteurised milk. When milk is allowed to stand, the fat and some of the protein rises to the top as **cream** and if this is removed, **skimmed milk** or **separated milk** is left according to the method used. If the milk stands for several days or if the enzyme *rennin,* commonly known as rennet, is added to it, coagulation occurs. The solid portion or **curd** consists mainly of casein with some lactose and minerals and the liquid part, or **whey,** of water, lactose and mineral salts. The adding of rennet, which is used in the making of junket and cheese is an enzyme action and the result is known as **clotting** but the **coagulation** which takes place on standing is caused by bacteria which oxidise the lactose to lactic acid and this precipitates the caseinogen as *casein.* The effect of boiling milk is to coagulate the lactalbumin, caseinogen being unaffected. This coagulated lactalbumin forms the "skin" on the surface of the milk and the steam from the boiling water underneath exerts sufficient pressure to cause the milk to boil over unless the heating is stopped in time. Cream is the separated fat of the milk but it also contains a certain amount of casein, lactose and minerals.

Milk, containing as it does, all the essential foodstuffs, mineral salts, most vitamins and water, is the natural and most perfect food for infants. For growing children and adults, however, solid food is necessary to provide bulk and roughage and for them milk alone contains too much water. A comparison of human milk with cow's milk shows that cow's milk contains more proteins, minerals and fat but less carbohydrate than human milk. If, therefore, breast feeding is unsatisfactory or undesirable for an infant, cow's milk can be used instead provided that sugar and water are added. There are, of course, several proprietary brands of dried milk suitable for infants from which the liquid can be made up.

Percentage Composition of Milk

	Human	Cow
Protein	1·7	3·5
Carbohydrate	6·4	4·5
Fat	3·3	4·0
Mineral Matter	0·2	0·7
Water	88·4	87·3

Butter

In the making of butter, cream is separated from milk and churned, eighteen pints of milk being required to make one pound of butter. After separation, it is allowed to "ripen" and this is brought about by bacteria which determine the flavour of the butter. Vitamins A and D from the milk are retained to a large extent in the butter which is almost entirely fat with a trace of protein. A small quantity of salt is added in salted butter.

Margarine

Margarine is made from both vegetable and animal fats and oils, the latter being "hardened" by hydrogenation. Specified amounts of vitamins A and D must be added to margarine in this country by law. Many margarines on the market have a stated percentage of butter added.

Yogurt

Yogurt is a form of curdled milk with a fairly high protein content. It is more easily digested than natural milk. It is prepared by heating and treatment with a culture of bacteria known as *Bacillus bulgaricus.* It is rather acid in taste but it can be flavoured with other substances.

Butter-milk

This is left after the churning of cream in the making of butter. It contains casein and lactose in percentages approaching those in milk, a little less mineral matter and, of course, but little fat.

Cheese

Cheese is prepared from milk by adding rennet which causes it to curdle. The curd is allowed to ripen by its own contained ferments or by the addition of bacterial cultures. Different kinds of cheese are produced by different cultures. Most English cheeses like Cheddar, Cheshire and Stilton, those from Canada and New Zealand, Dutch cheeses and Gruyère from Switzerland have the whey removed by pressure and are known as "hard cheeses". In "cream cheeses" and French cheeses like Brie and Camembert the whey is retained and these are "soft cheeses". The addition of bacteria and different methods of processing influence the flavour of the cheese. Roquefort cheese is made from sheep's milk and the blue colour, as is the case with Gorgonzola and Stilton, is due to a mould *Penicillium roqueforti.*

Cheese is a valuable food because the casein and fat are in a highly concentrated form, there being about $33\frac{1}{3}$ per cent. of each, the other $\frac{1}{3}$ being water. But lactose and minerals, largely calcium, and vitamins A and B_2 are also present.

Eggs

The shells of eggs are chiefly composed of calcium carbonate, though traces of the phosphate are also present. These shells are porous to permit the entry of air to the embryo chick and the exit of the carbon dioxide it produces. The white of the egg is composed largely of *ovalbumin,* together with a trace of fat and a large quantity of water. The yolk is suspended in the white by two twisted cords called chalazae and both white and yolk are enclosed in membranes. The yolk consists of a higher percentage of protein than the white, a high percentage of fat and a lower water content than the white. The proteins are ovalbumin and a chromoprotein called *vitellin* which contains phosphorus. Both yolk and white contain minerals, the most important being calcium and iron. Vitamins are also present, A and D occurring in the yolk.

Eggs thus form an excellent item of diet and they can be eaten raw if beaten up, preferably mixed with milk to make them more palatable. All methods of cooking coagulate the proteins and have no ill effects on their value as a food. Apart from hen's eggs, those of other poultry, duck's and in some countries geese are used. Hard roes of herrings contain their eggs and the luxury food caviar is composed of sturgeon's eggs. Mayonnaise contains egg yolk mixed with vinegar. Eggs are also used in the making of omelettes, custards and puddings and these therefore have nutritive value apart from that provided by the other ingredients. White of egg is used for glazing pastry and in the making of meringues.

Beverages and Other Drinks

Tea

Tea consists of the dried leaves of a shrub grown in India, Ceylon, China, Japan and Sumatra. The leaves are first allowed to wither, then they are broken up by rolling after which the broken leaves are covered with a damp material and allowed to ferment. Finally they are dried in hot air, a process called "firing". This produces black tea. If the fermentation process is omitted, green tea is obtained. Blending of different varieties is often done before sale.

There is no nutritive value in tea but it is a mild stimulant because it contains an alkaloid, *caffeine,* which acts on the central nervous system. A very small amount of vitamin B_2 is also present. The making of tea by adding boiling water forms an *infusion* in which the caffeine is dissolved together with *tannin.* The flavour is due to *theine* and this is spoiled by excess tannin, which develops if the tea is allowed to stand for a long time. A further disadvantage of tannin is that it makes proteins indigestible: it should not, therefore, be taken with meat. China tea contains no tannin.

Coffee

Coffee is obtained from coffee beans which are the seeds of the coffee tree and this grows in moist tropical countries such as Brazil, Jamaica and Kenya. The beans are roasted and then ground after which they should be kept in an air-tight container as the roasting develops the flavour and aroma and these are quickly lost on exposure to air. *Chicory,* obtained from the root of the wild endive, is sometimes added to coffee. To make coffee boiling water can be added and the mixture well stirred or the boiling water can be percolated through the coffee. Instant coffee powders are made from coffee beans and in some cases from percolated coffee which has been desiccated. Coffee contains less caffeine but more tannin than tea and its aroma is due to volatile oils. It is a stronger stimulant than tea, having a greater effect on the central nervous system.

Cocoa and Chocolate

Cocoa is prepared from the beans of the cocoa tree which grows in West Africa, Brazil and elsewhere. After roasting, the beans are ground and some of the fat called *cocoa-butter* removed. The residue is then ground to powder. In addition to a small alkaloid content of *theobromine* cocoa contains some protein and carbohydrate as well as fat and vitamins A, B_1 and B_2 together with calcium and iron. Chocolate is cocoa to which sugar and cocoa-butter have been added in plain chocolate and also milk in milk chocolate. Both cocoa and chocolate have some food value but not a great deal is obtained from them taken in the form of drinks owing to the small quantities used.

Mineral Waters and Fruit Drinks

Fruit drinks, provided that they are made from fresh fruit and contain sugar, have some food value and contain vitamin C. Synthetic products have no food value apart from any sugar which they contain. They are often made effervescent by dissolving carbon dioxide under pressure.

Alcoholic Drinks

These have little food value but they do produce some energy by virtue of any sugar which is contained in them. They contain **ethyl alcohol** and this is made by the fermentation of sugar by yeast which contains the enzyme complex called zymase which effects the reaction and carbon dioxide is given off as in bread-making

$$C_6H_{12}O_6 = 2C_2H_5OH + 2CO_2$$

Alcohol is rapidly absorbed into the blood and it is a poison. In small quantities it has a stimulating effect on the body, increasing the rate of heart beat and producing a feeling of well-being but in increasing amounts it impairs judgement, produces over-confidence and a lowering of caution in speech and behaviour and eventually a complete lack of mental and muscular co-ordination. It is because of these effects that the driving of a car under the influence of drink is quite rightly a serious offence, for a car can be a lethal weapon. There can be no logical objection to the compulsory breathalyser, blood and urine tests by the police if they have reason to suspect that a driver is in this condition. The reduction in the number of accidents after its introduction is ample proof of its efficiency. Under the present law the maximum permissible amount of alcohol in the blood of a driver is 80 mg. per 100 ml. of blood. It is therefore wise to keep the alcoholic intake down to a minimum or omit it altogether when driving a car. Apart from this, moderate drinking is comparatively harmless and a pleasant social habit but long continued immoderate consumption of alcohol results in cirrhosis of the liver, degeneration of the central nervous system and other diseases and it is the heavy drinker whose body suffers. Alcoholism is a serious problem which is worse in some countries than it is in others and, regrettably, amongst the young.

[*Methyl alcohol* (CH_3COOH) is very much more poisonous than ethyl alcohol. It is obtained from the pyroligneous acid which is produced by the destructive distillation of wood and is known as *wood spirit. Methylated spirit* consists of ethyl alcohol to which methyl alcohol and naphtha have been added. A violet dye is also added to make it unfit for human consumption. It has very serious effects on the body if drunk, neuritis and blindness being but two.]

Beer

In the making of beer, barley is soaked in water and allowed to begin to germinate in order to convert the stored starch into sugar. The germinating barley is called *malt.* The germination is stopped after a time by applying heat and yeast is added in large vats. To the alcoholic liquor produced hops and other substances are added to give a flavour. The alcohol content is between 3 and 7 per cent. Stout and similar drinks are similar but have a higher percentage of alcohol.

Cider

The source of sugar for fermentation in the making of cider is contained in cider apples and the alcohol content is up to 6 per cent.

Wines

Wines are obtained by the fermentation of the glucose and fructose in grapes. The addition of yeast is unnecessary as it is present in the so-called "bloom" on the outside of the fruit. This comes into action when the grapes have been crushed. Red and white wines are obtained from red and white grapes respectively and contain up to 10 per cent. of alcohol. Port and sherry are "fortified" by the addition of extra alcohol, giving a proportion of slightly over 20 per cent.

Spirits

These contain much higher percentage of alcohol than beers and wines and their different flavours are due to various aromatic compounds. The concentration of alcohol may be expressed as a percentage or by the number of degrees *under proof.* **Proof Spirit** is 50 per cent. alcohol (49·28 per cent. by weight = 57·1 per cent. by volume) and the term was originally derived from the fact that when an alcohol of this concentration was poured on to gunpowder and a light applied, the gunpowder burned whereas the higher water content in weaker solutions prevented this. Thus if a drink is labelled 20° under proof it means that 120 parts of the liquid are equivalent to 100 parts of proof spirit whereas 10° over proof would mean that 90 parts of the liquid are equivalent to 100 parts of proof spirit.

Whisky is made by the fermentation of malt but the process is carried further than is the case with beer thereby giving a higher proportion of alcohol. This liquid is distilled to about 43 per cent. by volume and it is then about 75° proof.

Brandy is made by the distillation of wines and contains about 43 per cent. of alcohol.

Gin, with an alcohol content of about 37 per cent. has aromatic liquids such as oil of juniper added to the fermented malt before it is distilled.

Rum is made from Molasses. This is a concentrated syrup from sugar cane from which no more sugar can be crystallised out. After dilution this is fermented and distilled. It contains around 43 per cent. of alcohol.

Vodka is obtained from rye by distillation. Its alcoholic content is about 38 per cent, and it is 65·5° proof.

Liqueurs have a high alcohol content of about 50 per cent. and various flavours are added.

Condiments and Spices

The effect of the addition of these foods is to aid digestion by virtue of the fact that they increase the secretion of saliva and gastric juice by making the food appetising. They are extracted from different organs in various plants. Vinegar is a dilute solution of acetic acid obtained by the action of bacteria on malt, cider or wine. It contains 5–6 per cent. of acetic acid.

THE PRESENTATION OF FOOD AND COOKING

The desire to eat arises when the body is in need of food to provide the calories of which it is getting short. If the food has an attractive appearance, taste and smell, the appetite will be aroused and saliva will be secreted into the mouth and gastric juice into the stomach. The smell of bacon frying first thing in the morning when the stomach is empty is a good illustration of this conditioned reflex. Some of the foods we eat, however, may have little nutritional value and may be indigestible. We eat them because they are "tasty". Although we eat some raw, others need to be cooked to make them more palatable and more easily digested, though in some instances this process may make them more indigestible. The way in which food is served and the conditions in which we eat also help in the value we derive from our meals.

Cooking meat breaks up the muscle fibres and softens the connective tissue and liquefies the fat. In the case of vegetables, the cellulose cell walls and the envelopes of the starch grains are split open so that digestive enzymes are able to come in contact with the starch. It has also been claimed that cooking destroys any infective bacteria or other parasites in the food but the temperatures at which food is cooked are neither high enough nor sufficiently prolonged to effect such sterilisation with any degree of certainty. If there is any doubt whatever of the fitness of food for human consumption, it should not be eaten in any case, even if cooked. The disadvantage of cooking is that vitamin C is destroyed by the heat though other vitamins suffer to a smaller extent.

There are five methods of cooking:—

(i) Roasting and grilling
(ii) Boiling
(iii) Steaming including pressure cooking
(iv) Stewing
(v) Frying

These can be carried out in utensils made of aluminium, copper (generally coated with tin) or enamelled iron. Neither aluminium nor copper is easily corroded but the most resistant is enamel, the only disadvantage of which is its liability to chip if carelessly handled. Needless to say all utensils used in the preparation of food and in cooking should be scrupulously clean.

Roasting and Grilling

Whether roasted in an open tin or in a covered one, there is a slight loss of water and salts from the meat but in a closed tin the loss of the latter is greater and that of water less. The open roast takes longer and water loss is by evaporation; shrinkage is also greater. Grilling is similar to roasting in an open tin and is well suited to the cooking of fish, steak, bacon and other meats which are more digestible than when fried. Nutritive juices which seep out of the meat are not lost and can be used in the making of gravy.

Boiling

In boiling, the temperature of the water is, of course, restricted to its boiling point but the meat does not reach this temperature because it is a bad conductor of heat. In any case it is better if the water is below boiling point to avoid making the meat tough. Juices escape from the meat and there is a loss of water and salts causing shrinkage. Similar conditions apply to the cooking of fish. Salt is lost as it dissolves out of the food but this loss is made good by the salt used in cooking. Eggs are most digestible when lightly boiled, a process which coagulates the protein. In the case of green vegetables, the calcium content may be considerably increased if they are cooked in hard water as the hardness is due to calcium salts.

Steaming

This is similar in effect to boiling but evaporation of water from the surface of the meat does not take place though there is a loss of weight due to the exudation of juices. In fish this is less owing to their becoming saturated with water. In pressure cooking there is less chance of beneficial juices being lost as only a small quantity of water is used and the temperature is much higher. Consequently the cooking time is very much reduced.

Stewing

In this method the meat is kept in water which is just simmering. Evaporation of water is very much reduced and the meat cooks in its own juices.

Frying

Meat, fish, bacon, eggs and potatoes can be cooked in hot fat, the temperature of which is much higher than that of boiling water. There is considerable loss of water but cooking is quick. Butter, lard and dripping are the most commonly used fats. Fried food, though very palatable, is somewhat indigestible.

MEALS

The number and size of the meals eaten during the day should be regulated according to one's age and activities as these determine the body's requirements. Failure to provide adequate amounts of food containing all the necessary nutrient substances and vitamins will result in *malnutrition*. Over indulgence, however, can lead to *obesity* and ill health. To start the day there is nothing better than a cooked breakfast (a thing of the past for many people!)

as one's efficiency at work is thereby definitely increased. Cereals, smoked fish, egg and bacon, kidneys and sausages all provide protein and fat while cereals, bread or toast and marmalade supply carbohydrate.

Growing children need proteins and vitamins but the meals should not be too large. Eating too many sweets is one of the causes of dental decay because the bacteria causing caries thrive on the deposits left behind between the teeth. Milk is a good inclusion as it provides calcium and vitamin B_2; hence the provision of "school milk". On reaching adolescence, protein is of great importance. In adult age, a good breakfast, a light luncheon and a cooked meal in the evening is the best arrangement to fit in with modern every-day life (tea can hardly be regarded as a meal), but customs must vary according to working conditions and in some countries must differ according to the climate. In old age calorie requirements are less and adequate protein should be taken. Care should be taken to see that diet is not neglected and that adequate food is provided. Pregnant women and nursing mothers need special diets with a good supply of calcium and vitamins, particularly vitamin C which should be increased by about a third; vitamin D, too, must be increased considerably. There must be adequate amounts of fats and proteins and of fruit and vegetables and sugar to provide for the additional nutrition of the developing child. Special diets are necessary for invalids during illness and convalescence and in certain diseases, as for example in diabetes mellitus, when the carbohydrate content must be carefully regulated.

PRACTICAL WORK

Meat

Put a few shreds of lean meat in a test-tube, add a little water and shake. Then apply one of the protein tests given at the end of Chapter VII (Biochemistry). Test also for carbohydrate and fat to see if any is present.

Vegetables and Fruit

1. Test some green vegetables such as cabbage for protein, starch and fat. Before testing for starch it will be necessary to remove the chlorophyll from the leaf. To do this, first dip the leaf in boiling water and then soak it in hot alcohol, kept boiling on a water-bath, until the leaf is white. Then apply the iodine test.

2. Soak a broad bean or pea in water for 24 hours to soften it. Then remove the seed coat, chop up the rest of the seed which consists of the embryo with two large lobes or cotyledons. Then

test for starch and protein. Follow up these tests with one for fats.

3. Test some uncooked potato and carrot for protein, carbohydrate and fat.

4. Apply protein, starch, sugar and fats tests to portions of fruits such as apple, tomato, grape and orange.

Flour

Put a little flour on a piece of muslin, tie it up to form a bag and immerse it in a small trough of water. Squeeze the flour thoroughly with the fingers until a sticky mass remains. This is the protein gluten. The washings make the water in the trough milky and this then contains starch and albumen. Verify by applying the following tests:—

1. Apply a protein test to the gluten.
2. Add iodine to some of the liquid.
3. Filter a little of the milky liquid and apply a protein test to see if any is present.
4. Place a small quantity of the residue on the filter paper with a drop of water on a microscopical slide. Put on a cover slip and examine the shape and size of the starch grains.

Bread

1. Add some iodine to a few crumbs of bread.
2. Perform a protein test with a few crumbs of bread.
3. Test for fat in bread.

Wheat and Other Cereals

1. Examine a slide of a L.S. of a wheat or maize grain. Note the embryo or "germ" at the more pointed end and the large endosperm occupying most of the grain. This is composed chiefly

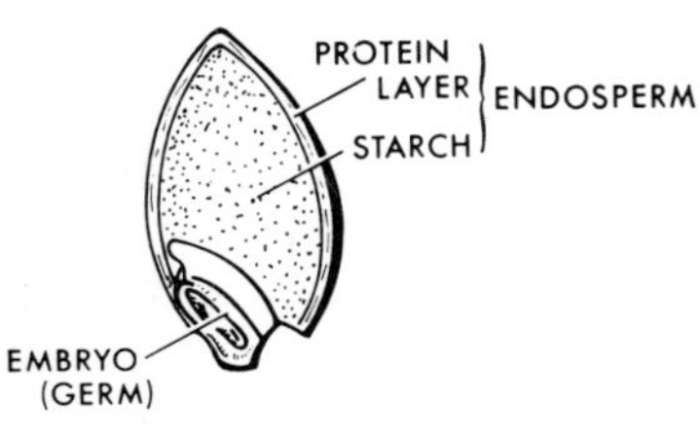

Fig. 23.1. **L.S. Wheat grain.**

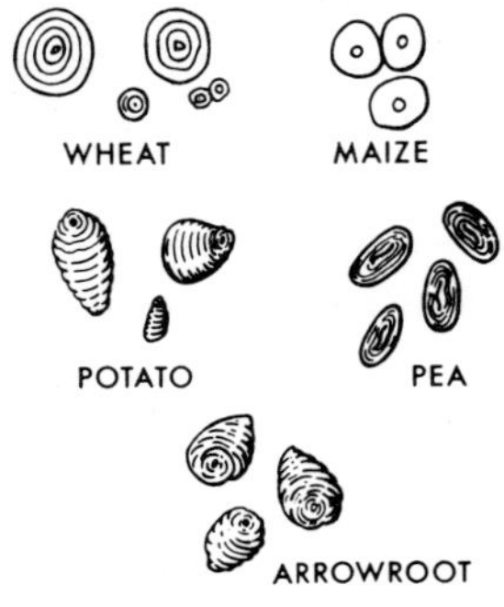

Fig. 23.2. **Starch grains.**

of starch with a layer of protein, the aleurone layer, round the edge. The whole is enclosed in an outer sheath which is the fused seed-coat or testa and fruit wall or pericarp.

2. Cut a grain of wheat or maize longitudinally at right angles to the broad surface and add a few drops of iodine to the cut surface.

3. Examine slides of starch grains from potato, wheat, pea or bean, maize and rice and compare their structure, shape and size. Note the series of concentric or eccentric layers of starch in each grain enclosed in a cellulose envelope.

4. Test for protein, starch and fat in a sample of oatmeal.

Milk

Specific Gravity:

Find the specific gravity of a sample of milk by means of an hydrometer or lactometer. It should be between 1·028 and 1·034.

Milk an Emulsion:

Place a drop of milk on a slide, cover and examine under the microscope. Suspended droplets of fat will be seen.

Separation on Standing:

1. Examine some milk which has been allowed to stand in a test-tube for several hours. Note the layer of cream on the top. Measure the depth of this and of the lower layer and thus estimate approximately the proportion of cream.

2. Remove some of the cream by means of a pipette and transfer it to a watch-glass. Add Sudan III or "osmic acid".

3. Examine some milk which has been standing in a beaker for several days. Note that coagulation has taken place, a thick curd forming on top and a watery liquid, the whey, beneath.

Action of Rennet:

Heat some milk in a beaker to body temperature (37° C), add some rennet, stir and allow it to cool. Note that the milk sets into a solid. This is called junket and is the curd. Break up the curd with a spatula when a watery whey will separate out from it.

Composition of Curds and Whey:

1. Remove some of the curd and place it in a test-tube. Perform a protein test. The positive result obtained is due to casein.

2. Pour some of the whey into a test-tube and perform Fehling's test. The sugar indicated is lactose.

3. Boil another sample of the whey. Coagulation takes place indicating protein. This is albumin.

Cheese

Test a sample of cheese for protein, carbohydrate and fat.

Eggs

1. Break the shell of a hen's egg and pour the contents into an evaporating basin. Separate the yolk from the white and transfer it to another basin. Test the white and the yolk for protein. Coagulation on heating will occur if protein is present.

2. Test the yolk for fat.

3. Put a few small pieces of the shell into a test-tube and cover it with dilute hydrochloric acid. Effervescence takes place. Pour the gas evolved into lime-water in another test-tube, taking care that no liquid runs over. Then shake the lime-water. It turns milky showing the gas to be carbon dioxide and this indicates the presence of carbonate.

Tea

1. To show the presence of tannin:

Make an infusion of tea in a beaker in the usual way by adding boiling water to tea leaves. Pour some of the infusion into a test-tube dilute with water if the colour is too dark and add a few drops of ferric chloride solution. A deep blue colouration is produced, showing the presence of tannin.

To show the effect of protein:

2. Allow some tea infusion to stand for about half an hour or so. Pour some uncooked white of an egg into a test-tube and add some of the tea. Note the coagulation of the protein.

Coffee

Test for tannin in coffee, diluted if necessary, by the ferric chloride test given above.

Fruit Drinks

Test for sugar in given samples of various fruit drinks by means of Fehling's solution.

Alcohol

1. Make up a solution of glucose by dissolving about 50 gm. in 250 ml. of water in a 500 ml. flask. Add a small quantity of yeast and connect the flask to a Dreschel bottle containing some lime water. Leave the apparatus in a warm place for several hours. Bubbles of carbon dioxide will be given off as shown by the fact that the lime water turns milky.

When the action has stopped, decant some of the liquid into

a beaker and filter it. Pour the filtrate into a distillation flask and distil it, using a Liebig's condenser.

Test some ethyl alcohol from the bottle in the laboratory by adding a few crystals of potassium dichromate and a few ml. of dilute sulphuric acid and heating the mixture. A characteristic fruity smell of acetaldehyde will be given off. Now repeat this test with the distillate.

2. Distil a small quantity of beer and perform the aldehyde test on the distillate.

Proof Spirit:

Apply a light to alcohols of different concentrations, 90 per dent., 57 per cent. and below 57 per cent. by volume, and note which of them burns.

Vitamin Tests

1. Vitamin A

Take a sample of milk or a raw egg and add a drop or two of a saturated solution of antimony trichloride in chloroform. Note the bright blue colour produced.

2. Vitamin C

To a small quantity of lemon juice add a drop of an aqueous solution of dichlorophenol indophenol (DPIP) in a test-tube. The pale blue solution will probably at first turn pink but then fades and becomes colourless owing to the reducing action of the vitamin C. If a standard solution of the DPIP is used it is possible to calculate the actual number of milligrams of vitamin C present.

3. **To compare the Amount of Vitamin C in Different Fruit Juices**

Using a 2 ml. pipette, suck up this volume of lemon juice and run it into a large test-tube. Then, using a graduated pipette fitted with a teat, draw up the solution of DPIP and run it into the fruit juice drop by drop, counting the drops necessary to cause the colour to fade. Repeat the experiment with orange juice or other fruit juice and compare results. This will indicate the relative amounts of vitamin C present in the different juices.

Chapter XXIV

THE PRESERVATION AND PROTECTION OF FOOD

Contaminated food can be the cause not merely of digestive troubles but of poisoning and disease. It is therefore of paramount importance that food should be adequately protected from its source right up to the time it is eaten. Food "goes bad" or decomposes as the result of the action of micro-organisms and moulds and these organisms thrive in air if moisture is present and the temperature suitable for their growth and multiplication. Food preservation, therefore, relies on providing conditions which are unfavourable to them, namely exclusion of air, dehydration, high temperatures or refrigeration. The causative agents of putrefaction and disease will be studied in Chapter XXV.

Food hygiene is under the control of the Ministry of Agriculture, Fisheries and Food together with the Department of Health and Social Security and Local Authorities are responsible for the enforcement of the laws concerned. The most important laws are the Food Regulations Act and the Food and Drugs Act. In 1974 6,227 cases of food poisoning were reported in England and Wales and it has been estimated that figures in excess of 28,000 man-hours are lost by it every month.

METHODS OF FOOD PRESERVATION

Canning and Bottling

In canning, the food which must be absolutely fresh, is put into clean tins which are then covered except for a small hole in each lid. The tin is then heated to drive out the air. The hole is then sealed after which the tin is heated to a high temperature to kill any micro-organisms or spores which may be present. If this is improperly done the food will decompose and gases will be given off which will cause the tin to bulge. Canning is a certain way of preserving food as there is no possible way in which bacteria can enter and cause contamination. An excellent example in support of this statement was provided by canned foods found on one of Captain Scott's bases on his last, fatal Antarctic Expedition in 1910. They were discovered over forty years later in 1956 and the food in the tins was found to be in excellent condition. Admittedly the cans had been subjected to the extremely low temperatures of the Antarctic during that time. This method of food preservation is applicable to meat, fish, vegetables, fruits and condensed milk.

Bottling, like canning, is highly effective for fruit preservation. In fact, the only difference is in the container and the consequent difference of method. The bottles, like the cans, must be clean and sterile at the outset. The fruit is put into the bottles, sterilised by heat and then covered with boiling syrup. The glass covers are provided with rubber washers and tested by lifting to ensure a completely air-tight seal. They are then kept in place by screw-on metal tops. There are modifications of this method but the underlying principle is the same.

Dehydration, Salting and Smoking

Preservation by removal of the greater part of the moisture is possible for certain foods such as milk and fruits like prunes (dried plums), apricots, raisins and sultanas (dried grapes). The deficiency of moisture prevents the development of micro-organisms but putrefaction will set in if the dried food becomes moist. There is some vitamin loss by this method.

Salting is a method suitable for meat and fish as the addition of salt in sufficient quantity withdraws water from them. It also kills bacteria and this property is improved by a further addition of saltpetre (potassium nitrate).

Smoking is a method which is used for bacon and some fish, the food being subjected to smoke from burning wood, and this seals the surface preventing bacterial entry to some extent.

Refrigeration

This method inhibits the action of bacteria but it does not kill them. Consequently refrigerated food will soon decompose after removal from the refrigerator, particularly if it is frequently returned to the refrigerator and taken out, at any rate in the case of some foods. Meat, fish, vegetables and, in fact, most foods can be preserved by this method. If the temperature is around Freezing Point, meat is said to be *chilled* but if it is kept at -10°C (14°F) or less it is said to be *frozen*. Frozen meat and fish will keep fresh for a year or longer but chilled meat begins to decompose after about a month. Refrigerated meat must be thawed before it is cooked.

Deep Freezing

Food preserved by this method must be absolutely fresh and must be frozen quickly at -28°C (-18°F) as slow freezing not only affects the texture and flavour but it can also raise the temperature of the food already in the deep freeze. Ice crystals form within the structure of the food and the faster it is frozen the smaller are these crystals and consequently the less the distortion of the structure. But food with a natural high water content can never be successfully frozen. The storage temperature is – 18°C (0°F). Opinions differ as to whether deep freezing affects the flavour of the food.

PROTECTION OF FOOD

Protection of food *at source* is effected by rigid laws concerning the use of healthy animals, the construction of slaughter houses and the slaughtering of animals, the provision of hygienic conditions in food factories and the personal hygiene of all workers who handle food.

During storage, apart from protection by refrigeration both on the large scale and in the home, food must be protected against contamination by insects such as flies and against vermin.

Protection *during transport*, when the risk may be high if the distance is considerable, is made possible by the use of suitable food containers, refrigeration during the journey and suitable clothing and personal hygiene of the workers who handle the food.

Milk

Milk is an important vehicle for the transmission of disease organisms because it contains the food substances on which they thrive. Tuberculosis, diphtheria, scarlet fever and typhoid fever can all be acquired by drinking infected milk. Strict laws therefore govern the production, storage and distribution of clean, healthy milk.

First and foremost the cows must be free of disease and the udders and neighbouring parts must be washed before milking. Cowsheds in which the animals are milked must be kept scrupulously clean and must be well ventilated. All utensils, buckets and churns and the mechanical milkers if used, must be sterilised with boiling water after use and thoroughly washed before use. Finally the cowmen responsible for the milking must themselves be free from disease and they should wear protective overalls.

Milk is graded according to its origin and method of treatment. **Tuberculin Tested** milk is obtained from herds which are regularly inspected for tuberculosis by a veterinary surgeon, certified by him as being free of the disease and kept separate from other herds of cows. The milk must be sold and labelled as "Tuberculin Tested Milk" and if bottling is done at source it should also be labelled as "Certified Milk". This is the safest kind of milk to drink.

Accredited herds are not submitted to the tuberculin test but they are inspected by veterinary surgeons and must be free from disease. As cows may suffer from tuberculosis without evidence of it, tuberculin tested milk is obviously safer. These two types of *raw milk*, as it is called, are suitable for sale without further treatment and no raw milk may be sold unless it is from one or other of theses two sources.

Sterilisation by heat ensures further safety from pathogenic bacteria. **Sterilised milk** is milk which has been heated to boiling and bottled in hermetically sealed bottles. There is some impairment of taste. **Pasteurisation** is the method used for the bulk of the milk which is sold. In this process, the milk is either heated to 63° to 65°C (145°–150°F) for at least thirty minutes and immediately cooled or it is heated to 72°C (161°F) for at least $\frac{1}{4}$ minute and immediately cooled. In both cases the milk must be cooled to 13°C (55°F). This is sold as **Pasteurised milk**. Its only real disadvantage is that some of the vitamin C content is destroyed.

The result of bacterial action, apart from the transmission of infectious diseases, may be **food poisoning** caused by the production of toxins by bacteria such as *Micrococcus pyogenes*. These take effect in a matter of hours causing vomiting, diarrhoea and pain. One very serious form of food poisoning is *botulism*, obtained from infected meat. Paralysis of the respiratory system occurs in this disease which is nearly always fatal. The cause is *Bacillus* (*Clostridium*) *botulinus* which produces the toxin. The spores are very resistant and bacteria may therefore develop from them after preservative measures have been taken.

Bacteria can enter food by the visits of insects such as the housefly which shows no discrimination in its landing grounds and is quite likely to settle on a manure heap or other refuse and then fly on to human food. Rats and mice can easily infect food if they come in contact with it. Cooks and kitchen-hands who fail to pay attention to personal hygiene before handling food can also convey infection to it. Carriers, people who transmit diseases by harbouring the infective agents without showing any symptoms of the disease, are sources of considerable danger in this respect. Diseases which can be carried in this way are typhoid fever, paratyphoid fever and dysentery.

It is common sense, therefore, that food should be protected from flies and other insects in shops as well as in the home and from other animals by storing it in places to which they cannot have access.

Chapter XXV

MICRO-ORGANISMS AND INFECTIOUS DISEASES

Excluding natural causes and apart from wars, accidents and starvation due to famine, the chief cause of death is disease. Unfortunately much of this came with civilisation. There are many diseases, several of them mentioned at the end of the chapters on the various systems of the body, to which no attributable cause may yet have been discovered and which cannot be conveyed to other people by the person suffering from them. But there is a very large number of **communicable** or **infectious disease** known to be caused by **micro-organisms** which gain entry into the body and it is this type of disease and the organisms—germs or microbes as they are popularly called—which cause them to which we shall now turn our attention.

In the latter part of the 17th century a Dutch draper, Anthony van Leeuwenhoek, who was interested in the making and use of lenses succeeded in magnifying drops of dirty water and scrapings from teeth up to some 270 times with a simple microscope and so discovered the existence of microbes. These were, in fact, **bacteria** though he described them as tiny animalcules.

THE CAUSES OF COMMUNICABLE DISEASES

Although it had been known for a very long time that some diseases could be "caught" it was not until the second half of the 19th century that the germ-origin of these diseases became known. For this the world is indebted to **Louis Pasteur** who was born in 1822 in the small town of Dole, some 50 km. south-west of Dijon in the French Jura. The house is now a museum where some interesting specimens and records of his experiments can be seen. Pasteur held a Degree in Chemistry and at first his interest lay in crystallography but later turned to the study of fermentation. This arose from an investigation into the souring of wines which were threatening the industry in France. Formerly it had been believed that organisms seen in fermenting and putrefying material had developed from that material *i.e.* that they had arisen spontaneously and the German chemist Liebig maintained that fermentation was simply a chemical change. Pasteur doubted the veracity of both these facts and suspected that living organisms entered the material from the air and caused it to ferment. He therefore heated the wines up to 55°C to kill the organisms and found that these wines did not turn sour. The outcome of this was the saving of the French wine industry. Pasteur also experimented with milk and obtained similar results. The term *pasteurisation* for this method of milk preservation is, of course, named after him.

His classic experiment which overthrew once and for all the idea of spontaneous generation is well known. Suspecting that the infection causing putrefaction came from the air, he set about to verify it. He took a series of flasks with necks drawn out as shown in Fig. 25.1, each flask containing broth of some kind. One in the museum at Dole is labelled in his own handwriting "Bouillon de poule du Aout 1883". The contents of the flasks

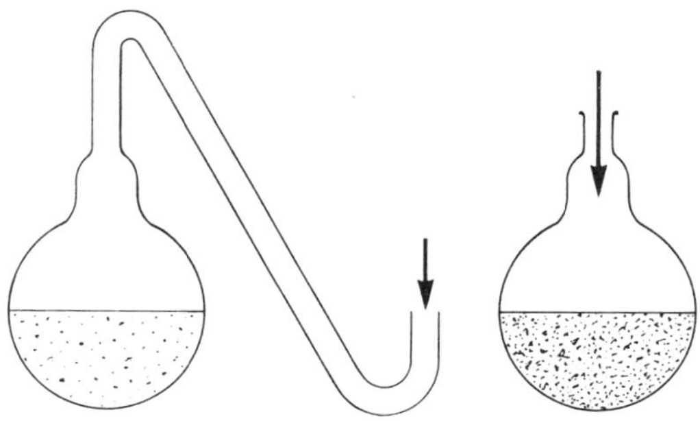

Fig. 25.1. **Pasteur's flasks.**

were heated to boiling with the object of killing any micro-organisms which might be present. None could enter the broth from the air owing to the swan-like shape of the necks of the flasks and the liquids remained fresh. The one referred above which the author saw in 1960 and which he was given to understand was original looked perfectly fresh. Another seen in this country about ten years after the experiment had been performed just before the last war, looked equally sterile. Pasteur then broke off the necks of some of the flasks after which fermentation and putrefaction soon set in. The two which the author examined, Pasteur's and the later repetition of his experiment, looked perfectly revolting!

Thus this famous chemist proved that putrefaction was due to micro-organisms and that the air was the source of infection. He next turned his attention to disease, suspecting that communicable disease might be caused in a similar manner. He first investigated a disease affecting silkworms, another disease threatening an industry in France. Then he investigated fowl cholera and in the course of his experiments accidentally injected hens from an old culture in which the activity of the germs had been lessened by age. Normally this disease is fatal but the injected hens survived. Furthermore they were unaffected when injected with the lethal doses. Thus Pasteur had discovered a method of **immunisation**. He repeated the experiment with anthrax, using sheep, which are normally affected by the disease, and the results were equally successful.

The next problem which faced Pasteur was whether or not these results applied to human beings and he investigated the disease known as *rabies* or *hydrophobia*. This is a disease which affects dogs, turning them mad and which is easily transmitted to human beings if bitten or even licked by infected dogs. Though he was unable to find any microbes which were the cause of the disease, he was convinced that they existed (it is now known that the disease is caused by a virus) and he discovered that it was possible to immunise dogs against rabies by injection of rabies vaccine as immunising injections are now called. Fortunately a human case came to hand in which he was able to try it out. The patient was a young boy from Alsace who had been bitten by a rabid dog. Pasteur injected him with rabies vaccine and the boy did not develop the disease. Pasteur had saved his life. Thus the **germ-origin of disease** was established and methods of prevention attained. After much fierce opposition and criticism, Pasteur's conclusions, supported by a great deal of other evidence, were eventually accepted throughout the medical world. The idea of spontaneous generation was finally abandoned and the *law of biogenesis*—living things can come only from living things—took its place. Thus Pasteur laid the foundations for the science of **bacteriology** and **immunology.** He died in 1895 and was buried in the crypt of the Pasteur Institute in Paris, a hive of research founded in his honour.

The real father of Bacteriology, however, was a German doctor named **Robert Koch** who was born near Hanover in 1843, for it was he who, with great skill, developed the techniques for the study of bacteria. Strangely enough, he knew nothing of the work of Pasteur and his first study was of anthrax. After demonstrating the presence of rod-shaped microbes in the blood of animals suffering from this disease, he prepared a culture of the germs and injected some into a mouse which quickly developed anthrax. Further cultures were prepared from the animal and injected into others. This was done several times until it became indisputably certain that the cause of the disease was the rod-shaped microbe. He also showed that they could form resistant spores. Koch then turned his attention to tuberculosis and again, in 1882, he was able to isolate the organism and show that the disease was caused by it. This was possible because he had devised methods of culturing and staining the germs to make them visible under the microscope. He was also the discoverer of the germ of cholera.

Let us now enter the field of **microbiology** and make a study of the microbes or germs which cause disease and which are collectively known as **pathogens**. The organisms concerned are *bacteria, protozoa, richettsiae* and *fungi*.

BACTERIA

Though, like fungi, bacteria could with some justification be placed in a separate Kingdom, since they show certain similarities to and differences from both plants and animals, they are, in fact, classified as plants. Until the advent of the electron microscope, the study of their structure was extremely difficult owing to their minute size. The smallest are about $0{\cdot}5\mu$* in diameter and the largest around 20μ in length. With the exception of a few very large ones, they cannot be seen with the ordinary high power of the microscope and a 2 mm (1 /12 in.) oil-immersion (O.I.) objective is necessary. Each cell is composed of cytoplasm enclosed in a cytoplasmic membrane and although they appear to be non-nucleated, the electron microscope reveals a ring, sometimes incomplete, in the cytoplasm and this contains D.N.A. It is therefore regarded as a nucleus. The cytoplasm contains both R.N.A. and D.N.A. and the whole is enclosed in a cell wall. A few develop **capsules** outside this wall, the function of which is protection. The majority are colourless and require specific staining to render them visible. Bacteria can be placed in one or other of two main groups known as **Gram-positive** and **Gram-negative**. They are so-called according as to whether or not they retain a stain called Gram's stain (gentian violet and iodine) after treatment with it.

Classification is as follows:—

Bacilli—rod-shaped.

Cocci—spherical.

- **Micrococci**—occurring singly.
- **Diplococci**—grouping in pairs.
- **Staphylococci**—grouping in bunches.
- **Streptococci**—grouping in chains.
- **Sarcina**—grouping in three dimensions of space forming "packets" like minute cubes.

Spirilla—twisted, cork-screw-like.

Spirochaetes—spiral.

Vibrios—comma-shaped.

* μ (pronounced *mew*) = micron = $0{\cdot}001$ mm.

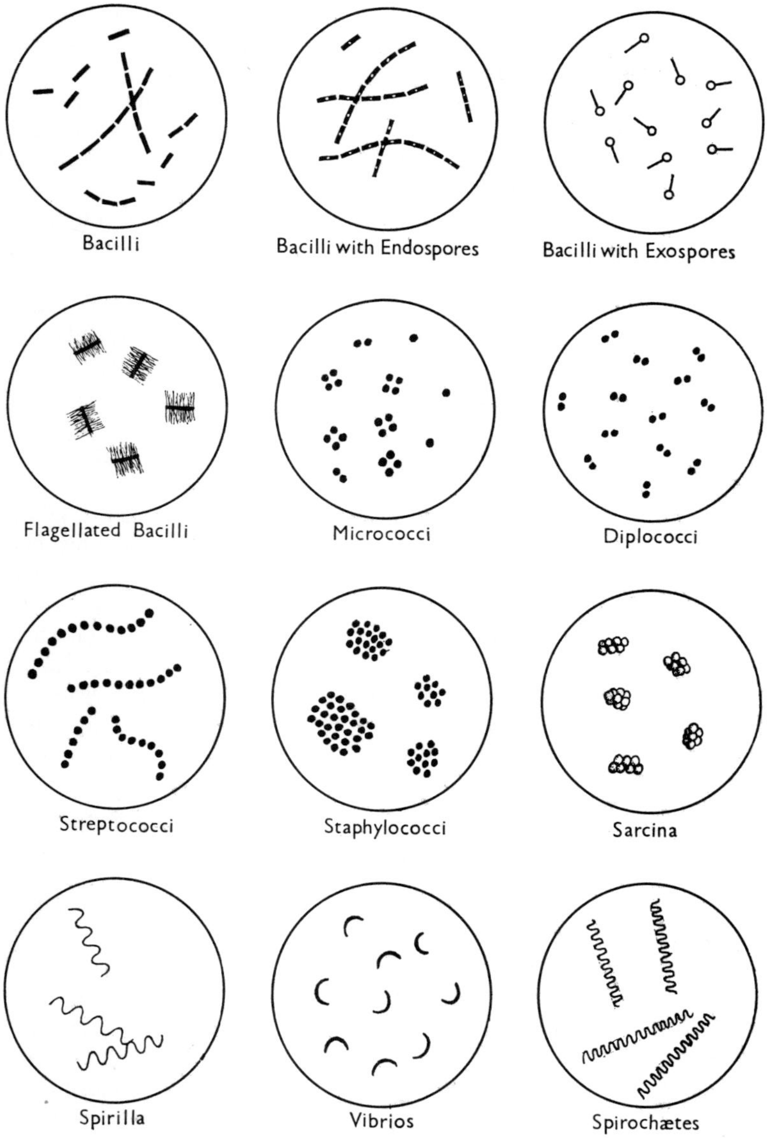

Fig. 25.2. **Forms of bacteria.**
(*From Wallis—Practical Biology.*)

Some bear long whip-like protoplasmic threads called *flagella* and are therefore motile, but some are motile though devoid of flagella. Reproduction is by binary fission, that is to say the bacterium grows, the nucleus is replicated and then the cell splits transversely into two. This, under very favourable conditions can take place every twenty minutes to half-an-hour and a single bacterium dividing once every thirty minutes could give rise to 2^{47} (over 140,000,000 million) in twenty-four hours. Fortunately such favourable conditions never arise owing to the lack of food and unfavourable conditions produced by their own metabolism. As was stated in Chapter XIX, conjugation between bacteria has been observed under the electron microscope. Some bacteria produce spherical **spores** which are very resistant forms able to withstand adverse conditions such as high temperatures and desiccation. These spores may develop within the bacterium (*endospores*) or externally (*exospores*) but only one arises from each.

The majority of bacteria thrive at temperatures favourable to other living organisms and the human pathogenic bacteria have an optimum temperature of 37°C, that of the body, and are killed by higher temperatures such as 55°–70°C in a few minutes but the spores are more resistant. Some respire solely in free oxygen in the air of their surroundings (*obligatory aerobes*), others respire solely without access to free oxygen (*obligatory anaerobes*) while yet others are able to use both methods of respiration (*facultative anaerobes*). Many are killed by ultra-violet light and therefore by exposure to sunlight. Dehydration is also fatal to them as are high degrees of acidity and alkalinity in many cases and subjection to certain chemical substances known as disinfectants.

The vast majority of bacteria are *saprophytic,* obtaining their nutrition from non-living organic matter and some are even able to obtain their energy from inorganic materials and these all play their part in processes which affect man such as the putrefaction of his food and the souring of milk and other processes which are beneficial to him like the ripening of cheese, the making of vinegar, the treatment of sewage to render it harmless and the decay of dead plant and animal remains into nitrates which are returned to the soil where they provide nitrogen for plants serving as food for man. *B. coli* which lives normally in the intestine is a further example of this. There is, in fact, a normal population of saprophytic bacteria living symbiotically in the human body. *Symbiosis* is an association between two organisms which live together to their mutual benefit. But a minority of all known bacteria are *parasitic,* about 10 per cent. being responsible for communicable

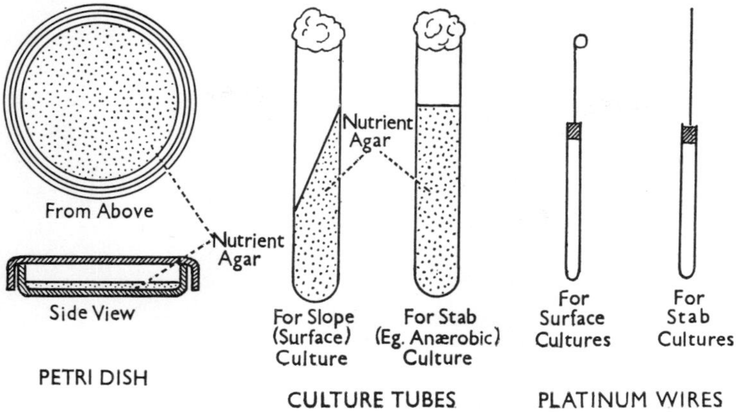

Fig. 25.3. **Bacteriological apparatus.**
(From Wallis—Practical Biology.)

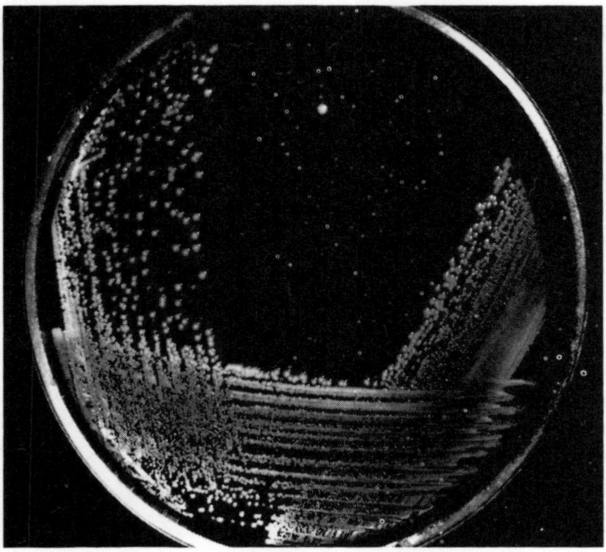

Fig. 25.4. **Staphylococi growing on nutrient agar in a Petri dish.**
(By Courtesy of the Wright-Fleming Institute of Microbiology, St. Mary's Hospital, Paddington, London.)

diseases in man and it is these **pathogenic bacteria** and the diseases they cause which concern us in this chapter.

Bacteria can be cultured in the laboratory on **culture media** made from a jelly-like substance, *agar-agar,* obtained from certain

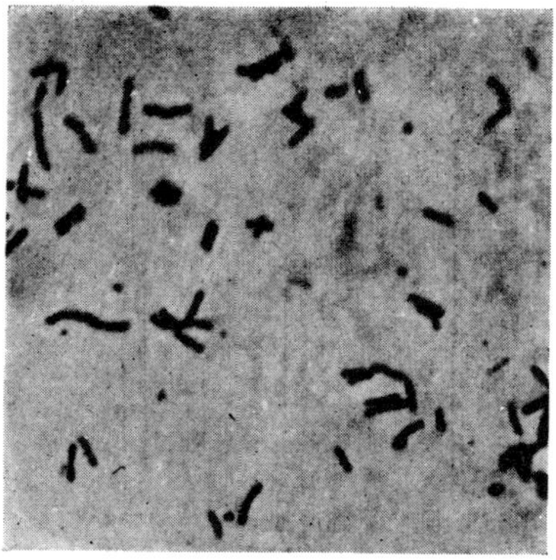

B. TUBERCULOSIS X1000

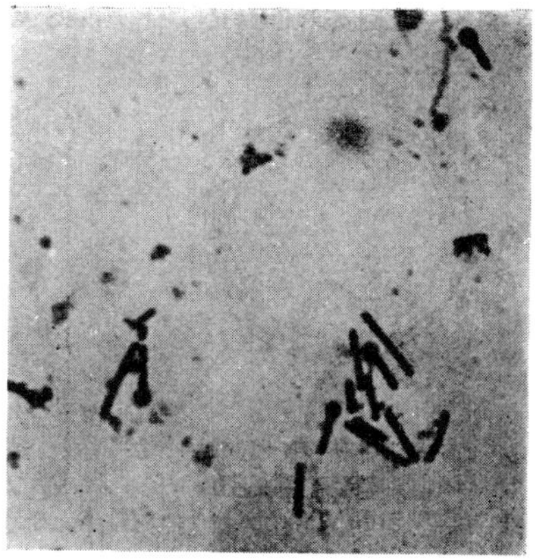

B. TETANUS X5000

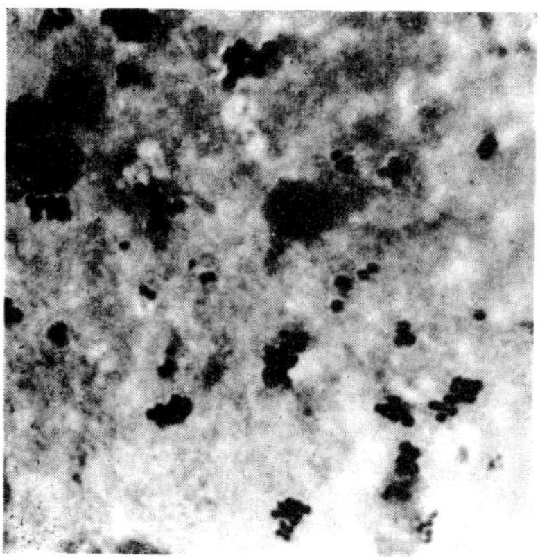

STAPHYLOCOCCI IN PUS X1000

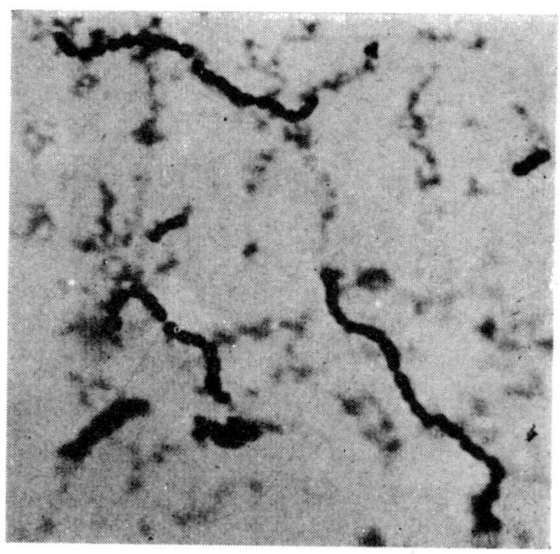

STREPTOCOCCI IN PUS X1500

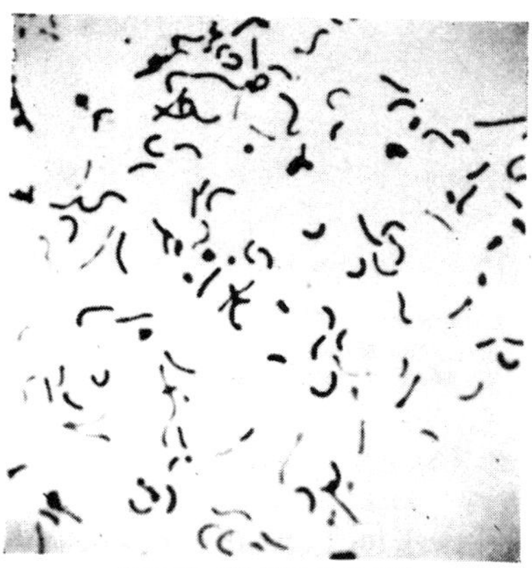

VIBRIO CHOLERÆ X5000

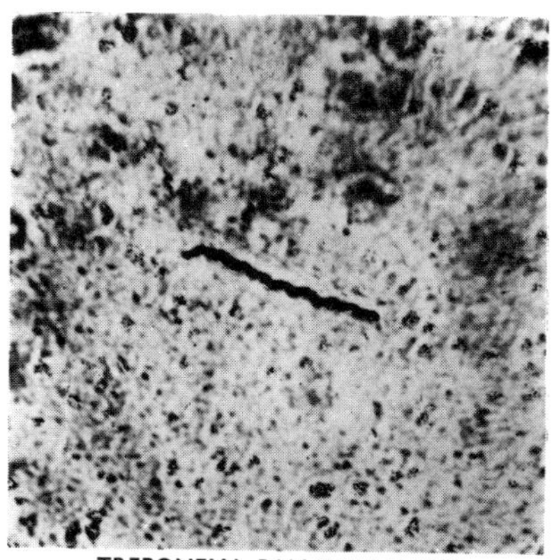

TREPONEMA PALLIDUM X3500

FIG. 25.5. **Photomicrographs of pathogenic bacteria.**

(By Courtesy of Dr. A. J. Duggan director of the Burroughs Wellcome Medical Museum.)

oriental seaweeds to which nutrient substances have been added. This is done in sterile *Petri dishes* or *bacteriological test-tubes*. When the medium has been inoculated the Petri dish or tube is kept at body temperature in an *incubator*. The bacteria multiply and produce colonies which are visible to the naked eye (see the practical work at the end of this chapter). These colonies are characteristic in shape, form, and colour for different species of bacteria.

A few examples of diseases caused by bacteria are given below. Tuberculosis (*B. tuberculosis*), diphtheria (*Corynebacterium* diphtheriae*), typhoid (enteric) fever (*B. typhosus* or *Salmonella* typhi*), bacilliary dysentery (*Shigella* dysenteriae*), pneumonia (*Diploccocus pneumoniae*), meningitis (*Meningococcus*), gonorrhoea (*Gonococcus*), lockjaw (*B. tetanus* or *Clostridium tetani*), anthrax (*B. anthracis*), erysipelas (*Streptococcus pyogenes*), cholera (*Vibrio cholerae*), syphilis (*Treponema*† *pallidum*), relapsing fever (*Borrelia*† *recurrentis*) and leprosy (*Mycobacterium leprae*).

Boils are caused by *Staphylococcus aureus*. One of the most serious menaces in our hospitals is the airborne *Staphyloccus aureus* which spreads infection rapidly from ward to ward and into the operating theatre. Many people are innocent carriers who only suffer when their resistance is lowered by illness. Furthermore strains resistant to antibiotics and drugs have developed and control therefore becomes increasingly difficult in spite of scrupulous cleanliness.

Food poisoning is caused by various bacilli of the *Salmonella Group* (so-called after a certain Dr. Salmon who made a study of them) and botulism by *Clostridium botulinum*.

PROTOZOA

Protozoa are minute non-cellular animals of which the common Amoeba, usually the first organism to be examined by the student of biology, is an example. Each is composed of a small mass of protoplasm in which lies a nucleus and, with some exceptions such as Amoeba, this is enclosed in a pellicle. Some have flagella or cilia for locomotion.

Most Protozoa are free-living but a few are pathogenic and they are the cause of many tropical diseases of which the following are examples:—

Malaria (*Plasmodium*), sleeping sickness (*Trypanosoma*)—which must not be confused with sleepy sickness—and amoebic dysentery (*Entamoeba hystolytica*) which is different from bacillary dysentery.

* A bacilliary genus.

† A spirochaete.

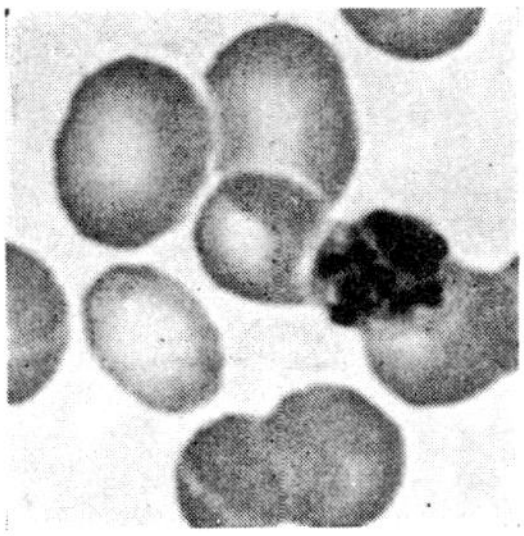

PLASMODIUM MALARIÆ
SCHIZONTS IN BLOOD
X 1000

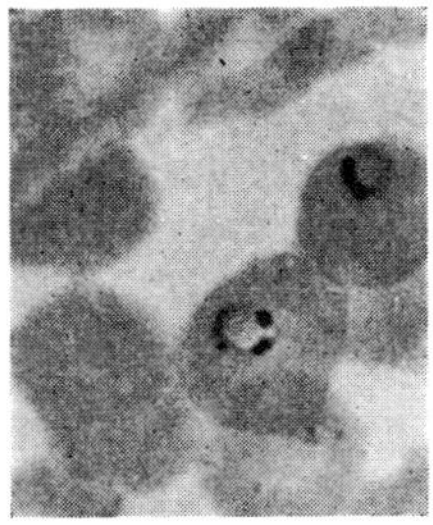

P. FALCIPARUM
SIGNET RING STAGE IN BLOOD
X 1000

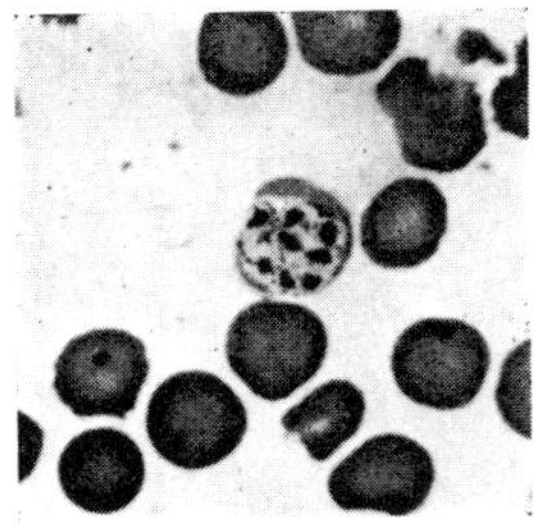

P. VIVAX
SCHIZONTS IN BLOOD
X 1000

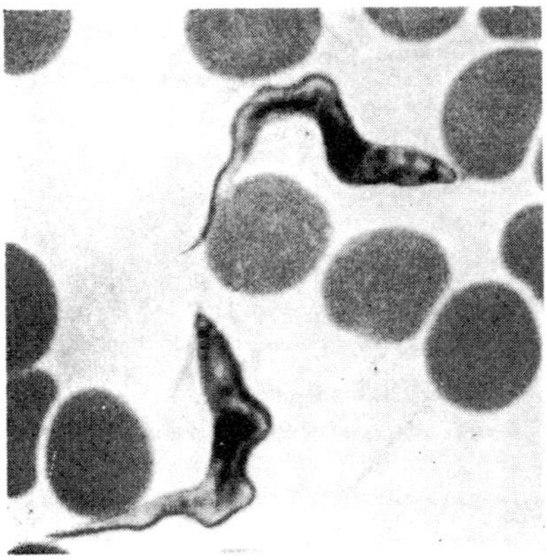

TRYPANOSOMA RHODESIENSE IN BLOOD
X1000

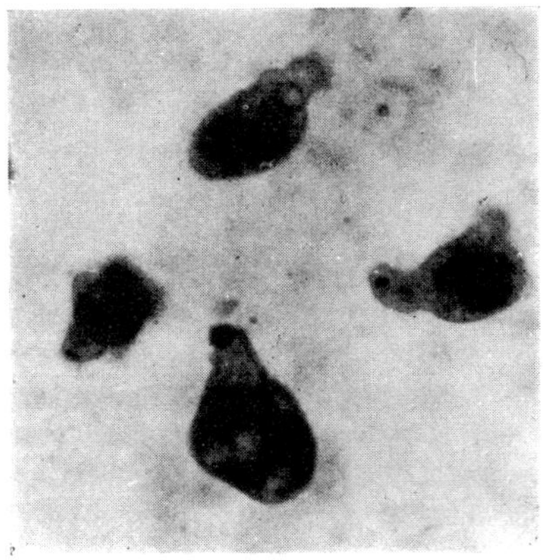

ENTAMŒBA HISTOLYTICA IN BLOOD
X1000

Fig. 25.6. **Photomicrographs of pathogenic protozoa.**
(*By courtesy of Mr. C. James Webb, London School of Hygiene and Tropical Medicine.*)

A Protozoon closely related to the last mentioned is *Entamoeba coli*, a normal harmless inhabitant of the human intestine where it lives saprophytically on bacteria and acts as a scavenger.

VIRUSES

The name virus, which has its origin from the Latin *virus* meaning a poison or slime, arose from the fact that when certain infected material was passed through the finest of porcelain filters which retain bacteria, the liquid which passed through retained powers of infection in spite of the lack of any visible organisms. The existence of viruses has been known for a long time but they are so extremely minute that it was not until the electron microscope was invented that it was possible to see them for the first time. The virus of small pox, one of the largest, is some 250 mμ* but that of polio, which is one of the smallest is only about 25 mμ. These extraordinary particles which are called **virons** and not cells can exist in what appears to be a non-living crystalline form yet they exhibit characteristics of life within living cells. In crystalline form they show a regular shape which varies with the virus. In living cells, and only in living cells of the host, they are seen to reproduce. In fact the view is held that they themselves are not capable of reproduction but are replicated by the cells in which they exist. They are composed of a protein coat which encloses a core of nucleic acid, in some cases D.N.A. and in others R.N.A. Viruses may well be on the border-line between the living world and the non-living.

They are killed by ultra-violet light. It is interesting to find, and this will give some idea of their extreme minuteness, that some bacteria are infected by viruses and these viruses are known as **bacteriophages**. They are rather tadpole-like in shape with hexagonal heads and narrow tails and their method of attack has actually been demonstrated under the electron microscope. Having invaded the bacterium, the virus uses the latter's protoplasm for its own reproduction. On release from the victim's cell by the splitting of its wall, a process called **lysis**, the new virus particles enter other

* mμ (millimicron) = 10^{-6}mm.

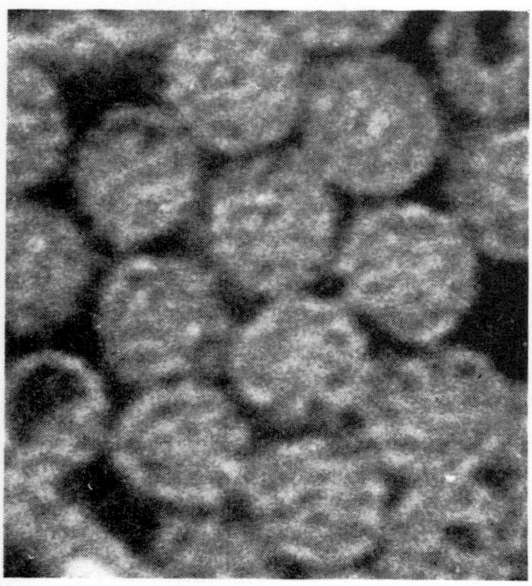

POLIOMYELITIS X 500,000

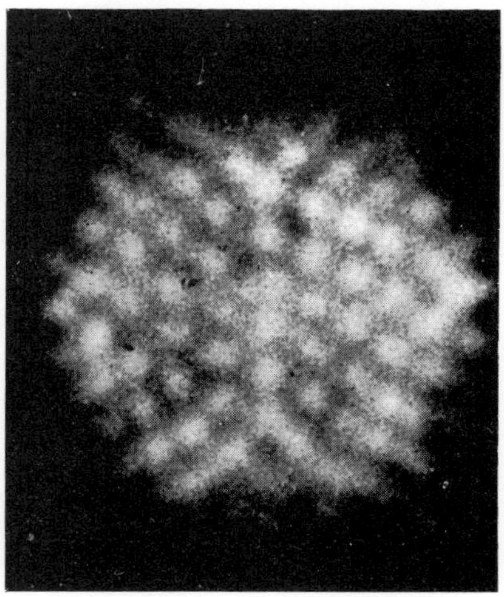

ADENOVIRUS X 450,000

Fig. 25.7. **Electronmicrographs of viruses.**
(By courtesy of Dr. R. W. Horne, St. Johns Innes Institute, Norwich.)

bacteria. No method of virus culture such as is used for bacteria has as yet been devised.

A few examples of diseases caused by viruses which affect man are the common cold, influenza, measles, mumps, chicken pox, smallpox, shingles, poliomyelitis, hepatitis, rabies and yellow fever. Inflammation of the respiratory tubes is caused by *adenovirus*. Foot and mouth disease which occurs in man's domestic animals is another viral disease, causing great havoc, spreading rapidly and extensively and fatal to them. Man is not normally subject to it though there has certainly been one exception to this at least and in this instance it did not prove fatal. A virus was discovered a few years ago which is the cause of cancer in monkeys but so far there is no conclusive evidence that carcinoma in man is a viral infection. Plants are also affected by virus diseases. Examples are leaf curl and mosaic disease, the latter being the first viral infection to be discovered as such. Both these diseases affect crops.

Viral infections are very easily transmitted and stringent precautions are necessary to prevent their spreading. The number of man-hours lost in industry by the common cold and influenza each year is very considerable indeed.

RICKETTSIAE

These micro-organisms, are smaller than the majority of bacteria but larger than viruses and are named after a bacteriologist, Dr. H. T. Ricketts who investigated the cause of typhus and who died while so doing. They are the cause of typhus, spotted fever and trench fever.

FUNGI

Microscopical forms of these non-green plants are the cause of several skin diseases and they are spread by spores. One of the commonest is ringworm of the scalp (Tinea capitis) which has nothing to do with worms and is caused by *Microsporon audouini*. Other species cause the disease in other parts of the body. Another quite frequently occurring infection is athlete's foot (Tinea pedis), caused by *Trichophyton rubrum.* Another fungus, *Candia*, causes thrush, an infection of the mucous membrane of the mouth and vulvo-vaginitis, an inflammation of the female reproductive tract.

METHODS OF INFECTION

There are various ways in which pathogens are able to enter the body and so cause disease, by physical contact and human carriers, in food and water, through wounds, by insect bites, from dust and dirt and by infected droplets in the air. Animals, too, are often carriers of diseases to man.

Personal Contact

Some diseases, particularly venereal diseases and skin diseases are contracted by actual physical contact or by contact with some article with which contact has been made by an infected person. Examples of the latter are spoons and forks, glasses, crockery (especially if it is cracked), towels, bedding, clothing and toilet seats. The importance of hand-washing after going to the toilet cannot be over emphasised. Venereal diseases have been dealt with at length at the end of Chapter XVII and this should be studied by all readers.

Anyone who has been in company with an infected person may also contract some diseases and he or she is referred to as a **contact**.

Human Carriers

Human carriers are people who harbour pathogens from contact with a disease or from previously suffering from it themselves, without actually suffering from it at the time and who can therefore infect others. They are usually perfectly innocent of their danger to the community and are difficult to trace. Typhoid fever, diphtheria, smallpox and cholera are examples of diseases which can be transmitted in this way.

Food

Infected particles of dust may settle on food or in milk. Food may be contaminated by flies or it may be touched by people with dirty hands or hands contaminated after defaecation or micturition because of failure to wash after performing these processes.

Water

Infection by water may be through contamination at source, by sewage or by workers who are carriers. Typhoid fever, cholera and dysentery can be contracted through water, and the eating of oysters from beds near the discharge of sewage has been found to be the origin of a typhoid epidemic. Just before the last World War there was a serious epidemic of typhoid in Croydon in which thirty-four people died. It was eventually traced to the reservoir from which the town drew its water supply. The contamination was found to originate from one of the men who were working on a nearby well; he was a carrier and habitually micturated on the ground. The bacteria found their way into the water of the reservoir. In 1964 an outbreak occurred in Aberdeen. This was traced to tinned meat from a factory in Argentina which used untreated water from a river to cool its hot tins. A year earlier there had been an epidemic in the beautiful mountain village of Zermatt at the foot of the Matterhorn in Switzerland and this was the result of a broken water main. But one of the worst epidemics occurred in Mexico in 1972 when 3,577 cases were recorded with a fatality rate of 13·5 per cent. at the outset. The overall percentage for the year was 3·6 per cent. In 1973 it fell to 1·9 per cent. and in 1974 to nil.

Wounds and Insect Bites

If the skin is pricked, cut or grazed, bacteria can gain ingress to the blood and cause an infection by entry into the damaged capillaries. Tetanus and anthrax can be contracted in this manner. Insect bites can also be responsible for the entry of pathogens if the insects have been in contact with infected dirt. Dust and any form of dirt commonly harbour infective agents and will therefore bring about infection if they enter a wound of any kind, just as they will if taken with food.

Droplet Infection

Anyone suffering from the common cold, influenza, diphtheria, tuberculosis and other diseases of the respiratory tract pass the pathogens into the air in minute droplets of moisture which are exhaled in the breath, and even when speaking. Sneezing and coughing shoot these droplets out with considerable force. Others may breathe in these droplets from the air or they may settle on food or in drinks. Alternatively they may settle on the ground as in the disgusting habit of spitting where they dry up amongst the dust and so get conveyed to other people. This **air-borne infection** as it is called is very easily contracted.

Animal Carriers

These are called **vectors** and include insects, rats, mice and domestic animals. The malarial parasite is carried by a particular mosquito, the female *Anopheles*, that of sleeping sickness is carried by the tsetse fly, rabies virus by dogs, and the typhus virus by lice. Rabies can also be transmitted by other animals such as mice, rabbits, foxes and wolves and even domestic animals can be infected. During the 1939–45 World War, the disease was spread into Western Europe by wolves which were fleeing behind the German Lines as the Russian troops advanced westward. Unfortunately the story does not end there for as recently as 1974 and 1975 the disease has been rife and out of control in France but fortunately so far no-one has died during the present outbreak.

If these conditions continue it is quite possible for the virus to be brought to this country by bats. Our laws are very strict concerning the import of domestic animals from abroad and all dogs must be put in quarantine for twelve months. But pets have been smuggled in, for example in lorries, and there was a case here in which a woman was bitten by a rabid dog from West Germany which had been in quarantine here and evidently contracted the disease while there. The law has now been tightened up and by a new Act in 1974 dogs in quarantine are kept completely apart. Much heavier fines can be imposed on anyone smuggling animals into the country, fines are unlimited and terms of imprisonment can be given. There have, in fact, been many successful prosecutions. The woman in question above was treated with rabies vaccine and did not contract the disease. The responsibility of the common house-fly in transmitting disease has already been mentioned. Bubonic plague, such as in the Great Plague of 1665 when over 25 per cent. of the population of London died, is caused by a bacillus (*Pasteurella pestis*), which is carried by the rat flea.

When a disease breaks out, spreads rapidly and affects a large number of the population in a country it is said to be **epidemic**. If it spreads to other countries throughout the world it is said to be **pandemic** and if a particular disease constantly occurs in a particular region, such as bubonic plague or malaria, it is said to be **endemic**. On the other hand, isolated cases of diseases which break out without any clear evidence of their origin are said to be **sporadic**.

HOW PATHOGENS CAUSE DISEASE

Once pathogenic bacteria have entered the body, they multiply and produce poisonous substances of a protein nature called **toxins** and these are responsible for the symptoms of the particular disease. The time elapsing between the entry of the pathogens and the appearance of the first symptoms is called the **incubation period**. These toxins may be *exotoxins* which are liberated from the bacteria into the body or *endotoxins* which remain in the bacteria until the latter disintegrate.

HOW THE BODY REACTS TO INFECTION

When pathogens enter the blood certain general symptoms more or less common to all develop, such as a general feeling of being "off-colour", loss of appetitite and high temperature which we call *fever*. In addition symptoms arise which are peculiar to each particular disease and caused by the toxins produced by the specific pathogens concerned. Examples of these are spots, rashes, localised pain and so on and it is these specific symptoms which make a doctor's diagnosis possible. Some diseases follow a definite course which, once diagnosis has been made, can be predicted with some certainty.

Once the pathogens have become established the toxins cause the body's defences to come into action and this is done in various ways. **Antigens** effect the production of **antibodies** which are specific to each pathogen. These are proteins and they are as follows:—

Antitoxins which neutralise the toxins so that they cease to take effect.

Agglutinins which cause bacteria to clump together, thus effectively preventing their activity.

Opsonins which affect the pathogens and prepare them for attack by the leucocytes (see below).

Bacteriocides which kill the bacteria.

Interferon is a protein produced in cells infected by viruses, the presence of which causes its production by the cells. It inhibits further viral replication in the cells. Our knowledge of interferon is still in its elementary stages but research into its possible production and therapeutic use continues.

In addition certain leucocytes ingest and destroy pathogenic bacteria. This is known as **phagocytosis** and the white corpuscles are then known as **phagocytes**. If the bacteria are extremely virulent this may not come about, the phagocytes appearing to be repelled and the bacteria are able to continue unhampered. The phagocytes are often killed when infection is kept localised and a thick, whitish, yellowish or green fluid consisting of the dead phagocytes together with dead tissue, is formed. This is **pus**. When the infection is kept localised and sealed off, *inflammation* sets in and if the sealing is complete an **abscess** containing pus develops. This can occur almost anywhere in the body, for example on the root of a tooth or in the intestine. Minute abscesses are known as **pustules**. Though it may cause a great deal of pain, the formation of an abscess is a good sign because it indicates that the infection is being kept localised and dealt with. The alternative is entry of the pathogens into the blood stream which causes more general infection known as blood-poisoning or **septicaemia**. When toxins from an abscess pass into the blood, the blood poisoning is referred to as **toxaemia**. The infection may be overcome, of course, by the antibodies in the blood.

The power pathogens have in producing disease is known as their **virulence**. This depends on their ability to enter the cells or body fluids and once there to multiply. For example *B. typhosus* can enter only through the wall of the alimentary tract. This virulence can vary in the same pathogen and it is thought that rapid transfer from host to host increases it and that it is probably the result of mutations. Because of this, the virulence of particular pathogens often varies over a period of years.

In disease, once the infection has been overcome, **recovery** takes place but if the body defences are unsuccessful in their war on the invaders death may ensue.

IMMUNITY

Man is immune to many diseases which affect the lower animals and this is called **natural immunity**. As the result of an infection he may obtain immunity from a particular disease in a subsequent invasion by its pathogens because the capability of providing the specific antibodies remain in his blood. This is **naturally acquired immunity** and it lasts for a certain length of time. Often a child who has had measles, for example, does not contract it on a future occasion when other children with whom he comes in contact are infectious. **Inherited immunity** is sometimes obtained by a baby from its mother though this lasts for only a few months. But immunity can also be created by artificial means. This **artificially acquired immunity** is of two kinds. In one, **active immunity**, the body produces its own antibodies and this is brought about by injection of what are called **vaccines** composed of either the dead disease organisms or attenuated (weakened) forms of them. This method which is used, for example, in anti-typhoid and anti-diphtheria inoculation is slow in developing but long lasting and is therefore suitable as a preventative measure if a person is liable to be exposed to the risk of infection. In **passive immunity** the body does not itself produce antibodies. The serum of an animal containing them because it has been injected with pathogens or toxins thereby producing the antibodies, is injected into the person concerned. The advantage of this method is that it is quick-acting and is therefore a suitable one when a person has already been exposed to infection but the immunity is short-lasting. However there is normally adequate time for the body to produce its own antibodies to continue the immunity. It is used in the case of diphtheria and, as is often the case the horse is the animal used to supply the serum. This is because horses are not liable to contract many diseases which affect the human subject and high concentrations of antibodies can be obtained by injecting them with large doses of pathogens or toxins. These toxins are neutralised to make them non-toxic and they are then known as **toxoids**.

PREVENTION OF INFECTION

Protection against infection by various methods of prevention is called **prophylaxis**. It is clearly the duty of every infected person and of anyone who has been in contact with infection to take all possible measures to avoid passing it on to other members of the community. For this reason infected cases must be **isolated** and contacts with infection must avoid coming into contact with other healthy people for a certain period of time known as **quarantine** until the risk of infection is over. Quarantine is the maximum incubation time for a particular pathogen and varies with different diseases. As a general measure the avoidance of overcrowding and adequate ventilation go a long way in preventing the spread of communicable diseases.

Certain infectious diseases such as smallpox, diphtheria, scarlet fever, typhoid fever and poliomyelitis are **notifiable diseases**. This means that a medical practitioner must by law inform the Community Physician (formerly known as the Medical Officer of Health) in his District immediately he diagnoses or suspects a case. Measures can then be taken to trace the cause and prevent the spread of the disease.

Immunisation by inoculation with appropriate vaccines is another way in which members of the community can be protected against infection. This method may be considered to have arisen in the first instance from the practice of vaccination against smallpox and this was over fifty years before Pasteur discovered the germ origin of disease. It all began with a Gloucestershire doctor named **Edward Jenner** who was born in 1748 and who died a year after Pasteur was born.

Variola or **smallpox** was a very grave and prevalent disease affecting large numbers of the community in Jenner's time and he observed that milkers on farms who contracted **vaccinia** or **cowpox**, a similar but much milder disease affecting cattle and to which human beings are subject, rarely became infected with smallpox or, if they were, they suffered very mildly when there were outbreaks of the disease. He noticed that sores similar to those on the udders of cows with cowpox appeared on the hands of the milkers and he suspected that cowpox provided protection against smallpox. In 1796, with considerable risk, he took a very bold step to test the idea. He took some of the discharge from a cowpox sore on the arm of a milkmaid and transferred it into the arm of a young boy. This produced a sore on the boy's arm.

A few weeks later Jenner injected him with the discharge from a smallpox sore and the boy did not contract the disease. Several repetitions with others producing equally successful results convinced this country doctor that he had found a method of immunisation against smallpox. It is known as **vaccination** from the Latin *vacca* (= cow) but the term vaccine is now used for injections used for immunisation against other diseases. After the considerable opposition to vaccination had died down it became widely practised not only in this country but abroad and the disease was almost wiped out here by 1880.

Figures from Sweden are interesting. Deaths from smallpox in that country in 1774 were 2,049 per million. Compulsory vaccination was introduced in 1816 and in the years 1812–1821 the death-rate from this disease had been reduced to 133. By the time compulsory vaccination had had a chance to take full effect the number was still further reduced to 1 per million. In 1898 vaccination was made compulsory in this country but the law was rescinded when the National Health Service started as cases were so very rare and these were attributable to overseas travellers. Almost 80 per cent. of cases in the world occur in Asia, mainly in India, Indonesia and Pakistan. Some countries, our own included, insist on vaccination of visitors from certain countries abroad before they are allowed to enter the country. If anyone comes in contact with a suspected or confirmed case of smallpox or a carrier or if there is an outbreak of the disease in their district, common sense dictates that he should immediately seek vaccination.

Later Jenner found that fluid from the cowpox sore on the arm of one human being was as effective as that from a cow and so he used this method. It is now known, of course, that Jenner's success was due to the development of antibodies in the blood of his patients produced as the result of cowpox antigens and the use of calf lymph vaccine has replaced this method of vaccination.

The term **vaccine** is used generally for immunising inoculation material which is prepared from pure cultures of bacteria and stocks of these vaccines are kept. Injection is either intravenous or intramuscular and, with the possible exception of the polio vaccine, oral vaccines have not proved successful. In the case of this disease it is usual to give intramuscular injections of the *Salk vaccine* or oral doses of the *Sabin vaccine* as they are called (see below). A jet-injector is now used when large numbers have to be vaccinated, as in epidemics of cholera and smallpox. Using this instrument it is possible to vaccinate up to 1,000 persons an hour.

Susceptibility to diphtheria can be ascertained by the *Schick test* in which a minute dose of the toxin is injected into the skin.

If the person is susceptible to the infection, local inflammation occurs where the injection was made whereas those with resistance to it show no reaction. The former should be given an injection of anti-diphtheritic serum. Children are more liable than adults to contract the disease though, as a result of mass immunisation, it has been almost eradicated.

As some diseases can be contracted from food and water both must be protected from source to consumer and the methods used for food have already been given in Chapter XXIV. Water will be considered in a later chapter. Meanwhile it should be stated that any water under suspicion should be avoided if possible; otherwise it should be boiled before use.

As far as protection from droplet infection is concerned, avoidance of overcrowding and proper and sufficient ventilation are extremely important. It should be unnecessary to add that it is essential to avoid infected droplets getting into the air when coughing or sneezing by the use of a handkerchief. Nor should it be necessary to mention that, apart from its being a disgusting and very objectionable habit, the act of spitting can be a serious source of droplet infection as the dried sputum gets into dust and thus spreads disease.

Bites of insects and other animals as well as cuts and grazes should always be washed with a suitable disinfectant to kill any infective bacteria which could get in.

In conclusion, it should always be remembered that **prophylaxis** (prevention of disease) is better than having to employ **therapeutic** (curative) treatment, and a great deal can be done to attain this end by the maintenance of a healthy life in healthy surroundings with plenty of fresh air and an adequate and suitable diet. One should always breathe through the nose and not the mouth as infective germs are trapped in the nose and so cannot gain entry into the bronchial passages. In 1973–4 over 319,000,000 working days were lost due to illness, the largest single cause being bronchitis.

CHARACTERISTICS OF A FEW COMMON DISEASES CAUSED BY MICRO-ORGANISMS

Several of these have already been described at the end of the chapters dealing with the various systems of the body. The notes which follow give further details.

General characteristics, such as fever due to high temperature, headache, shivering, and in some cases rash, which are common to most diseases have already been mentioned.

Chickenpox (Varicella)

A mild disease, mostly affecting children. *Cause:* a virus. *Mode of Infection:* contact. *Incubation Period:* 17–20 days. *Infective Period:* about 14 days; until last scab falls off. *Quarantine Period:* 20 days. *Symptoms:* rash of pimples which become small blisters forming scabs. These fall off.

Diphtheria

A serious disease affecting children which may lead to complications. *Cause: B. diphtheriae (Corynebacterium diphtheriae). Mode of Infection:* contact, carriers. *Incubation Period:* 2–3 days. *Infective Period:* up to 14 days. *Quarantine Period:* 7 days. *Symptoms:* sore throat, loss of appetite, headache. Toxins can cause paralysis of nerves (post-diphtheritic neuritis) and the palate and heart may be affected. Injection of anti-toxins as early as possible is essential to prevent this. Identification of the disease is by taking a swab of the throat which is examined bacteriologically. The *Schick Test* (pp. 341 seq.) determines susceptibility and all susceptible cases should be immunised by immediate injection of a toxoid, anti-diphtheritic vaccine. As a result of mass immunisation of children introduced in 1960, epidemics are now rare. In 1940 the number of cases was 45,000 with 2,400 deaths but in 1966 only 20 cases were notified and in 1974 no deaths were recorded.

German Measles (Rubella)

A mild infection of children. *Cause:* a virus. *Mode of Infection:* droplet by contact. *Incubation Period:* 14–21 days. *Infective Period:* 7 days. *Quarantine Period:* 7 days. *Symptoms:* pink rash, swollen glands. If it occurs during pregnancy, damage of the foetus may occur.

Measles (Morbilla)

A very infectious disease of children which may lead to complications. *Cause:* a virus. *Mode of Infection:* contact. *Incubation Period:* 14 days after the period of rash. *Quarantine Period:* 16 days. *Symptoms:* cold in the head, cough, redness of eyes, dark rash. Epidemics occur and about 250,000 children have been known to contract it in a year. In 1961 the number was 750,000 and of these 152 died. Deaths in 1962 dropped to 39 but went up again in the following year to 127, dropping again in 1974 to 20 out of 109,567 cases which were notified. The virulence of the virus may vary. A successful immunising vaccine has recently been introduced which can be given to babies in the second year and to children up to the age of fifteen and it is hoped that this will reduce considerably the incidence of this highly infectious disease. A second attack of the disease is rare, naturally acquired immunity being the result of the first infection.

Mumps (Epidemic Parotitis)

A disease affecting children and occasionally adults. *Cause:* a virus. *Mode of Infection:* contact. *Incubation Period:* 21 days. *Infective Period:* until swelling of the glands subsides, usually about 21 days. *Quarantine Period:* 25 days. *Symptoms:* painful swelling of the parotid glands and this may spread to other salivary glands. In males the testes may be similarly affected and this in an adult may give rise to sterility.

Poliomyelitis

A very serious disease affecting children and young adults. *Cause:* a virus. (Fig. 25.7, p. 333.) *Mode of Infection:* contact or carriers. *Incubation Period:* normally 7–14 days but can be up to 3 weeks. *Infective Period:* 6 weeks. *Quarantine Period:* 21 days. *Symptoms:* fever, headache, vomiting and general irritability. Later, pain in muscles and paralysis may or may not occur. This is the result of infection of the motor neurons in the brain and spinal cord and may affect the limbs. If paralysis of the respiratory muscles occurs, it is essential that the patient be put into an *iron lung* to enable respiration to continue. Two different types of vaccine are in use. In one, the *Salk vaccine*, non-living virus material is used and the other, the *Sabin vaccine*, contains attenuated living virus. Both are most effective and the introduction of these vaccines caused a drop in the number of cases from 32,000 in 1956 to 23 in 1966 and to 5 in 1974. Two cases were recorded in Wales in 1975.

Scarlet Fever (Scarlatina)

This affects children mostly but adults can contract the disease which can lead to complications. *Cause:* a Streptococcus. *Mode of Infection:* contact. *Incubation Period:* 3 days. *Infective Period:* 28 days. *Quarantine Period:* 7 days. *Symptoms:* sore throat, vomiting and headache followed after about a couple of days by a scarlet rash which first appears on the chest and later spreads over the entire body. When it disappears after about 7 days the skin peels. The severity of the virulence has been very much reduced in recent years, the number of cases falling from well over 38,000 to a little over 23,000 in a period of ten years recently. But virulence can vary and there is always the danger of a more virulent form reappearing. 10,402 cases were notified in 1974.

Whooping Cough (Pertussis)

A disease of children which is particularly serious in infants. *Cause:* a Bacillus. *Mode of Infection:* droplet by contact but can be by carriers. *Incubation Period:* 10–14 days. *Infective Period:*

4–6 weeks. *Quarantine Period:* 14 days. *Symptoms:* cough which becomes spasmodic and paroxysmal and ends in a long inspiration which is the "whoop". There are effective vaccines for relieving the symptoms and for immunisation and the number of cases dropped by over 84,000 between 1956 and 1966. In 1974, 16,225 cases were notified.

The Common Cold

Cause: a large number of viruses. *Mode of Infection:* droplet. Chills and general dampness in Winter favour development. *Symptoms:* well-known *viz.* sneezing, watery discharge from nose, followed by yellow discharge, headache and general listlessness. Methods of immunisation so far unseccessful as there are so many viruses concerned. The disease spreads rapidly in schools, offices, trains, 'buses, theatres, cinemas, factories and, in fact, wherever large numbers of people are in contact. Symptoms can be eased by staying in bed for a day or two but so far there is no known cure.

Influenza

An extremely infective disease affecting all ages which often leads to complications. It occurs in epidemics. *Cause:* a virus. *Mode of Infection:* droplet by contact. *Incubation Period:* 1–3 days. *Infective Period:* up to 7 days. *Symptoms:* cough, sneezing, high temperature, headache, congestion and a general feeling of prostration. No really effective vaccine, prophylactic or immunising, has yet been discovered. Deaths in 1957 were 6,716, in 1960 3,672 and in 1974, they were 1,235. Immediately after the First World War a particularly virulent form of the disease which became pandemic broke out causing some 18,000,000 deaths. An enormous number of man-hours are lost in industry every year from influenza.

Typhoid Fever (Enteric Fever)

Cause: B. typhosus (Salmonella typhi). Mode of Infection: food or water, carriers. *Incubation Period:* 12–14 days. *Infective Period:* until bacteriological examination shows absence of the bacilli. *Symptoms:* headache, nose-bleeding, high temperature particularly at night, abdominal pains, diarrhoea or constipation, pink eruption on chest and abdomen. Effective vaccine known as T.A.B., which also immunises against the milder paratyphoid fever. Three epidemics have been described under droplet infection on pp. 335 seq. Only two deaths occurred in 1974.

Smallpox (Variola)

An extremely infectious and serious disease causing epidemics.

Cause: a virus. *Mode of Infection:* contact, droplet and infected articles. *Incubation Period:* 14 days. *Infective Period:* until all scabs have healed, usually about 4 weeks. *Quarantine Period:* 16 days. *Symptoms:* headache, vomiting, fever. Rash of small red pimples develops and these become fluid-filled pustules (these are the "pocks"). They dry up and become scabs. Preventive vaccination is essential in any area in which an epidemic occurs. Vaccination within 3 or 4 days after exposure to contacts confers immunity. Vaccination has already been discussed earlier in the chapter.

Tuberculosis

Cause: B. tuberculosis. (Fig. 25.5, p. 330.) *Mode of Infection:* contact with a case or use of articles used by them, dried sputum from an infected person in dust. Milk from cows suffering from Bovine tuberculosis affects mostly children. *Incubation Period:* 3–8 weeks. *Infective Period:* as long as the bacillus is found in the sputum. *Symptoms:* tubercules develop in the lungs in **pulmonary tuberculosis**, particularly at their apices, and this results in their becoming more solid. There is a dry cough and great weakness develops. Mass X-rays have enabled cases to be diagnosed in their early stages and treatment can then be given. Cure can be effected. The vaccine contains attenuated bacteria and is known as the B.C.G. vaccine, so-called from bacillus and the first letters of the surnames of the two French bacteriologists who first produced it. The number of notified cases fell by over thirteen and a half thousand in the ten years succeeding 1956 and deaths by around 50 per cent. over the same period, from over 4,000 to 2,000 and these figures are small compared with those at the beginning of the century.

Other parts of the body can be infected apart from the lungs such as bones, joints and glands. In 1974, 10,687 cases were notified of which 2,577 were other than respiratory. A survey carried out by the Medical Research Council has shown that the risk of tuberculosis infection is halving every five years.

Pneumonia

There are various types of pneumonia. **Pneumococcal pneumonia** is a serious disease affecting the lungs. *Cause: Pneumococcus. Mode of Infection:* droplet. *Incubation Period:* may be extremely rapid or 2–6 days. *Symptoms:* pain in chest, shivering, high temperature, cough, high respiration rate and shallow breathing, thick brown sputum coughed up. If one lung only is affected it is called **lobar pneumonia** but if both lungs are infected in patches it is known as **bronchial pneumonia**. Abscesses may develop in the lungs and heart failure may occur. Antibiotics administered in time will effect a cure. Pneumonia may also be caused by viruses.

Lockjaw (Tetanus)

Cause: B tetanus (Clostridium tetani), (Fig. 25.5, p. 330.) *Mode of Infection:* through wounds. *Incubation Period:* 3–21 days. *Symptoms:* stiffness of jaw muscles making the opening of the mouth difficult or impossible. This may spread to the muscles of the back and abdomen. Treatment by antitoxin possible but difficult and immunisation by injection of toxoid is essential in all cases where there is risk of infection.

Meningitis (Cerebro-spinal meningitis)

Cause: Meningococcus. Mode of Infection: contact or droplet. *Symptoms:* headache, stiff neck, rash of red spots. Formerly the death-rate was high but the use of antibiotics now brings about a cure in most cases.

Cholera

Cause: Vibrio cholerae. (Fig. 25.5, p. 330.) *Mode of Infection:* water, food, contact. *Incubation Period:* 2–6 days. *Symptoms:* sudden severe diarrhoea of watery fluid containing mucus causing dehydration of the body, vomiting, cramp. Can be fatal in two days. A disease of the Far East but cases have been reported in Europe, the disease having been contracted by contact infection. Immunisation by injection of vaccine containing dead vibrios.

Rabies

Cause: a virus. *Mode of Infection:* bite or lick of a dog or other animal suffering from the disease. *Incubation Period:* 1 to 2 weeks or even several months. The disease is always fatal to human beings but prevention of its development after being bitten by a rabid dog or from other source of infection is successful if immunising vaccine is given before the symptoms develop. Unfortunately rabies infected dogs do not always reveal the symptoms for several months. *Symptoms:* extremely painful spasm in throat when trying to swallow causing fear of drinking. Hence alternative name of *hydrophobia* (fear of water). (Two people died in this country in 1975 from rabies contracted when abroad and three died in Spain, one man from the bite of a rabid dog and two women from scratches by a rabid cat.)

Veneral Diseases (S.T.D.)

These have already been dealt with in some detail at the end of Chapter XVII (pp. 197–200).

STERILISATION, ANTISEPTICS AND DISINFECTANTS

Sterilisation is the process whereby anything is made germ-free. This can be brought about by antiseptics, disinfectants, dry heat, boiling, steam at atmospheric pressure and steam under increased pressure.

Antiseptics and Disinfectants

Strictly speaking an **antiseptic** is a substance which inhibits any further multiplication of infective bacteria and a **disinfectant** is a substance or process which kills infective organisms. The two terms are often confused but many disinfectants are also antiseptics; fresh air and direct sunlight also serve as disinfectants.

One of the first chemical disinfectants or **germicides** to be used was *carbolic acid* or *phenol* (C_6H_5OH) and this was in the treatment of sewage. Until the second half of the 19th century surgical operations were always beset with the danger of the wounds becoming *septic,* a condition known as *hospital gangrene,* and amputations were invariably fatal. This caused many surgeons to avoid performing operations which today are commonplace. Hospitals were often unpleasant and dirty places and were usually overcrowded. Surgeons kept old clothes for surgical operations when they were no longer fit for ordinary wear, just as we now keep them for gardening or cleaning the car. It was no wonder that infection was rife. But a contemporary of Pasteur put an end to all this with the introduction of antiseptic surgery. He was **Joseph Lister** who was born in Essex in 1827. After serving as a House Surgeon at Edinburgh Royal Infirmary he was appointed Professor of Surgery at Glasgow University in 1860.

Learning of Pasteur's discoveries, he realised the similarity between putrefaction and wound sepsis and considered that microorganisms could be the cause of the latter. He therefore devised a method of killing them without harm being done to the tissues of the patient and the substance he used was carbolic acid. This was also used to sterilise the instruments and the hands of the surgeons and nurses. The experiment resulted in complete success and in 1867 the first records of his **antiseptic surgery** were published. Unfortunately carbolic acid in high concentration is harmful to tissues so Lister turned to weaker solutions which he found to be equally efficacious. Antiseptic surgery in which the germs were killed to avoid infection was later replaced by **aseptic surgery** in which they were killed beforehand and everything which could possibly infect the wound was sterilised and made germ-free. This method is that which is still in use to this day. Every instrument,

FIG. 25.8. **A modern operating theatre. The patient is anaesthetised.**
(By Courtesy of the World Health Organisation, Geneva.)

swab, dressing and bandage is sterilised and the surgeons and their various assistants wear sterilised gloves, overalls, caps and boots and masks over their mouths and nostrils. The skin around the area of incision is painted with antiseptic after any hair has been removed. What chance have bacteria of causing infection during surgery under these conditions?

Lister made himself unpopular by insisting on absolute cleanliness throughout the hospital. Anything which could harbour dirt had to be removed and walls, floors and furniture scrubbed with carbolic acid. He realised that dirt harboured germs and germs caused disease and he was indeed right as the result of his alleged fanaticism proved without question. Needless to say there was much opposition to Lister's methods and demands, as is always the case when something new is discovered and put into practice, but this was eventually overcome. He died in 1912 after receiving a peerage and the O.M.

In addition to carbolic acid, chemical antiseptics and disinfectants in common use are *cresol, lysol, dettol* and various others under proprietary names all of which are effective in very dilute solution (about 1 per cent.). *Mercuric chloride (corrosive sublimate,* $HgCl_2$) is a very powerful one and can be used in as low a concentration is 0·1 per cent. but it cannot be used with metal instruments. *Mercury biniodide* is used for sterilising gauze, lint and cat-gut, *iodoform* is sometimes used for dressings, *silver nitrate* finds a use in eye lotion, *hydrogen peroxide* is used for cleansing wounds and abscesses and as an antiseptic mouthwash and *surgical spirit* for cleaning the skin. After thorough washing and scrubbing, the hands can be sterilised by insertion in a solution of *potassium permanganate. Formalin,* a 40 per cent. solution of the gas *formaldehyde* (H.CHO), can be used in various concentrations for many purposes.

Rooms in which there have been cases of air-borne infectious diseases can be sprayed with a 5 per cent. solution of formalin and the gas can be liberated by heating *paraformaldehyde,* which is a solid, or by adding formalin to potassium permanganate. *Sulphur dioxide* is sometimes used and this is obtained by heating what is called a sulphur candle. The other disinfectants referred to above are suitable for tables, floors, toilet seats, towels, sheets, handkerchiefs and so on or a solution of formalin can be used. It is important that excreta in some infectious diseases such as typhoid fever should be treated before being disposed of and formalin is a most suitable disinfectant for this.

Dry Heat

This is one of the best methods when it can be used. It is

done directly or in a hot air oven at a temperature of 150°C for an hour. This method is suitable for glassware such as petri dishes and bacteriological test-tubes and various instruments but is unsuitable for dressings, swabs, cotton wool and delicate materials which would be charred at this temperature. Books used by infected persons can be treated in this way though it causes damage to them and, unless they are of value, they are best destroyed.

Boiling Water

If this method is used it must be for a prolonged period and for some purposes it must be repeated after an interval of a day as spores are resistant to it and time must be allowed for them to develop into bacteria which are then killed by the second boiling. In fact, boiling on a third day will ensure the material's being absolutely sterile. This method is suitable for sputum pots, crockery, knives, forks and spoons and similar articles.

Steam at Atmospheric Pressure

Various liquids and culture media can be sterilised in glass vessels placed in an atmosphere of steam in a special piece of apparatus appropriately known as a **steamer** in which steam from boiling water is retained. There must be an outlet at the top to allow steam to escape for reasons of safety.

Steam at Increased Pressure

On a small scale this could be done in an ordinary pressure-cooker but in hospitals and laboratories a much larger closed boiler called an **autoclave** is used. The steam being under pressure is at a much higher temperature and sterilisation is effected more quickly. It is used in hospitals for the sterilisation of surgical instruments, dressings etc. and is the most effective and useful method.

DISINFESTATION

Animal parasites and vectors such as bed bugs, lice and mosquitoes which infest or attack the body are best killed by a toxic agent such as **D.D.T.*** † This process is called **disinfestation** and the substances used are known as **insecticides**. The use of D.D.T. for this purpose was first discovered during the last war when it was found to be effective in preventing the spread of typhus and it is widely used in killing mosquitoes which carry the malarial parasite. It is toxic to domestic animals and care is needed to be exercised in its use. Many new strains of insects have arisen which are D.D.T. resistant and in these cases other insecticides must be used. Animals

*But see p. 404. † Dichloro-diphenyl-trichloroethane.

such as rats can be killed by such gaseous poisons as **hydrocyanic acid (prussic acid** HCN) and **chlorine** but as they are extremely poisonous and dangerous to use, extreme care is necessary.

PRACTICAL WORK
PATHOGENIC BACTERIA

Examine prepared stained microscopical preparations of **pathogenic bacteria** under the 2 mm. O.I. objective as follows:—

First see that the coverslip on the slide is clean and free from fingermarks. Use a microscope lamp and adjust the mirror and sub-stage condenser to give maximum illumination. Then put a drop of immersion oil on the coverslip. Keep the body-tube in a vertical position and place the slide on the stage. Carefully lower the objective into the oil and by gentle slow focussing with the fine adjustment the bacteria will be brought into view.

Suitable specimens for examination are:—
B. tuberculosis, B. anthracis, B. (*Corynebacterium*) *diphtheriae, B. tetanus, Diplococcus pneumoniae, Staphylococcus pyogenes, Streptococcus pyogenes, Treponema pallidum, Vibrio cholerae.* Look for endospores in *B. anthracis* and for exospores in *B. tetanus.*

To Prepare Cultures of Bacteria

Sterilise some Petri dishes by heating in an oven at 150°–200°C for half an hour or boil in water for an hour. Melt some nutrient agar by immersing the tube or phial in hot water, first loosening the cap of the phial. Then pour the contents into a sterile Petri dish and quickly replace the cover. This will serve as a *control* for the subsequent experiments. It will remain sterile as long as the cover is left on. No colonies should appear in this dish. The inoculated agar in the dishes in the experiments which follow should be left for a few days at room temperature and then examined for bacterial colonies.

Bacteria in Air

Remove the cover from a dish containing nutrient agar. Leave the agar exposed to the air for twenty minutes to half an hour and then replace the cover. Distinguish any fungal growths from bacterial colonies when examining a few days later.

Bacteria in River or Pond Water

1. Sterilise a small pipette by boiling in water for $\frac{1}{4}$ hour and then, by means of it, add a few drops of river (or pond) water to some melted nutrient agar. Pour the inoculated agar into a Petri dish and examine later;

2. Repeat the above experiment with a sample of the same water which has been boiled for $\frac{1}{4}$ hour and compare the result with that from the unboiled specimen.

Bacteria in Soil and Manure

Sterilise a pair of forceps by heating the tips in the bunsen flame. Then pick up a few small particles of soil and add them to agar in a Petri dish. Sterilise the forceps again and pick up a few particles of manure. Add them to agar in another Petri dish. Examine both dishes later.

Bacteria in Milk

Using the method described above for water test for the presence of bacteria in (i) Pasteurised milk (ii) unpasteurised milk (if available) (iii) milk which has been allowed to stand exposed to the air for a few days (iv) milk which has been boiled once and (v) milk which has been boiled on three consecutive days (this should ensure that any spores or bacteria from spores have been killed). Compare results.

Bacteria on the Skin

Pour some nutrient agar into a Petri dish, replace the cover and allow it to set. Then remove the cover and rub your finger over the surface and quickly replace the cover.

Bacteria in the Mouth

Gently scrape the surface of your teeth with a small knife or scalpel which has been sterilised by boiling in water for $\frac{1}{4}$ hour and transfer the scrapings to a drop of water on a clean slide. Stain with methylene blue, cover and examine with a 2 mm. O.I. objective. Various saprophytic bacteria will be seen.

PATHOGENIC PROTOZOA

The following specimens must be examined under the high power.

The Malarial Parasite, *Plasmodium*

Examine a stained preparation of **plasmodium in the blood of man**. Various stages may be seen. They may appear in an amoeboid form, as a vacuolated signet ring-shaped trophozoite with the nucleus to one side or as a mulberry-like mass of cells within the red blood corpuscles. Again there may be numbers of separate cells (merozoites or schizozoites) in the plasma which have been liberated from corpuscles. Later stages may also be seen, microgametocytes with large nuclei and large megagametocytes with small nuclei. The number of stages seen will vary from slide to slide.

The Sleeping Sickness Parasite, *Trypanosoma*

Examine a slide of **human blood containing trypanosoma**. The parasites are spindle-shaped and will be seen in the plasma. Each has a large nucleus, an undulating membrane along the greater part of its length on one side and a long flagellum.

The Parasite of Amoebic Dysentery, *Entamoeba histolytica*

Examine a microscopical preparation of **Entamoeba histolytica**. Note the irregularly shaped amoeboid organisms, each with a nucleus.

PATHOGENIC FUNGI

The Parasite of Ringworm, *Microsporon audouini*

Examine a slide of **Microsporon audouini** under the high power. Note the thread-like mycelium and round spores of the fungus.

VIRUSES

The optical microscope is not of any help here but the Bioset entitled "The Virus" has some good electron micrographs of different types.

Chapter XXVI

ANTIBIOTICS, DRUGS AND ANAESTHETICS

ANTIBIOTICS

Antibiotics are substances produced by living organisms, usually fungi, which are used to kill pathogenic bacteria (*bacteriocidal*) or to inhibit their growth (*bacteriostatic*). It will be seen that there is no one antibiotic for all infections and it should be noted that some patients are allergic to certain antibiotics and serious side reactions may develop.

In 1929 **Sir Alexander Fleming**, working in the pathological laboratory of St. Mary's Hospital, Paddington in London, noticed that a bluish-green mould was growing amongst bacterial colonies which he was culturing in a Petri dish. Such contamination arose from spores of the fungus entering the dish before it was covered with its lid. But Fleming also observed something which proved to be of extreme importance, namely that in the vicinity of the fungal growth there were no bacteria. This aroused his curiosity and he was eventually able to show that the mould had produced

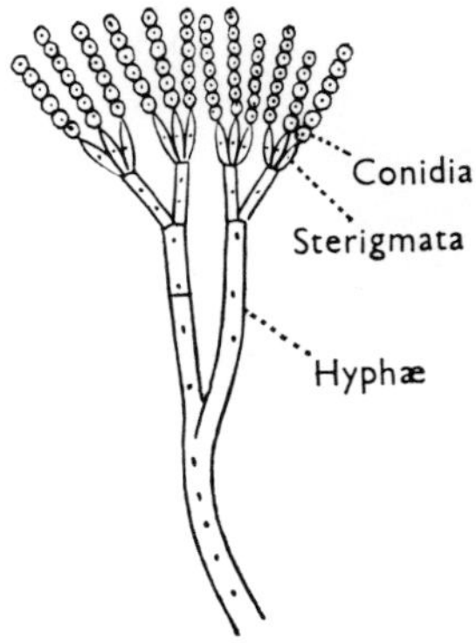

Fig. 26.1. **Penicillium.**
(From Wallis—Practical Biology.)

a substance which inhibited bacterial growth. The mould was *Penicillium notatum* and he called the substance **penicillin**. This was the first antibiotic to be discovered. Isolation of penicillin, however, was left to **Sir Howard Florey** and **Prof. Chain** working at Oxford and others in the U.S.A. some ten years later. It was found to be non-toxic to human beings and animals but toxic to most Gram-positive bacteria such as B. anthracis, B. tetanus, Streptococcus aureus, Staphylococcus pyogenes and the Gram-negative Meningococcus and Pneumococcus. Thereafter penicillin came into use in the treatment of wounds during the 1939–45 war. It is obtainable both from *P. notatum* and *P. chryosogenum*, and it has been synthesised but some people suffer side-effects from its use. For minor infections it can be taken orally; otherwise it is administered by

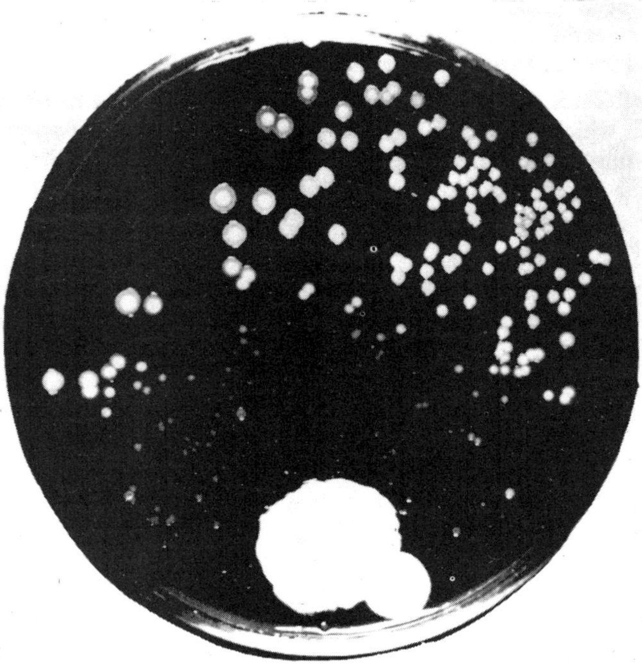

Fig. 26.2. **Sir Alexander Fleming's original culture plate. Showing the inhibition of bacterial growth around the colony of pencillium at the bottom of the photograph.**
(By courtesy of the Wright-Fleming Institute of Microbiology, St. Mary's Hospital, Paddington, London.)

injection. Antibiotic treatment should be given only when absolutely necessary as resistance develops in the patients and it eventually becomes less effective or useless. Unfortunately strains of penicillin-resistant bacteria have appeared as mutations. In fact Fleming came across penicillin-resistant bacteria. D.N.A. passes across in conjugation. While other bacteria are killed off by penicillin the resistant forms multiply and eventually take over.

Several new antibiotics have since been discovered. One of these is **streptomycin** which has proved to be particularly active against *B. tuberculosis.* It also affects *Brucella abortus*, the cause of *contagious abortion* in cattle from the milk of which man can be infected and contract *undulant fever.* Many other Gram-negative bacteria which are resistant to penicillin are killed by streptomycin. Unfortunately prolonged use of it can cause deafness. **Chloramphenicol** is a most effective antibiotic against *B. typhosus*, the typhus Rickettsia, certain viruses and bacteria in the Salmonella group like B. typhosus and those causing food-poisoning together with other Gram-negative bacteria. **Chlorotetracycline** has proved itself of value in venereal disease, against Rickettsia, sittacosis, a virus disease of parrots which man sometimes contracts from them, and various Gram-negative bacteria resistant to penicillin. Further antibiotics which have been discovered are **erythromycin** which is effective against *C. diphtheriae*, gonococcus and staphylococci, **novobiocin** which acts against *gonococcus*, *meningococcus* and *H. influenzae.* Other antibiotics such as **nystatin** and **griseofulvin** are specific against certain fungal infections such as thrush and ringworm respectively. Recently a new antibiotic called **auroxil**, a semi-synthetic penicillin, has been discovered which has proved successful in effecting relief in bronchitis. It can also be used in the treatment of a number of other bacterial infections.

DRUGS

Chemotherapy or treatment of disease by chemical agents or **drugs** can be said to date from 1910 when **Paul Erhlich**, a worker under Robert Koch, discovered an arsenical derivative, *salvarsan*, to be a cure for syphilis. As the discovery was his six hundred and sixth experiment it became known as "606". Since that time a large number of drugs of various kinds have been discovered, and are still being discovered, which have a curative effect in certain diseases. Salvarsan proved to have toxic effects and has now been replaced by penicillin in the treatment of syphilis.

In 1935 a German chemist named **Domagk** discovered that a dye called *prontosil* had a bacteriostatic effect on streptococci. Further research proved that this was due to its being converted into **sulphanilamide**. This is one of a group of drugs known as **sulphonamides**. Many of these *sulphur drugs* such as sulphamerazine, sulphaguanidine and sulphacetamide, have been found to possess therapeutic benefit in certain cases. They have been proved to be active against various streptococci, staphylococci causing septicaemia, meningococcus, pneumococcus and gonococcus. Side effects may be experienced from their administration, however. In some instances their use has been replaced by penicillin and other antibiotics, of which there are several with different specific uses as already seen.

Some drugs such as *opium, cocaine, morphine, atropine, heroin, strychnine, Indian hemp* and *barbiturates* are extremely poisonous and their use is very strictly controlled under the **Misuse of Drugs Act**. Many of them are organic nitrogeneous bases forming salts with acids known as **alkaloids**. They are mostly obtained from various plants, for example morphine and codeine come from the opium poppy, strychnine from the seeds of the nux vomica, cocaine from the leaves of the coca plant and atropine from Belladonna. Caffeine, theine, quinine and *nicotine* are also alkaloids. Quinine is used as a therapeutic measure in malaria though others are now used such as *mepacrine* and *paludrine* which provide immunity from the symptoms of the disease.

Drugs are also used in the relief of pain, *cocaine, morphine, benzamine* and *aspirin* (*acetyl-salicylic acid*) being examples, and these are known as **analgesics**. *Benzedrine* has the opposite effect in that it is stimulating in action. Its use, except under medical advice and control, can be dangerous. In fact, drugs should not be taken except under medical supervision owing to the danger of overdose and the development of the drug habit leading to drug addiction. This is a very serious and unfortunate habit, regrettably on the increase. It brings nothing but misery and depravity and ultimately causes death. **Drug-addiction**, or absolute dependence on drugs, is a habit which develops quickly and which becomes almost or completely impossible to break. Many of these drugs such as *Marijuana* (*cannabis* or *hashish*, also called "pot") is a so-called *soft drug* because it is considered to be non-addictive whereas *LSD* (an abbreviation for lysergic acid diethylamide), *heroin* and *cocaine* on which those who take them become dependent are referred to as *hard drugs*. But it is extremely unwise and dangerous to experiment with the soft drugs as the results produced may so easily tempt the user to try the hard drugs. It should also be noted that *barbiturates*, often prescribed as sleeping pills, can become addictive and *withdrawal symptoms* when the user ceases to take the drug or even lessens the intake are extremely serious, consisting of hallucinations, fits, increased pulse rate and may eventually be fatal. Cannabis is smoked in "reefers" and has a relaxing effect but this can lead to depression and anxiety though there are no withdrawal symptoms. Evidence has been produced which shows that heavy smoking of Marijuana can reduce the amount of testosterone in males to such a low level that, if prolonged, they could become impotent or unfertile. It has also been claimed that there is evidence that it could adversely affect the genes and so possibly have ill effects on future generations and that there is reason to believe that it could lead to cancer. On the other hand LSD, usually taken on a lump of sugar, produces the most odd hallucinations resulting in actions which are sometimes fatal. However, this, too, has no withdrawal symptoms. Heroin, on the other hand, which has to be injected produces a pleasant feeling of relaxation but its continued use makes it quite impossible to lead a normal life and the craving for the drug will lead its addicts to go to any lengths to obtain supplies. The hard drugs eventually bring about depression and delusion and may even lead to violence and complete lack of self control. It is nothing short of tragedy that drug addiction is on the increase and it has become a serious problem which has got to be tackled in a big way. The number of recorded drug addicts in the U.K. in 1974 was 1,980, an increase of 9 per cent. on 1973, and the most common age was 23–26. Fortunately the number under 21 has been declining in the last four years.

ANAESTHETICS

Connected with the matter of disease, since surgical operations have sometimes to be performed, is the subject of **anaesthesia** which will be of interest to students.

The use of modern anaesthetics dates from 1846 when **nitrous oxide** (N_2O) and **ether** ($C_2H_5.O.C_2H_5$) were used in America for the first time in tooth extractions. A year later **Sir James Young Simpson,** an Edinburgh surgeon, discovered the anaesthetic property of **chloroform** ($CHCl_3$) and used it in midwifery. Nowadays nitrous oxide or a mixture of this gas with oxygen provides sufficient anaesthesia for minor operations and has the advantage that the patient recovers quickly from it. But for major operations a **general anaesthetic** providing a much more prolonged state of unconsciousness is necessary. Ether and chloroform were formerly used but nowadays **Halothane**, which is midway in potency between the very powerful chloroform and the weaker ether, is the most popular inhalational anaesthetic used. It is comparatively safe and has largely ousted ether while chloroform is very seldom used. **Trichlorethylene** is used in obstetrics.

Prior to the administration of the general anaesthetic the patient is usually given an intravenous injection of **Pentothal** or **Brietal** to induce rapid loss of consciousness and this may be sufficient for short operations. A new steroid based intravenous drug, **Althesin,** has been introduced which has similar uses and is popular. Anaesthesia is then maintained by nitrous oxide and oxygen, either carrying a percentage of Halothane or analgesia is provided by powerful analgesic drugs such as phenoperidine or pethidine, both administered intravenously.

Should muscular relaxation be required, long-acting non-depolarising drugs such as **Pavulon** or **tubarine** are given intravenously and the patient ventilated artificially until the end of the operation when the muscle relaxants are reversed by **neostigmine**. Short-acting depolarising drugs may be used for tracheal intubation or short procedures.

Spinal anaesthesia is not popular in Britain. **Extra**- or **Epi-Durals**, drugs injected into the epi-dural space just outside the spinal canal, being considered much safer. They are used in obstetrics as these pudenal nerve blocks allow pain-free labour, the mother being conscious and co-operative. Drugs used are **lignocaine** and the recently introduced **marcaine (bupivacaine**) which has a high safety margin.

Chapter XXVII

VECTORS AND OTHER PARASITES AFFECTING MAN

As we have already learned, parasites may be transmitted to man through food, water and other agencies but some require the assistance of another organism or **vector** for this purpose. These may be purely *mechanical vectors* like the house-fly which carries the parasites without taking any part in their life-history or they may be *biological vectors* within which part of the life-cycle of the parasite takes place and which are essential for the completion of the parasite's life-cycle.

RATS

There are two chief varieties of rat, the black rat, *Rattus rattus* and the brown or sewer rat, *R. norvegicus.* These animals have very pointed heads, the black rat having a more pointed snout than the brown rat. The tails also differ. In the black rat it is scaly and slender and at least as long as its body whereas in the brown rat the tail is hairy and is shorter than the body.

These animals not only spread disease, but they contaminate and destroy food. All food should therefore be kept in places or receptacles to which they cannot gain access. Bubonic plague is caused by a bacillus which is carried by a rat flea. The prevention of rodents from boarding or leaving ships in port is mainly responsible for the elimination of this fatal disease from this country and from many others. Circular rat-guards on the ropes and cables of moored ships are a familiar sight in harbours. Another disease with a high mortality rate but which is, fortunately, more or less restricted to workers in damp conditions, is Weil's disease. It is caused by a spirochaete which is excreted in the urine of infected rats. In this disease suffering is serious and consists of severe jaundice, haemorrhage and kidney failure. Rats can be destroyed by the setting of poison or caught and killed by rat-catchers working for sanitary authorities. This is essential as the animals breed extremely rapidly, one pair of rats being able to give rise to 200 offspring in a single year and they can do £10–£15,000,000 worth of damage in the same period.

THE HOUSE FLY

Most insects have two pairs of wings but the house fly (*Musca domestica*) in common with other flies is a two-winged insect, the posterior pair of wings being modified into dumb-bell shaped balancers. Mosquitoes which carry the malarial parasite and tsetse flies which convey the parasite of sleeping sickness are also *Diptera* or two-winged insects.

There are three pairs of legs, covered with hairs and ending in pads. The mouth parts consist of a protrusible tube called a

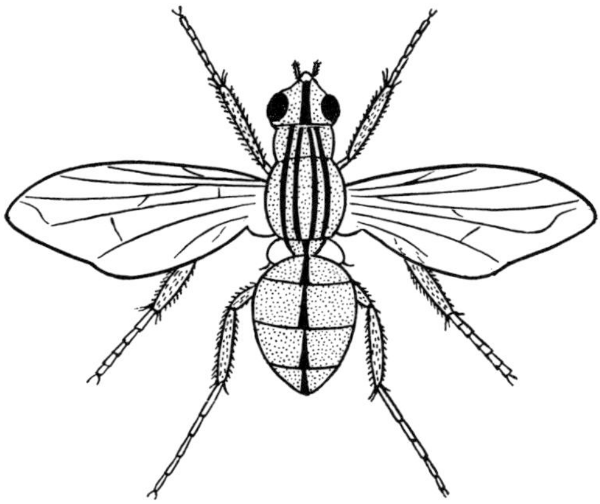

Fig. 27.1. **House fly.**
(*From Wallis—Practical Biology.*)

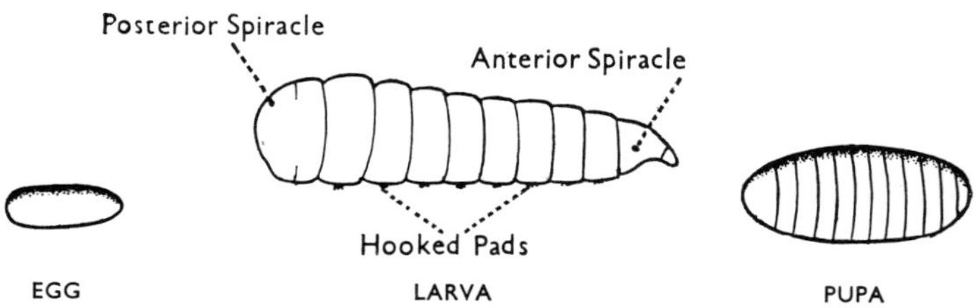

Fig. 27.2. **House fly, egg, larva and pupa.**
(*From Wallis—Practical Biology.*)

proboscis through which the food is sucked after it has been rendered into liquid form by saliva containing enzymes which the fly spreads over it. Regurgitation and defaecation often take place on the food and flies may alight on human food after previously resting on filth of any kind. Flies are therefore mechanical vectors which can transmit diseases such as dysentery, typhoid fever, paratyphoid fever and cholera. The female fly lays batches of small white elongated eggs, each about 1 mm. long, just below the surface of the food by means of an oviposter. In a matter of hours or days, according to temperature conditions, limbless segmented grubs, the larvae, often referred to as *maggots*, hatch out and feed on the organic material in which the eggs were laid. After three moults enabling growth to take place, these develop into brown barrel-shaped resting pupae from which the imago or fly emerges after about a week.

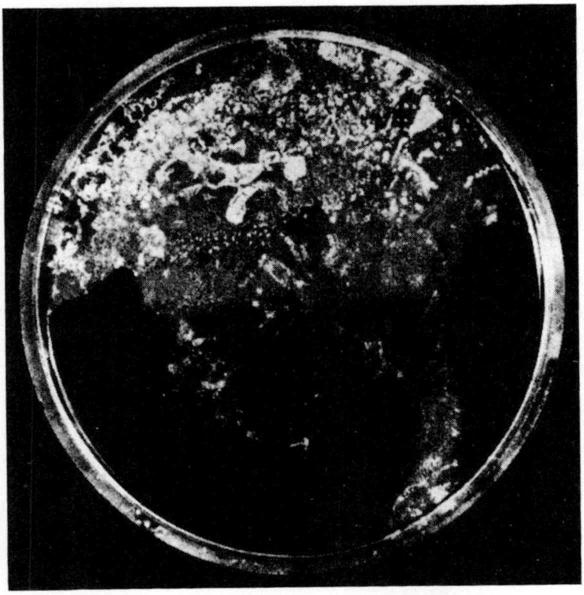

Fig. 27.3. **Bacteria growing on agar plate over which a house fly was allowed to crawl.** *In the author's laboratory.*

Measures must clearly be taken to protect man against the diseases for which this mechanical vector, the house fly, can be responsible. Food must be kept in a larder or refrigerator and in the former it should be kept covered. Milk must always be kept covered. The breeding of flies must be kept down to a minimum by the burning of all refuse as far as possible and by the placing of unburnable waste in properly covered dustbins. Refuse dumps and manure heaps should be kept well away from dwelling houses. It always amazes the author that in many continental countries huge manure heaps are kept right outside farmer's houses and that domestic animals are also kept under the same roof. He remembers driving through a village in Bavaria in which every house in the single street had a manure heap outside. He took a photograph and was asked by one of the inhabitants, with apparent annoyance, what it was he found to photograph in the village! It would be interesting to know what the fly population is in a place like that. Certainly the house fly is a menace to health.

THE MOSQUITO

The ordinary mosquito, *Culex pipiens*, which is found in this country and many others, particularly in the vicinity of lakes, does little harm beyond causing a slight local infection and considerable irritation but the females of the *Anopheline* variety in tropical countries are the carriers of the malarial parasite. It had been known for a long time that malaria occurred in swampy districts—it was prevalent in such areas as the fens in England in the 17th and 18th centuries—and it was thought to be caused by the bad air in these regions; hence the name *malaria.* The actual parasite, *Plasmodium*, was first discovered by **Alphonse Lavaran** in 1880 but it was **Sir Patrick Manson** who suggested that it was carried by mosquitoes and **Sir Ronald Ross** who first demonstrated that part of the parasite's life-history took place in the "stomach" of the mosquito. This was in 1898.

It is only the female *Anopheles* which is responsible, as she needs blood in connection with the production of eggs whereas the male feeds on plant juices. She has long legs, a long slender abdomen, long delicate wings and a long unretractable proboscis. The insect comes to rest on the skin at a sharp angle with the head pointing downwards and the abdomen upwards whereas the

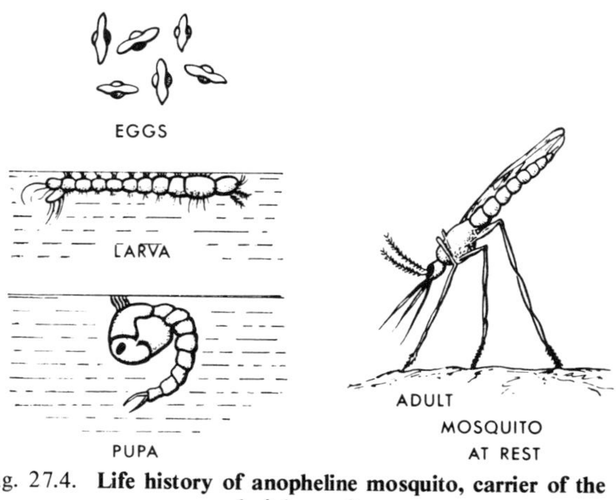

Fig. 27.4. **Life history of anopheline mosquito, carrier of the malarial parasite.**

gnat settles with the body more or less parallel with the skin when at rest. The tiny eggs are laid on the surface of stagnant water and the larvae hatch out in two or three days. These are active and wriggle about in the water. As they are air-breathers, they frequently come to the surface. There are several moults and after ten days or so the pupa is formed. This is also aquatic and air-breathing and is shaped rather like a large comma, Unlike most pupae it is motile though it does not feed. After a further four or five days the imago emerges and as soon as its wings are dry it flies away. The time for this complete metamorphosis varies according to climate conditions.

If the female mosquito "bites" a human being (it really pierces the skin with its sharp proboscis), it sucks blood. Should the human subject already be infected with Plasmodium in the gametocyte stage, this form of the parasite will, on entering the vector, develop into gametes and the sexual cycle of its life-history takes place. The sporozoites into which it develops reach its salivary glands and these will be injected into the next victim which she "bites". The asexual cycle then begins again and fever occurs as the result of this, the red corpuscles being broken down in the process. This is malaria.

Clearly a war on these vectors is a war on malaria and measures can be taken to reduce the mosquito population. Drainage of stagnant water or, if this is impossible, the pouring of oil on its surface will prevent the larvae and pupae from breathing and thus kill the organisms. Alternatively the introduction of fish which eat the aquatic stages will have a similar effect. The spraying of breeding grounds with D.D.T. and the use of insecticides on clothes have also proved effective.The use of mosquito nets is necessary to protect the human subject against being "bitten" at night when the insects are particularly active. Drugs such as quinine, mepacrine and paludrine can be used as prophylactics and of these, paludrine kills the parasite in all stages except when the parasite is in the liver sinusoids. On infection the parasite first invades this organ before entering the general circulation.

THE LOUSE

This insect is an *ectoparasite* living on the body of man, piercing the skin, sucking blood and causing considerable irritation. *Pediculus humanus capitis*, the head louse, lodges on the hair of the head, *Pediculus humanus corporis*, the body louse, on the body and clothing and the minute *Pediculus humanus pubis* on hair in the pubic region. This may be acquired as a result of sexual intercourse, though not necessarily so. The head louse is about 3 cm. long, somewhat oval in shape, flattened dorso-ventrally, dark in colour and wingless. The three pairs of legs are provided with claws for

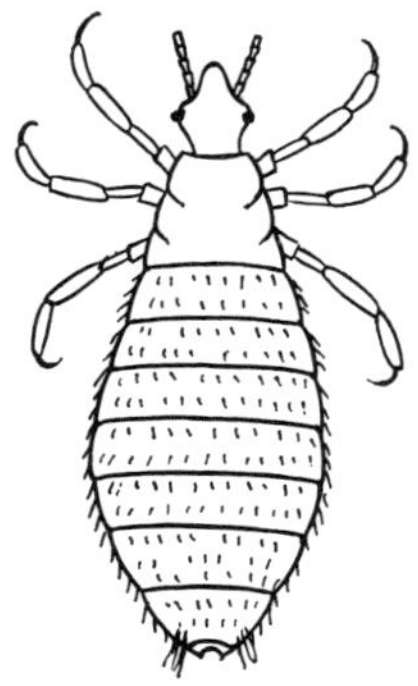

Fig. 27.5. **Head louse.**
(From Wallis—Practical Biology.)

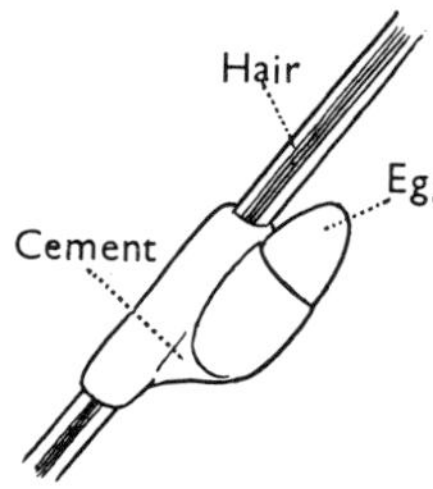

Fig. 27.6. **Egg of louse on hair.**
(From Wallis—Practical Biology.)

attachment to the hair and the proboscis is retractable. The female lays small, ovoid yellowish-white eggs, referred to as "nits" which are joined to the hair by cement. The larva which hatches out after about a week is a *nymph*, a miniature form of the adult. After three or four moults covering about as many weeks during which growth takes place, the adult is formed. There is no pupal stage.

As these insects thrive in dirt they can be the vectors of diseases such as typhus. Infection is by contact and once installed lice are tenaceous visitors which seldom leave their hosts. There are many special preparations for delousing which kill eggs and adult lice but the eggs must be removed by means of a fine tooth-comb or by cutting off the hair. D.D.T. is effective against lice, particularly on clothing, Absolute cleanliness and attention to personal hygiene is the obvious answer to the spread of this infection.

THE FLEA

The species of flea which attacks man is not a vector of disease but considerable irritation is experienced as the result of its visits. The insect pierces the skin and sucks blood. Other species such as the rat flea will, however, be dangerous if it carries the bacillus of bubonic plague of which it is the vector.

The human flea, *Pulex irritans*, is pale brown and its body wall, which is compressed laterally, is of tough texture. The insect is wingless but the three pairs of legs are well developed, the hind legs being longer than the others. This enables it to jump considerable distances. Minute white eggs are laid in the dust in crevices in dwelling houses and elsewhere. The larvae which hatch out a week or so later are legless, elongated creatures, also white except for the head which is brown. About a fortnight later a soft white pupa, rather like the imago but enclosed in a silken cocoon appears and it remains in this stage for a further fortnight when the imago emerges.

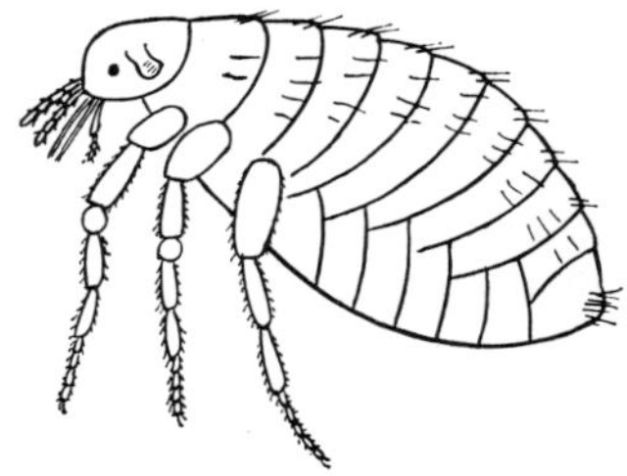

Fig. 27.7. **Human flea.**
(From Wallis—Practical Biology.)

Cleanliness of houses and such places and the closing up of any crevices in which fleas can lay their eggs are the means by which infestation by these insects can be avoided. Should this occur, the room must be fumigated.

THE BUG

The bed bug, *Climax lectularius,* is a common species affecting man. It is a night visitant rather than a parasite, being nocturnal in its appearance and remaining hidden by day. When it reaches man it sucks his blood.

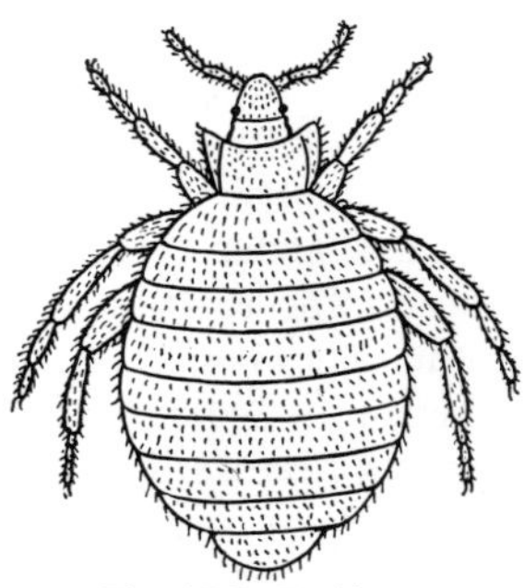

Fig. 27.8. **Bed bug.**
(From Wallis—Practical Biology.)

The ovoid body is dark brown, flattened dorso-ventrally and covered with short hairs and it is about 5 mm. in length. Bugs have a most unpleasant smell due to the secretion of the *stink glands* on the thorax. Batches of eggs are laid in crevices in walls, woodwork, furniture and particularly in beds. The larvae are nymphs, being similar to the adults though not as dark in colour. There is no pupal stage. The parasite can exist for long periods without food and once installed are difficult to eradicate. Fumigation of rooms and treatment of furniture with D.D.T. are the measures which must be taken.

THE ITCH MITE

This is truly parasitic and is the cause of **scabies** or "itch". *Scarcoptes scabii,* the itch mite, is not an insect but a minute *arachnid,* a class to which spiders, ticks and scorpions also belong. The soft body, rather rounded or oval in shape, is only about 0·3–0·45 mm. in length in the female and even less in the male.

Fig. 27.9. **Itch mite.**

It bears four pairs of short legs but no wings. The female burrows into the skin and forms small channels in which it lays its eggs. Once she has done so she dies. The parts of the body attacked are those free of hair, for example between the fingers, and great irritation **(pruritus)** and heat result from this. The larvae which hatch from the eggs burrow deep into the skin and the irritation leads to scratching which in turn leads to inflammation and the formation of scabs. It takes about three weeks for the adults to mature after which they escape to the surface of the skin where they mate and after this the males die. Then, once again, the fertilised females burrow into the skin. Scabies can be acquired through sexual intercourse, though not necessarily so.

Fumigation may be necessary to rid clothes of the parasite for if the females fail to find a suitable host they can exist for quite a long time in these garments. Treatment of the skin is by application of an emulsion of benzyl benzoate.

ROUND WORMS

Ascaris Lumbricoides

The Round Worms are unsegmented and are known as *Nematodes. Ascaris lumbricoides* is found in the small intestine, mostly of children, and is therefore an *endoparasite.* It is cylindrical in shape, tapering towards each end, though less so anteriorly. It may attain a length of 15 cm. or more in the male and as much as 22 cm. or more in the female. The mouth, bounded by three lips, is found at the anterior end. The posterior end is curved in both sexes but more so in the male.

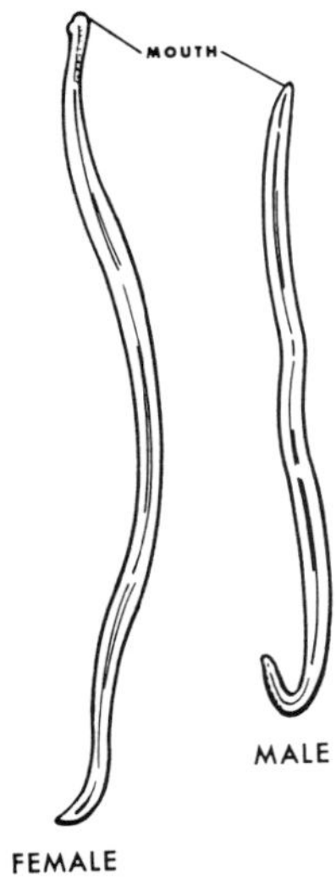

Fig. 27.10. **Round worm —Ascaris.**

The eggs are fertilised and laid in the intestine and the zygotes pass out with the faeces. Once outside they undergo maturation for some weeks and by this time they have reached an infective stage. Man may thus become infected through contaminated food or drink or from the hands. On reaching the intestine, the zygotes develop into larvae which make their way into the blood and this eventually carries them to the lungs. From the lungs they gain entry into the bronchi and trachea and so by the oesophagus to the stomach. On reaching the intestine the larvae develop into adult worms. Diarrhoea and, of course, an unusually large appetite since the worms feed on the host's food, are the only symptoms normally produced.

Trichinella Spiralis

This is a minute nematode reaching only about 0·3 cm. in length and is also an intestinal parasite. It causes the disease known as *trichinosis* which, fortunately, is rare here. The reason for this is that the parasite is contracted from contaminated pork in which the encysted, spirally coiled embryos reside. Proper inspection of meat and good sanitary conditions have made infection almost impossible.

Thread Worms

The thread worm, *Enterobius vermicularis* is found in the caecum and rectum and, like the other Nematodes, occurs mostly in children. It causes great anal irritation, and infection is through the mouth by contaminated fingers. The worm, like a piece of white thread, is only about 0·65 cm. long.

TAPEWORMS

Tapeworms, so-called from their being long, flat and white like a tape-measure, are parasites which require two entirely different hosts to complete their life cycle. There are many different species but the two mostly found in man are the pork tapeworm, *Taenia solium,* and the beef tapeworm, *Taenia saginata.* The adult lives coiled up in the intestine and may attain quite a considerable length, the pork tapeworm reaching perhaps 7 m. and the beef tapeworm considerably more. One which occurs in Eastern Europe and which has three hosts, a small crustacean, Cyclops, the pike and man, can reach over 9 m. in length!

T. solium consists of a minute bulbous head or **scolex** provided with a ring of hooks and four suckers by means of which it attaches itself to the wall of the intestine with great tenacity. Behind the scolex is a series of structures called **proglottides** (they should not be called segments as this term has a restricted meaning in biology). These are budded off from this region and they grow rapidly. Each proglottis at this end of the animal contains a complete set of both male and female organs when mature. As the parasite lives coiled up in the intestine, spermatozoa produced in one proglottis are able to fertilise ova in another. In the posterior proglottides the reproductive organs degenerate, apart from the uterus, as it is called, and this develops and branches to such an extent that it occupies most of the proglottis. It contains fertilised eggs. These proglottides become detached and pass out with the faeces. Good sanitation will avoid any further development for none will take place until the **six-hooked embryos** into which the fertilised ova develop enter the alternative host, the pig. Failing this they die after a time. In the soil the proglottides degenerate and a large number of embryos, each in a protective coat are set free. If eaten by a pig, the shell is dissolved and the liberated embryo bores its way into the blood vessels and is carried to the muscles where

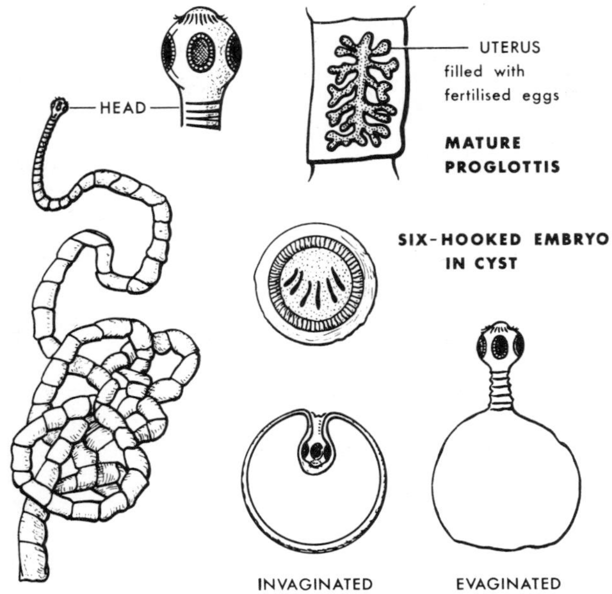

Fig. 27.11. **Tapeworm—life history.**

the hooks are lost and it develops into a **bladder worm**. This is composed of a small bladder with an invaginated scolex. No further development is possible in the pig but if this "measly" pork is eaten by man, once the bladder reaches the intestine the scolex evaginates and attaches itself to the intestinal wall, and the bladder is dissolved.

The life-cycle of *T. saginata* is similar to that of T. solium but the alternative host is the ox. In both cases man acquires the infection by eating raw or under-cooked meat which contains the bladder worm. Thanks to proper food hygiene including the careful inspection of meat and good sanitation, the tapeworm is almost extinct in this country but in certain eastern countries a very high percentage of the population, over 90 per cent., are infected though without apparent ill effect. The normal results of the presence of tapeworm are diarrhoea, voracious appetite and loss of weight due to inadequate nutrition.

There is another species of tapeworm, *T. echinococcus,* in which man may be the intermediate host instead of being the primary host in whom the sexual cycle takes place. The primary host is the dog and the worm is unusually small consisting of only three or four proglottides. The embryo on reaching man by his allowing an infected dog to lick his fingers or contaminate his food, forms an enormous bladder worm known as a **hydatid cyst** in the liver, lungs or brain. This may bud off a number of smaller cysts inside itself and the smallest contain invaginated scolices. Results may be serious as, for example, in the brain when a tumour may develop. Whenever possible hydatid cysts are removed by surgical means.

Treatment of normal infection by tapeworm consists of a period of starvation followed by the administration of an extract of the male shield fern, *Dryopteris*, and a purgative. This causes the tapeworm to be ejected on defaecation but if the scolex is not detached and discharged a new tapeworm will develop.

PRACTICAL WORK

House Fly

1. Examine the **eggs**, **larva**, **pupa** and **imago** of the **house fly** with a good hand lens.
2. Examine a slide of the **mouth-parts** of the **house fly** and note the extensible **proboscis**.
3. Examine a slide of one of the **legs** of the **house fly**.
4. Sterilise a Petri dish, melt some nutrient agar and pour it into the dish. When it has set, rub one or more legs removed from a fly over the surface of the agar, using sterilised forceps. Quickly replace the cover and examine a few days later for bacterial colonies. Some fungal growths may also appear.

Mosquito

1. Examine the **eggs**, **larva**, **pupa** and **imago** of the **mosquito**, either *Culex* or, if available, *Anopheles*, using a hand lens. These may be examined on slides or in a preserved mounted life-history specimen.
2. Examine a slide of the **mouth-parts of the mosquito** and note the long, unretractable **proboscis**.
3. Examine a slide of the **sporozoites of the malarial parasite** in the **salivary gland** of the *Anopheline* mosquito.

Louse, Flea and Bed Bug

Examine slides of the **louse**, **flea** and **bed bug** under a hand lens.

Itch Mite

Examine a slide of the **itch mite** under the microscope.

Round Worm

Examine preserved specimens of the male and female **round worm**, *Ascaris lumbricoides.*

Tapeworm

1. Examine a preserved specimen of a **tapeworm.**
2. Examine slides of (i) the **scolex** (ii) a **mature proglottis showing the sexual organs** (iii) a more **mature proglottis showing the branched uterus** and (iv) the **bladder worm**, under a dissecting microscope.

Thread Worm

Examine a slide of the **thread worm** *Enterobius vermicularis* under the microscope.

Chapter XXVIII

THE WATER SUPPLY AND ITS PROTECTION

How little do we realise the importance of water in our lives until there is a shortage of it! We drink about 1·5 litres ($\frac{1}{3}$ gal.) a day in various ways and use much more for washing, cooking and other purposes. Over and above our own personal requirements an enormous amount is used in industry and public services. In fact the average daily requirement in a town of say 80,000 inhabitants for all purposes may amount to as much as 2,500,000 or more gallons a day or over 136 litres (30 gal.) per head of the population and in some industrial districts this may be almost doubled. Not only must we have a sufficient supply of water in adequate quantity but the water must be physiologically and bacteriologically pure enough for drinking purposes.

Sources of Water

Water evaporates into the atmosphere from the huge surfaces of seas and oceans and returns to the land as rain, hail or snow. Rain water is the purest of the natural waters but it can become contaminated with dirt, particularly in industrial districts. What happens to the water after it has reached the earth depends on the nature of the soil. It may give rise to lakes and ponds or it may form rivers which ultimately take it back to the sea. This will be the case with soils impermeable to water such as those composed largely of clay. But if it is a sandy or chalky soil the water will drain through it until it reaches solid non-porous rock. In this way water collects underground and may become the source of wells and springs. The extent to which the water sinks depends, of course, on the arrangement and extent of the permeable and impermeable layers. Where there is an impermeable layer, water will lie on top of it whereas the water will soak through a permeable layer until it reaches the next impermeable one. Wells and springs arising from water above the uppermost impermeable layer are known as **shallow wells** and **springs** and as they may easily be contaminated or polluted they cannot be regarded as good sources of water. But the impermeable layers vary in position in different places and **deep wells** and **springs** which arise from water below the upper impermeable layer are less liable to contamination and are therefore safer. Furthermore there is less likelihood of their drying up when there is a shortage of rainfall as is the case with shallow ones. In some of these deep wells the pressure is sufficient

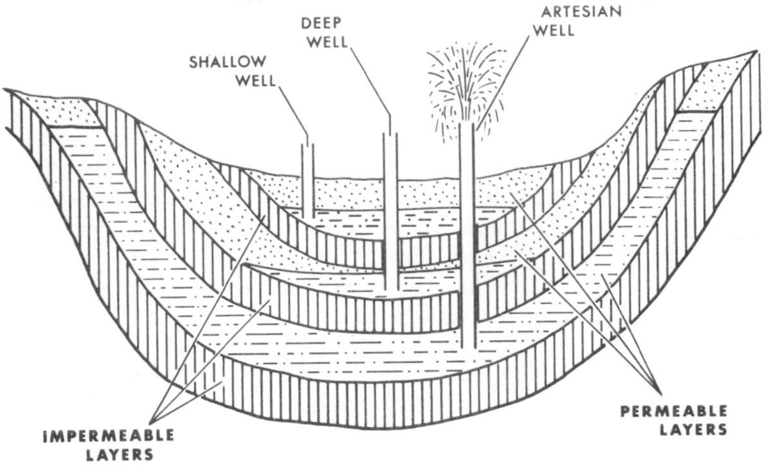

Fig. 28.1. **Shallow, deep and artesian wells.**

Fig. 28.2. **Lake Vrynwy, N. Wales. This shows how a reservoir can be constructed to harmonise with the surrounding countryside.**

(By courtesy of Liverpool Corporation.)

to force the water up to the surface and these are known as **artesian wells**. Spring and well water is often clear and sparkling as suspended matter has been filtered out of it in its passage through the soil.

Upland surface water in natural lakes is a source of water for some large towns and cities but in other cases man-made lakes or **reservoirs** made by damming suitable valleys serve as the source and these may be at considerable distances from the towns they supply. Birmingham for example, is supplied by water from beautifully landscaped artificial reservoirs in the Elan Valley and Liverpool from Lake Vyrnwy, both of which are in Wales, while Manchester's supply comes from Lake Thirlmere in Cumberland. London, on the other hand, has to obtain its water from the River Thames and its tributaries and many large man-made reservoirs have been built to provide storage for its water supply.

The region draining into a lake or reservoir is known as the **catchment area** and strict rules are enforced to control human habitation and animal grazing in these areas.

Purification of Water

In view of the enormous demands for this commodity, water must be conserved and up to six months supply should be stored if possible. But the supply must be free from pathogenic organisms, organic matter, poisonous salts and gases, dirt and other contaminants of any kind including radio-active matter. While upland surface water may be reasonably free of contamination, it is seldom that water for domestic use does not need some purification. Many of the village fountains in alpine countries provide clean, healthy water straight from the mountains but most water supplies have to be treated. This entails purification and regular chemical and bacteriological examination by public analysts. To some extent purification will take place in lakes and reservoirs, some suspended matter, including bacteria, settling to the bottom, organic matter being decomposed by saprophytic bacteria while some pathogenic bacteria fail to survive the sunlight to which they are exposed.

Next after this process of **sedimentation** all remaining suspended matter is removed by passing the water through specially constructed beds of sand called **filter beds**. These consist of a deep layer of sand, fine at the top and coarser beneath, on top of a layer of stones graded so that the smallest lie on top and the largest at the bottom. Algae grow in the upper part of the sand layer forming a jelly-like layer which helps to filter off pathogenic bacteria. This layer of sand needs to be changed periodically.

Finally the water is chlorinated, the gas being passed into the

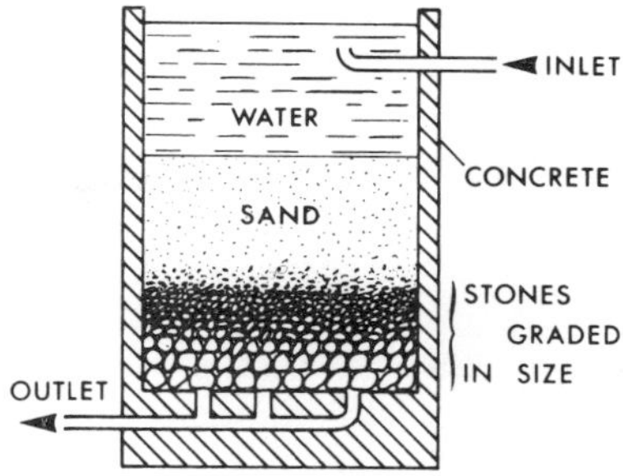

Fig. 28.3. **Filter bed.**

water in a regulated amount, 0·25 parts of chlorine per million parts of water, to kill any remaining pathogens. The water is then led to **service reservoirs** which must be at sufficient height above the region to be supplied to give an adequate head of water to supply the highest buildings in that region.

Distribution of Water

Water is carried from the service reservoirs by pipes buried under the soil and these give rise to the **mains** under the streets from which further pipes take water to houses and other buildings. There must be no possible chance of re-contamination from these pipes and glazed pipes must be used. Within the building the water is distributed to the various rooms by lead or copper pipes and when water is in constant supply the only need for storage in cisterns is for the heating system, baths and toilets.

Dissolved Substances in Water

Water is the commonest of all solvents and various mineral substances, according to the nature of the soil, may dissolve in it during its passage over or through the soil. Water lacking dissolved substances such as distilled water is "heavy" and flat to drink. Sometimes the mineral solutes have medicinal value. At Harrogate there are sulphur and chalybeate (iron containing) waters. At Bath the waters, which date from Roman times, are also chalybeate in the famous baths. Brine is found in the waters of Bad Gastein in Austria, Rheinfelden in Switzerland and Spa in Belgium (which is the origin of the term spa for such places), magnesium sulphate in Vichy water in France and those of Bad Reichenhall in Bavaria it has been claimed cure almost anything from rheumatism to impotence! Baden Baden in Germany has been known as a spa since Roman times.

Other mineral compounds, not of medicinal value but of domestic importance, are those which cause the **hardness of water**. A water is said to be soft if it readily forms a lather with soap. If it fails to do so it is said to be hard. Soap (sodium stearate) is soluble and the lather is due to surface tension. When the water is hard it is because the soap reacts with certain dissolved substances. Soft water gradually dissolves lead and therefore needs to be hardened somewhat to prevent this, for lead is a cumulative poison. This hardening is done before distribution so that a protective layer of an insoluble lead compound is formed on the inside of lead pipes which prevents the lead from dissolving in the water. It is claimed that the incidence of heart disease is higher in soft water districts than is the case in those in which it is harder.

The causes of hardness are magnesium and calcium compounds and water in chalky or limestone districts is therefore always hard. There are two kinds of hardness, *temporary hardness* due to calcium or magnesium bicarbonate and *permanent hardness* due to calcium or magnesium sulphate. Temporary hardness develops because the carbon dioxide in the soil air dissolves in the soil water and the carbonic acid so formed reacts with the chalk (calcium carbonate) to form calcium bicarbonate,

$$CO_2 + H_2O = H_2CO_3$$
$$CaCO_3 + H_2CO_3 = Ca(HCO_3)_2.$$

Permanent hardness arises from the direct solution of the salts from the soil.

Hardness has certain disadvantages. Both kinds of hardness require more soap to form a lather than does a soft water and this can prove expensive in laundries and very hard water is a distinct disadvantage in the home. Furthermore when temporarily hard water is boiled the calcium bicarbonate is decomposed into the insoluble carbonate and this is deposited as "fur" on the inside of pipes, boilers, kettles, etc. The result of this is a waste of fuel and time. Both of these defects are disadvantageous to the individual and in industry the additional costs may be considerable.

Temporary hardness may be removed by boiling

$$Ca(HCO_3)_2 = CaCO_3 + H_2O$$

or by adding lime

$$Ca(HCO_3)_2 + Ca(OH)_2 = 2\,CaCO_3 + 2H_2O.$$

This is known as *Clark's Process* and can be done on a large scale where boiling is obviously inconvenient.

Permanent hardness can be removed by the addition of washing soda (hydrated sodium carbonate—$Na_2CO_3.10H_2O$)

$$CaSO_4 + Na_2CO_3 = Na_2SO_4 + CaCO_3.$$

The sodium sulphate does no harm. Bath salts are sodium carbonate together with colouring matter and scent.

Both forms of hardness can be removed by the *permutit process* in which water is allowed to flow over permutit (sodium aluminium silicate). The sodium, being higher in the electrical chemical series than calcium replaces the latter forming calcium aluminium silicate. When this conversion is complete no more softening will occur until sodium aluminium silicate is restored and this is done by the addition of common salt when the sodium once again replaces the calcium. The calcium chloride formed as the result of this is run off. Domestic water softeners work on this principle and it is largely used in industry. Water pollution is dealt with on pp. 402 seq.

Chapter XXIX

HOUSING, HEATING, LIGHTING AND VENTILATION

Our health depends to a large extent on the houses in which we live, their siting, their construction and the manner in which they are heated, lighted and ventilated. As a great deal of our time is spent in working this also applies to the buildings in which we carry on our various occupations. Both are matters which concern the community as a whole but the house is the seat of family life in which we spend a great deal of our time.

HOUSING

Siting of Houses

It is seldom possible to choose the soil on which to build a house and so often in this country it has to be on clay which, being non-porous, is damp and cold. Gravel and sand are better but in either case provision must be made for good drainage and dampness must be avoided at all costs. Obviously the foundations on which a house or other building is to be constructed must be of sufficient strength to support it and for this purpose concrete is used once the ground has been excavated. The walls then stand on a bed of concrete. Walls made of concrete blocks are usually erected on a brick foundation. It is possible for dampness to rise through the walls from the ground and to avoid this a **damp-proof course** is constructed several inches above ground level. This is a layer of lead, slate, asphalt, plastic or other impervious material. In houses with wooden floors there is an air-space above the concrete and the floor joists are above the damp-proof course. Owing to the high cost of timber many modern houses and bungalows and blocks of flats have "solid floors" that is they are made entirely of concrete, at least on the ground floors. In this method there is no air-space and the damp-proof course is a thick polythene sheet under the concrete which is bent up so that it reaches beyond the damp-proof course in the walls. Alternatively a layer or damp-proof membrane of a bituminous liquid compound is used on top of the concrete. Next comes a two-inch layer of cement and sand called screed which is carefully levelled and smoothed, and this is covered with thermo-plastic tiles or wooden blocks.

Houses should not be built too close together and should be sited to give as much fresh air and sunlight as possible. The width of the streets is therefore related to the heights of the buildings along them. There is, unfortunately, frequent overcrowding in towns, for available space does not often allow for the building of "garden cities" and the construction of large tall blocks of flats, unsightly as they may be, seems to be the only way to provide sufficient accommodation for all who have to live there. London and other large cities have spread out into the countryside like octopi and unless this intrusion into our beautiful and natural heritage is stopped by building upwards instead of outwards, our valuable agricultural land and the country in which it is natural for us to relax will fast disappear.

Aspect of Houses

As to the aspect of houses, ideally the main living rooms and bedrooms should face the direction which provides the greatest amount of sunshine and, as far as this country is concerned, this means south. Other rooms such as kitchens, larders and so on which need to be kept cool can then be situated on the other side. Now that larders have been largely replaced by refrigerators and deep-freeze methods, this is perhaps of less importance and housewives, who spend so much of their lives in the kitchen, can have the benefits of sunshine and pleasant outlook.

Construction of Houses

Bricks, concrete, stone and wood are the materials used for

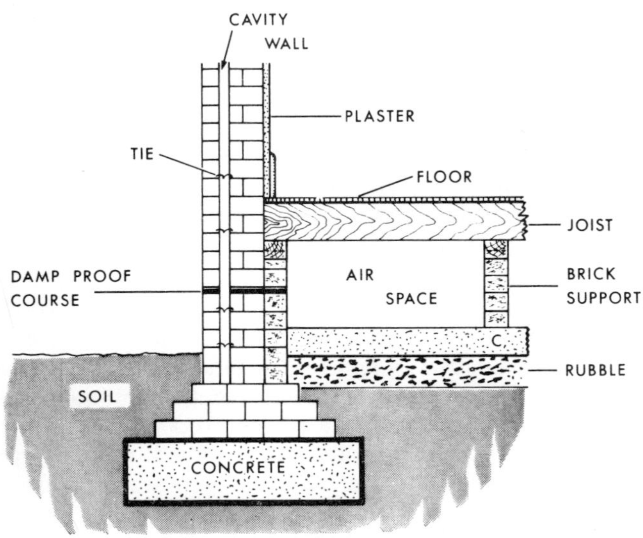

Fig. 29.1. **House foundation and damp-proof course.**
(wooden floors)

building houses in different countries and plastic materials such as fibreglass are now being experimented with. The materials used must keep the house free from damp, be insulated against loss of heat, be as sound-proof as possible and, of course, long lasting. Roofs must be waterproof and interlocking concrete tiles are now most commonly used in their construction. In old houses roofs were covered with slates. The roof timbers and trusses must be constructed to withstand the weight of the roof. Provision must be made for the drainage of water from roofs into gutters and so by pipes down to drains. Dampness through walls can be avoided by the building of **cavity walls**, that is two separate walls with an air cavity between them. This also has the advantage of improving insulation. Sometimes, as a further provision against damp and to improve the appearance, the outside walls are covered with cement and small stones sprayed on to it, an effect known as *rough cast.* This can be left as it is or a coloured paint can be mixed with it before spraying. When it is white it is now usually referred to as *Tyrolean finish.* In some cases stone-faced concrete blocks are used.

Partition walls which have no weight to support are single and made of brick or light-weight concrete blocks. This is covered with plaster, wooden laths being used to give it support, the method also used in ceiling construction in older houses. In lower priced modern houses plaster board is used for ceilings to keep down cost. Conventional plastering is used only in better quality houses. The wood used in the construction of floors, window frames and doors is sometimes affected by the fungus causing "dry rot" and by wood-worm in old houses. These should be treated immediately they are discovered as they are very liable to spread. As wood-worms are the larvae of wood beetles, the latter can fly to other woodwork including furniture where the trouble starts all over again. There are proprietary preparations which deal with this menace but they must be injected into the holes which the larvae bore and in which they live.

Houses should have a sufficient number of rooms and rooms of sufficient size to prevent overcrowding and the housing of large familes and of large numbers of families in inadequate accommodation is prejudicial to the good health of the families concerned and therefore of the community as a whole. Rooms must be well heated, lighted and ventilated and the house provided with proper cooking facilities, an adequate supply of clean, pure water, sufficient bathing and toilet accommodation and proper drainage for disposal of waste water and sewage.

Fortunately slum clearance is receiving attention throughout the country but it is necessarily a slow process and one which is not made any easier by the enormous increase in population in recent years. It has been estimated that eight million families will require new homes in the next twenty years. Much has already been done to improve working conditions in the building of new factories and offices. All these matters are covered by various Public Health and Planning Acts and while much is the concern of Town Planning Authorities, responsibility also rests on the shoulders of the individual in maintaining healthy conditions.

HEATING

Central heating, the fuel for which may be gas, oil, electricity or solid fuel, is steadily coming into much wider use in this country

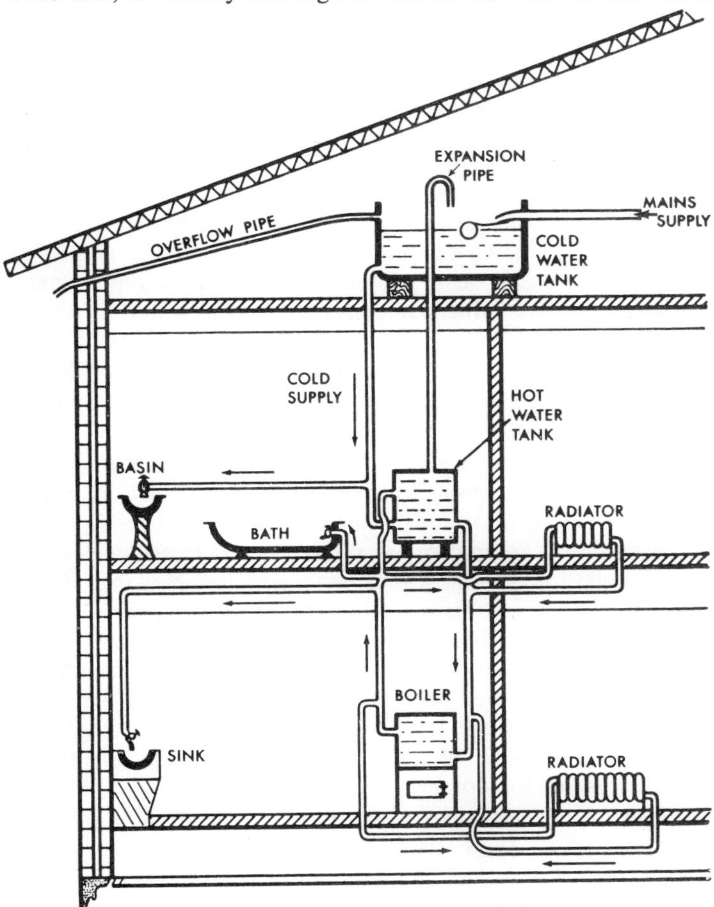

Fig. 29.2. **House water supply and central heating system.**

and in colder countries it is essential. It gives a uniform temperature, which can be regulated throughout the building whether it is a dwelling house, an office block or a factory. It can also be used to provide background heat which can be supplemented by other means. It is the only satisfactory method of heating large buildings and there is a great deal to be said in favour of its use in private houses. Heat is distributed by convection and the so-called radiators are really primarily convectors, only about 20 per cent. of the heat being provided by radiation and 80 per cent. by convection. Central heating can be combined with the hot water system of a house as can be seen in Fig. 29.2. This, however, has one disadvantage, namely that the radiators tend to cool down when large quantities of hot water are drawn off for baths etc. The outlet pipe from each radiator can be connected to the return pipe to the boiler instead of leading on to the next radiator. In large buildings the water is circulated under high pressure and in some factories steam is passed through the pipes instead of hot water.

If no central heating system has been installed water may be heated for the house by a special boiler in which anthracite or other solid fuel is burned or the boiler may be behind one of the open fires. There are also gas and electrical heaters for this purpose the latter being an immersion heater in the hot water tank. The temperature can be regulated by means of a thermostat. All work on the same principle, convection. Hot water has a lower density than cold water owing to the increase in volume of constant weight. Consequently the heated water rises and the colder water falls to take its place. Separate pipes enable a circulation to be set up and these supply the baths, toilet basins and kitchen.

Open fires burning solid fuel lose 50 per cent. of their heat up the chimney but they do give out a great deal of heat into the room, are cheerful to look at and assist ventilation. The burning of smokeless fuel, compulsory in some areas called *smokeless zones,* prevents pollution of the atmosphere and the blowing of smoke down the chimney into rooms which sometimes happens with some flues when the wind is in a certain direction. Modern open fires heat by radiation and by setting up convection currents in the air of the room. Closed heating stoves burning solid fuel heat by convection though some heat is given out by radiation. Two disadvantages of the use of solid fuel are the work it entails and the space required for storage.

Gas fires, like coal fires, are pleasant to look at and they are convenient to use. Heat, which can be regulated, is given out by the red hot radiants and modern gas fires have reached a high degree of efficiency. Some are self-igniting. Adequate flue provision to carry away the waste products of combustion and sufficient ventilation to supply the necessary oxygen are essential.

Electric fires and radiators are most convenient in use and highly efficient. Some designs can provide not only radiant heat from the red hot elements but convected heat as well. Electric radiators containing oil are also used and in all forms of electric heating the heat can be regulated within certain limits. Storage heaters work on the principle of heating up bricks to a fixed temperature by electricity which is then automatically switched off and heated units emit the stored heat. The current is again automatically switched on at certain fixed times and the bricks reheated. The rate for this method of heating is usually less than that charged for ordinary electrical heating. Under-floor heating and heating by elements behind the skirting boards and with ducts at the bottom of walls which allow warm air to enter the room are still further methods of electrical heating. Unfortunately heating by electricity is the most expensive method but cheap heating by any means is a thing of the past.

When **paraffin heaters** are used great care must be taken to see that the rooms are well ventilated as paraffin, like gas, requires oxygen for its combustion and produces carbon dioxide and water. These heaters are portable and are sometimes useful to give background heat when other means seem inadequate. Draughts must be avoided and great care exercised when using them owing to the danger of fire.

Heat for cooking can be provided by modern cookers burning solid fuel but this has the disadvantage of heating kitchens in the summer. These cookers also heat the water for the house. Cooking by gas or electricity is no doubt cleaner and of the two gas is more easily regulated.

Butane and **propane gas** provided in cylinders is used in caravans, boats and remote farms.

Solid fuel is sold by weight but gas and electricity have special units. On the metric system the unit of heat is the *calorie* but in this country at present it is the *British Thermal Unit* which is used. This is the amount of heat necessary to raise the temperature of 1 lb. of water 1° F. and as this is very small the *Therm*, which is 100,000 B.T.U., is the one used by the Gas Boards in calculating their charges. As the gas meters register the volume of gas consumed in cu. feet it is necessary for them to be converted to therms. The electrical unit is the *Kilowatt-hour* which is the expenditure of 1,000 watts for one hour, a watt being the product of the current in amperes and the voltage in volts. Electric meters record these units direct.

Solar Heating

The sun is the source of the world's energy and a method of heating houses and water by harnessing solar energy was invented in comparatively recent years. With the present high cost of fuel, more attention is now being paid to this and Japan and Israel seem to be well ahead of most other countries with research projects with Japan well in the lead. This is no doubt because there is neither coal nor oil in Japan. There are different types of solar collectors which consist of panels treated to absorb as much sunlight as possible and to lose as little of it as possible. In this country they must be placed on south facing roofs. They cover water compartments and the heated water is then stored in a hot water tank. This supplies the domestic hot water. One of the main problems is storage as to be fully effective the heat must be provided even when the sun is not shining and during the Winter months and there are various ways of coping with this. Space heating in the house as well as water has been done by solar energy but for whatever purpose it is to be used the cost of installation is at present high though upkeep is minimal. No doubt further research will eventually overcome the various problems for the fuel itself costs nothing and the ultimate saving of other fuels as sources of energy will be an enormous advantage throughout the world. It should also reduce very considerably domestic and industrial costs. Up to the present solar heating has been used mainly for swimming baths in this country as the temperature of the water is much lower than that required for other purposes.

Fire Prevention

An enormous amount of damage to and destruction of homes, factories and farms occurs every year due to fire and what is even more serious the loss of human life is quite considerable. Apart from deliberate arson much of this could be avoided if people took adequate precautions. The Home Office has issued a booklet entitled "Danger from Fire" to every household in the country. In this booklet not only are the dangers and risks from fire clearly stated but the measures which are vital for safety and the actions which must be taken in the event of fire are plainly given. Everyone should be conversant with this and every individual should know what action he should take in the event of this emergency.

Heat Insulation

Heat is conserved, as already stated, by the use of cavity walls, the air between the two layers of brick acting as an insulator. The insertion of insulating material (plastic foam) into the cavity is even more effective, a modern method which originated in Denmark. This can be blown into existing cavity walls but it must be done by experts otherwise it may cause damp and do more harm than good. In addition the roof can be insulated by the laying of glass fibre or other insulating material between the joists in the loft. Heat loss through windows can be considerable, particularly if the windows are large, and this can be overcome by double glazing in which the air between the two glass surfaces acts as an insulator. We are slowly coming round to it in this country but it has been an accepted practice in many countries abroad for a very long time.

LIGHTING

Good natural lighting of rooms is necessary to prevent eye-strain, to admit as much sunshine as possible as bacteria are killed by ultra-violet light and on account of the benefit good lighting confers on mental outlook. This is provided by having sufficient windows but ultra-violet rays do not pass through ordinary glass and the special glass which does admit them is very expensive. It is therefore necessary for windows to be open for the sunlight to have its full effect and, in any case, they should be open to provide ventilation. Large picture windows have the advantage of providing good lighting and give a room a pleasant outlook as they are usually fitted in rooms overlooking the garden or the countryside. Studios are lighted from above and a northern light is preferable. Whenever possible illumination in offices and schools and similar places should be from the left as most people are right-handed and a shadow is produced on the paper by the hand and arm. The reflection of bright sunlight from paper is, however, harmful to the eyes. The more window space there is the more cheerful is the room but the disadvantage lies in its making the room cold in the winter unless double glazing has been provided. Light walls and ceilings reflect more light than dark ones and therefore make a room lighter. They also give the impression of more space.

Artificial lighting by electricity is sometimes necessary even in daylight hours in factories, laboratories, hospitals and similar places in order to supply adequate illumination for the work in hand. Direct artificial lighting on to objects can produce eye-strain owing to the glare but is sometimes unavoidable. Lighting from the ceiling is desirable either by ordinary filament lamps arranged so that a diffuse light is reflected or by fluorescent strip lighting which gives excellent illumination without shadows and which is preferable. The intensity of illumination varies in different situations and insufficient or badly arranged lighting must be avoided. A great deal of energy from a lamp is lost as heat and this can be quite considerable as, for example, over an operating table in a hospital when the illumination provided must necessarily be very high and this can cause no small discomfort to those taking part.

Some electricity power stations depend on coal or oil, others on atomic energy and the electricity is carried to quite remote places through the *National Grid*.

Where electricity or coal gas is not available in some remote country districts, in caravans and camps and during some road repairs at night good lighting can be provided by butane or propane from cylinders or by paraffin pressure lamps, an excellent light being provided by an illuminated mantle protected inside a glass chimney.

The **illuminating power of a source of light, Luminous flux,** was originally measured as **candle-power**, a term derived from the method of measurement, using a standard candle. On the S.I. System it is measured in **lumens**, each unit being the light intensity of an International Candle*. Average daylight is about 500 C.P. (that of the sun is 3×10^{27} C.P.). A gas-filled electric lamp gives 1 C.P. for every $1\frac{1}{4}$ watts (approx.). Thus a 100 watt lamp has a candle power of $\frac{100}{1\frac{1}{4}} = 80$ C.P.

The **Intensity of Illumination of a Surface, Luminescence,** is the quantity of light falling on a unit area per second and is measured in **metre-candles**, on the British System in **foot-candles** and on the S.I. System in **lux** which is 1 lumen per square metre. The Intensity of Illumination is one foot-candle when the surface is one foot distant from a standard international candle source at right angles to the surface and this can be measured by a special meter. The intensity will vary inversely as the square of the distance from the source of light. Thus a 100 watt lamp (= 80 C.P.) will give an Intensity of Illumination at a distance of 5 ft. (= 1·524 m.) of $\frac{80}{5^2} =$ 3·2 ft.-candles

or $\frac{80}{1{\cdot}524^2} = 34{\cdot}5$ metre-candles.

In the average sitting room about 6 ft.-candles (= 64·2 metre-candles) is adequate intensity of illumination and it should be slightly more in a kitchen but the table in an operating theatre needs around 100 ft.-candles (= 1,070 metre-candles) (remember these readings are per unit area). In other words the illumination requirements vary according to the purpose for which they are needed.

* Though now measured by electrical means, the term is still used. An International Candle = 0·98 of the original English Standard Candle.

VENTILATION

Natural convection currents and winds keep the air out-of-doors in motion but the air in a room becomes stagnant unless means are adopted to keep it in circulation. The air out-of-doors is normally fresh and healthy though it can become polluted, particularly in towns and industrial districts, by smoke, sulphur dioxide and dust and in traffic congested streets, by carbon monoxide. Otherwise it retains its normal approximate composition of

Oxygen	20·9 per cent.
Nitrogen..	78·1 per cent.
Carbon dioxide..	0·03 per cent.

together with the inert gases (argon, neon, helium etc.) and varying amounts of water vapour.

Smoke, apart from making the atmosphere dirty, increases the possibility of fog, or "smog" as it is called when the causative agent is smoke. This is both unhealthy and creative of unsafe conditions. Smoke also decreases the amount of ultra-violet light which is so beneficial in germ destruction. The creation of "smokeless zones" has done a great deal to mitigate this nuisance. The provison of parks and other open-spaces in towns is clearly of the utmost importance.

The air in rooms is affected by the respiration and sweating of people occupying them. Some of the oxygen is removed and its place taken by carbon dioxide and the moisture content is increased. There is also danger of droplet infection. The air must therefore be continually changed if healthy conditions are to be maintained and this is ventilation.

It has been proved that the discomfort experienced in an ill-ventilated room is not brought about either by oxygen deficiency or excess of carbon dioxide but by increase in humidity from respiration and sweating and increase of temperature which cause the air to become stagnant. This produces a general feeling of lassitude and work and other activities suffer accordingly. The author once analysed the air in a classroom first thing in the morning before it had been used and again several hours later after it had been occupied by classes of twenty-five or more boys. In the meantime all the windows had been kept completely closed (which went very much against the grain but was necessary for the experiment) and the door opened as little as possible. The records showed the decrease in oxygen and increase in carbon dioxide were very small though the humidity and temperature had noticeably increased. For body comfort the temperature of a room should be around 15°C (60°F), the air must be in motion at about 0·6 to 0·9 metres

(2–3 ft.) per second and the Relative Humidity should not exceed 75 per cent. (relative humidity is the ratio of the actual mass of water vapour in a given volume of air to the mass of water vapour required to saturate an equal volume of air at the same temperature and is usually expressed as a percentage) and the higher the temperature the lower is the relative humidity.

In order to produce these conditions, particularly in an occupied room, ventilation is essential, the object being to replace the stale air by fresh air, and provision must be made for this when a building is constructed.

Ventilation may be natural due to convection currents set up by heating appliances which cause the heated air to rise and colder air to enter to take its place. This cold air enters through doors and the lower regions of open windows or by special ventilators set in walls at low levels about six inches or so above the floor. The hot air escapes up the chimney and out of the gaps at the tops of open windows or through ventilators high up in the walls. The open fire is an excellent means of bringing about this kind of ventilation but this is not always practicable or desirable. A few types of ventilators are shown in Fig. 29.3. In large halls like cinemas and theatres where large numbers of people are confined additional methods by mechanical means are necessary. Air is either forced into the building or extracted from it. In the former method the fresh air is pumped through ducts into the building and in the latter the air is withdrawn from it by extraction fans, fresh air entering to take its place. The two systems can be used in conjunction with each other and the air pumped in can be purified and heated before entering the hall. This air-conditioning has great advantages in giving protection against infection. It is important, however, that these currents should not be too rapid or they will create draughts which in turn will increase the rate of sweating and probably cause a chill.

It has been found that the air can be changed in a room three times an hour without causing draughts but the volume of air needed depends on the use to which the room is being put and the number of occupants. It is customary to express this in cu. ft. in this country and a minimum requirement is 600 cu. ft. (= 16,996 l.) per person per hour. Take a bedroom measuring 15 ft. by 12 ft. and 10 ft. high = (4·5 × 3·65 × 3·0 m.). The volume of air in the room is 1,800 cu. ft. (= 50,971 l.). Not more than three people should therefore sleep in the room. In sick rooms the minimum requirement should be increased by 50 per cent. at least and in this same room two patients at the most should therefore be accommodated. In hospital wards, 1,200

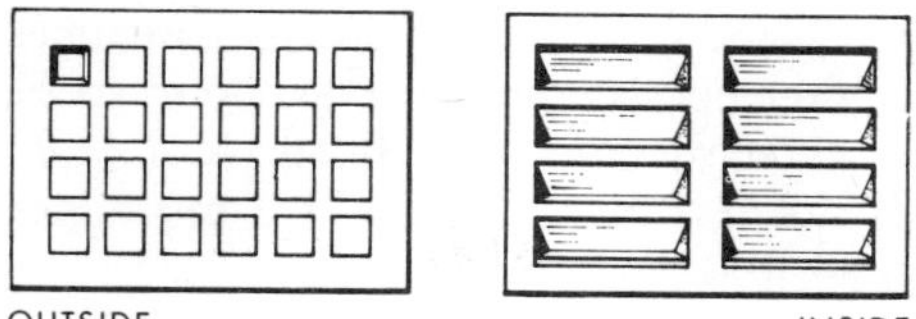

AIR BRICK

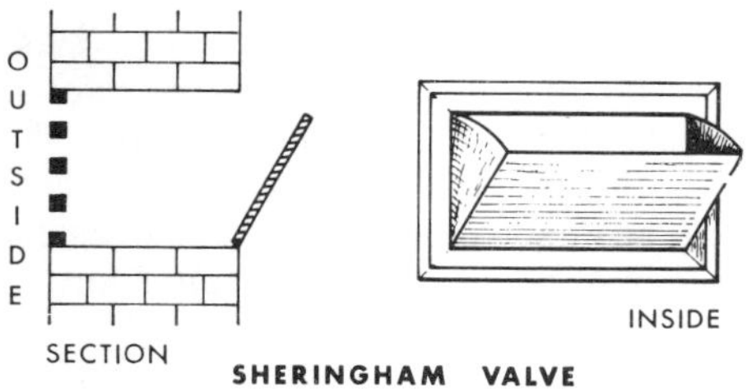

SHERINGHAM VALVE

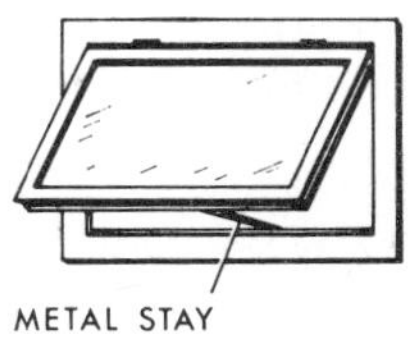

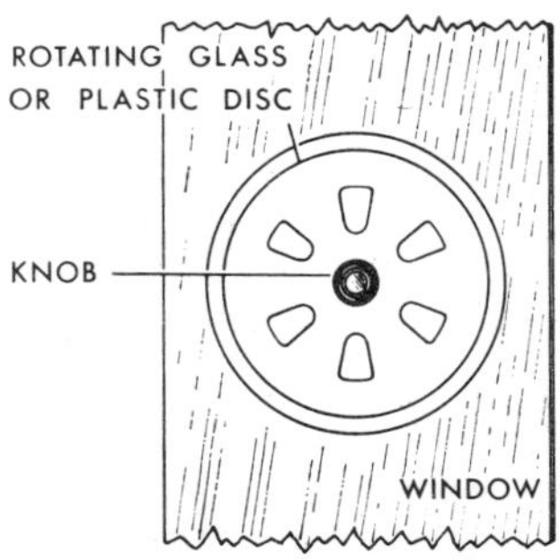

COOPER'S DISC

The disc on the inside of the window can be rotated so that the holes around its circumference correspond with similar holes in the window providing ventilation or moved so that the holes are closed.

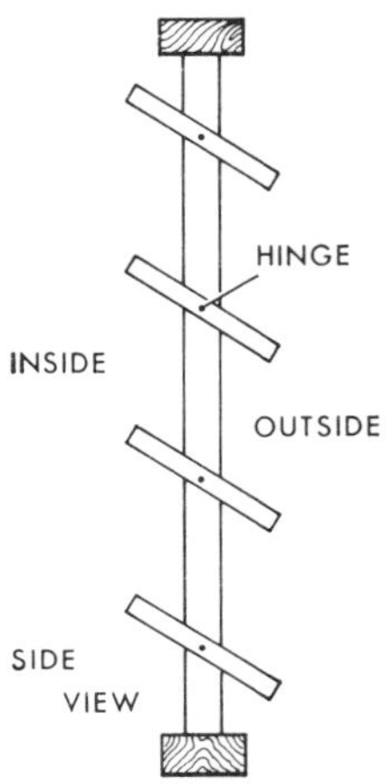

LOUVRE

Fig. 29.3. **Ventilators.**

cu. ft. (= 33,994 l.) per patient is the standard laid down. Sometimes the minimum ventilation requirement is expressed as floor space in square feet and in hospital wards this is 100 sq. ft. (= 93 sq. m.) per patient with beds not less than 5 ft. (= 1·5 m.) apart. Air pollution is considered on pp. 400 seq.

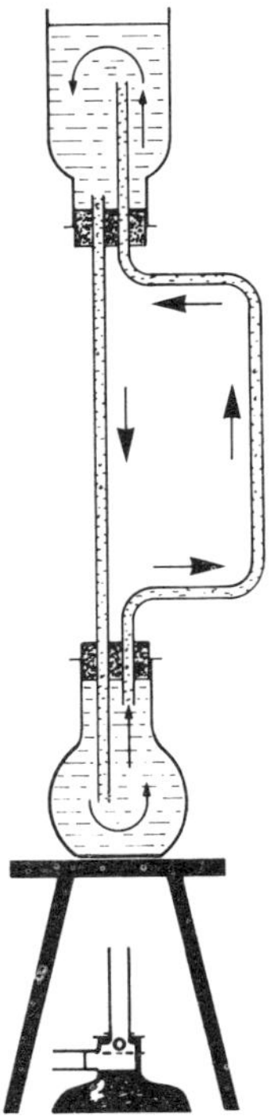

Fig. 29.4. **Convection apparatus.**

PRACTICAL WORK

Heating

1. Set up the apparatus shown in Fig. 29.4. Drop a crystal of potassium permanganate or other dye into the top vessel and stir. Heat the water in the flask and the convection currents set up will be made visible by the movement of the coloured liquid. Note the direction of the flow.

2. Hold a thermometer about 0·5 m. above and 0·5 m. away from a gas or electric fire and note the temperature. This is due to convection. Now hold the thermometer directly in front of the fire 0·5 m. away. The temperature recorded is due to radiation. Compare the two temperatures.

3. Hold a lighted candle at the bottom of an open doorway in a room in which the windows are closed. Note the direction in which the flame is blown. Now hold the candle near the top of the doorway and again note the direction of the flame. This experiment shows that air enters at the bottom and leaves at the top. It can be repeated at an open window.

Chapter XXX

DISPOSAL OF REFUSE AND SEWAGE

Refuse

If the plant and animal waste from eating and cooking and other sources is allowed to accumulate and is not frequently and regularly removed from the vicinity of human habitation disease is bound to occur. Other rubbish from houses, offices, schools, factories and so on must be disposed of and it will be appreciated that the amount of such refuse throughout the country is enormous. Waste food if it cannot be used for pigs and poultry should be burned whenever possible. Failing this it should be wrapped up in newspaper and put into a dustbin. On no account should it be put unwrapped into a waste receptacle as this encourages flies and makes it difficult to keep the inside of the bin clean. In Australia the refuse collectors have authority to refuse to take away any unwrapped refuse. The bin in which the refuse is stored prior to its removal by the Local Authority must be of galvanised iron or plastic and provided with a well fitting lid. Another type, used largely on the continent, consists of a large, strong, waterproof disposable bag which fits into a frame provided with a closely fitting metal lid. The collector simply removes the bag and its contents and another is put in its place. This type is available in this country but it is a more expensive method. Waste receptacles should be kept as far away from the house as possible but convenience of access must, of course, be considered.

Garden refuse should always be burned or used to make a compost heap. The Local Authorities' refuse vehicles must also be provided with proper covers. Useful materials such as scrap metal and paper are salvaged at the depot and sold and the remainder burned in an incinerator or put on a tip on waste land and covered with earth. The latter method should be avoided if possible as it encourages flies and rats.

Adequate covered refuse receptacles should be placed in all parking places on country roads. Motorists should use them and they should be emptied regularly and not allowed to spill over on to the ground. It is also desirable that toilet facilities within reasonable distances of each other should be available to motorists and that the countryside along our roads should not be fouled by excreta.

Sewage

All waste water and human excreta from buildings is carried

in glazed stoneware or cast iron pipes to sewage works where it is specially treated in order to render it completely harmless before it is discharged into rivers or the sea. This is known as the **water carriage system** but in some isolated country dwellings where there is no main drainage a different method called the **conservancy system** has to be used.

Conservancy System

One type, also suitable for caravans, boats and aeroplanes, is the **chemical closet** in which a special preparation containing formaldehyde or cresol is added to the water in the container. This causes the excreta to disintegrate and also acts as a deodorant. The contents can be buried or emptied into a **cess pit**. Large static caravans on permanent sites are often connected to mains water and drainage. Chemical closets can also be useful in parking places on country roads and on building sites and similar places where

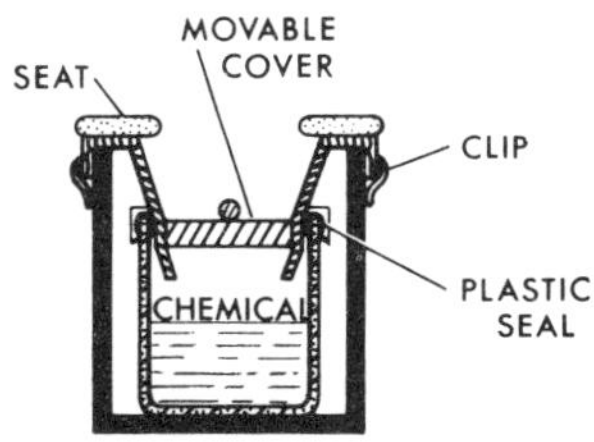

Fig. 30.1. **Chemical closet.**

mains drainage is not accessible. In one model the sewage container is enclosed in an outer one into which is fitted a tapering cover for the inner vessel ending in a circular hole provided with a movable lid. A proper seat and cover are fixed to the outer vessel. These closets can be made of metal protected by vitreous enamel or of a strong plastic.

In camp **latrines** trenches are dug on the leeward side of the camp and the excreta covered with earth or, alternatively, pails with seats and covers above them can be used. **Earth closets** are sometimes used in remote country dwellings, the earth in the container and covering the excreta causing it to disintegrate and acting as a deodorant.

A **septic tank** is used when a house or hamlet has a water supply but no sewers. It must be constructed of impervious material, have a tightly fitting cover and be well ventilated. It must be sited not less than fifty feet (15m.) from a dwelling and considerably more from a well or spring. Anaerobic bacteria decompose the excreta. Periodically the tank must be emptied and special motor vehicles fitted with suction pumps and hoses call and withdraw the contents into a tank on the vehicle which then carries away the sewage for disposal.

Water Carriage System

This system in which excreta and liquid waste is carried away by flowing water is by far the best method of sewage disposal. It makes use of pipes, drains, sewers and other appliances. Pipes carry the water into **drains** and drains into **sewers** which take it to **sewage works**.

The pipes taking waste water and sewage into drains must have a *water seal* known as a **trap** to prevent any foul gases passing back into the building. This consists of a pipe bent back on itself, water resting in the bend and acting as a seal. The pipes carrying

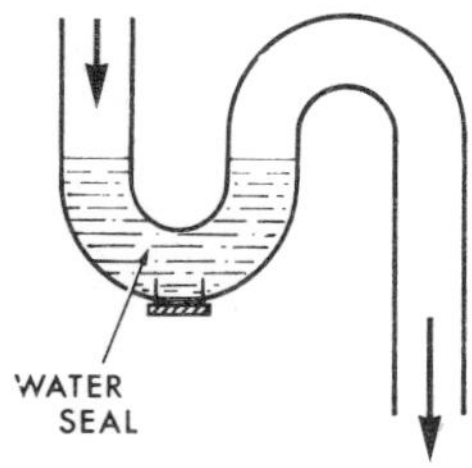

Fig. 30.2. **Trap.**

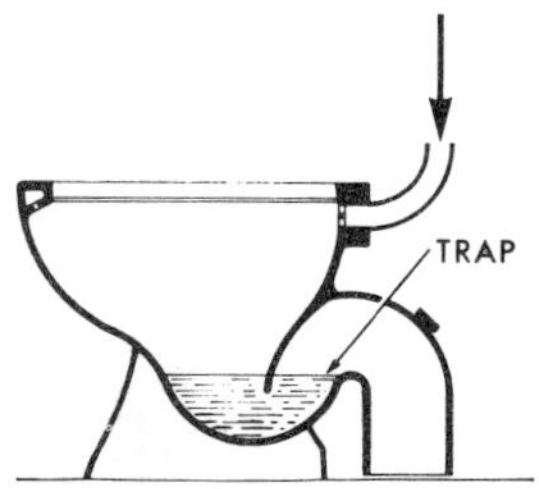

Fig. 30.3. **Water closet pedestal.**

water from baths, basins, sinks and gutters stop short of the drains in the open air and empty into a **gulley** above the grating or into a **gulley trap** below it, the entrance being covered with a removable grating to retain such things as leaves which could cause blockage. A water seal is situated beneath. All drains must slope gradually downwards all the way to the main sewer (or septic tank) so that there is a steady flow in that direction and **inspection chambers** giving access to the drains must be provided in case of any obstruction. Finally high **ventilation shafts** must be fitted into the drainage system to prevent the pressure of foul gases in sewers from breaking the water seals and so entering the building.

Pipes from toilets lead into a separate **soil-pipe** which runs down directly to the drain outside the building and extends upwards as a **ventilating shaft**. The top of this ventilating shaft is protected from birds by a wire cage. The trap forms part of the toilet pedestal. Water for flushing is contained in a cistern above the toilet, either

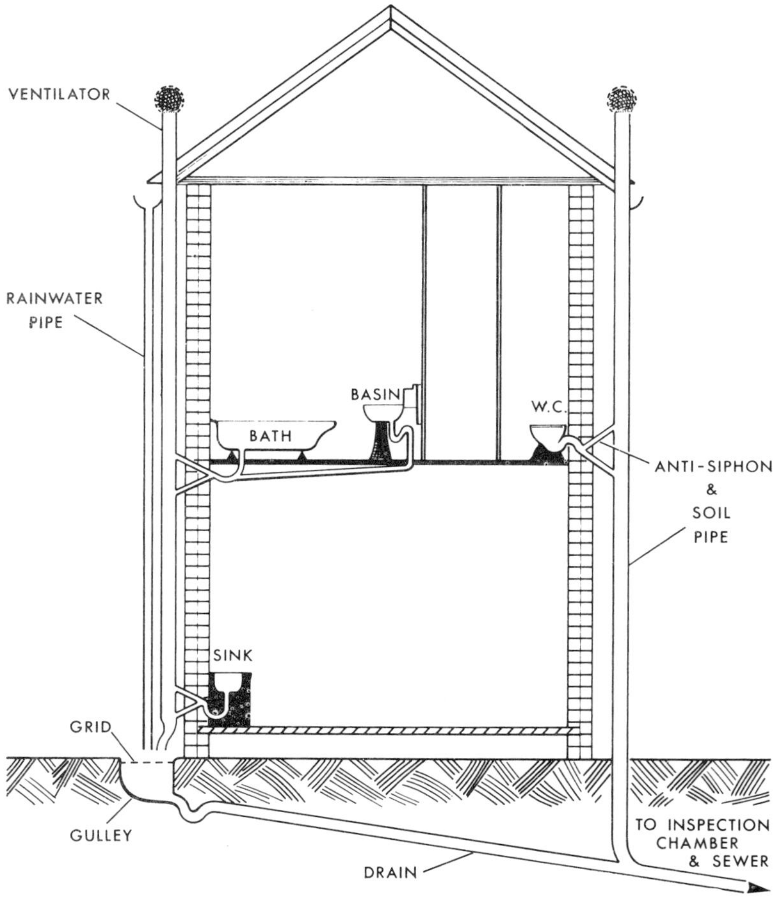

Fig. 30.4. **House drainage system.**

well above thus giving a good head of water or, in the modern low flush suites, just above it. In both cases the object is to give a powerful flush. A copper *ball valve* floating on the water in the cistern keeps the inlet closed. The action of the lever arm raises the outlet cover and allows the cistern to empty. With the ball-valve lying on the bottom, the inlet is opened and water enters to refill the cistern, the outlet being closed again on release of the lever arm. With the ball-valve floating on the water the inlet is again cut off on reaching a certain level. Cisterns must be fitted with overflow pipes which must not enter the drainage system.

Rain water from streets, which may be considerable after heavy rainfall, particularly in large towns and cities, is frequently carried in separate sewers from those which take the waste water and sewage from buildings and this can be taken directly into rivers or on to land.

TREATMENT OF SEWAGE

In 1831 an epidemic of cholera appeared almost for the first time in this country causing the death of more than 50,000 inhabitants. Just over twenty years later there was a further epidemic though less widespread. At that time it was the custom to discharge sewage from very insanitary *privies* into rivers which in some places, like London, provided water for the dwellings. Refuse was also dumped into rivers. When it was realised that there was an obvious connection between these practices and disease, Public Health Acts came into force which forbade them and which enforced the passage of excreta into sewers and the provision of healthy water supplies. It was not until nearly the end of the century, however, that really satisfactory methods of sewage disposal came into use.

The treatment of sewage at the sewage works takes place in four stages which are described below.

1. First the sewage is passed through metal *grids* to remove any solid matter such as paper, a process called **screening**. These grids are cleaned by mechanical rakes and the residue burned in incinerators or buried.

2. It then passes slowly through **detritus tanks** where grit and stones settle to the bottom.

3. From the detritus tanks it runs into **settlement tanks** where it is retained for several hours to allow suspended matter to settle to the bottom where it forms what is called **sludge**. This process of sedimentation can be speeded up by the addition of lime. After removal the sludge is either spread over land or dried and used

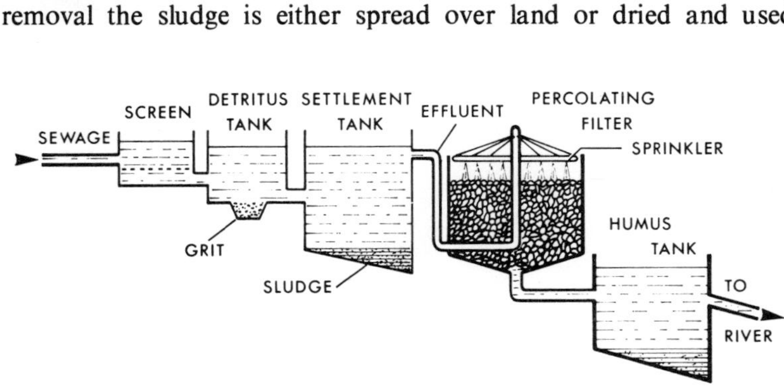

Fig. 30.5. **Sewage treatment.**

as manure. Methane or "marsh gas" (CH_4) is given off in the drying process and can be collected and used for driving machinery.

4. The liquid from the settlement tanks after sedimentation is known as **effluent**. As it still contains organic matter it is dangerous and requires further treatment.

One method, known as the **percolating filter method**, involves the pouring of the effluent on to coke or clinker in circular concrete tanks by means of slowly rotating perforated arms above the surface. The effluent filters through the coke or clinker where bacteria on its surface bring about decomposition. This is a biological action. The effluent which contains dead bacteria then runs into further **humus tanks** to enable these dead bacteria and algae, which are also present, to settle.

In the alternative **activated sludge method** the effluent is run into tanks through which compressed air is forced to provide oxygen for the bacteria which cause decomposition. It is left in the tanks for several hours and then run into a **settling tank** in which sludge settles to the bottom. The sludge, which contains active bacteria is disposed of by one of the methods described above though some of it may be pumped back into the first tank in this process to provide more bacteria for the first stage.

Finally the purified effluent is run into rivers or the sea. When discharged into the sea untreated sewage should be carried by pipes out to a sufficient distance beyond low tide level to prevent its being washed ashore by the tide. There have been complaints of this happening at some seaside resorts and this is not only extremely unpleasant to say the least, but detrimental to health.

Chapter XXXI

POLLUTION OF THE ENVIRONMENT

Pollution may be defined as the tainting or defiling of anything thereby destroying its purity. This is certainly taking place in our environment, in the air, on the land and in our waters, and it is occurring throughout the world. The whole subject is so vast that we can do little but touch on it here and some of the ways that it can take place have already been seen in earlier chapters. Man's health is being impaired by those activities of his which contaminate the environment and it is the bounden duty of governments to take measures to reduce and where possible eliminate pollution in their countries. As far as Great Britain is concerned this lies with the **Department of the Environment** coupled with other bodies. Oil pollution at sea and aircraft noise are the concern of the **Department of Trade**, agricultural chemicals such as pesticides come under the **Ministry of Agriculture, Fisheries and Food** and noise and fumes in factories are the responsibility of **H.M. Factory Inspectorate**. The **Secretary of State for the Environment** has a general overall co-ordinating role to perform and the whole subject of pollution in the United Kingdom and around its coasts is covered by the **Control of Pollution Act** of 1974.

The National Trust does a great deal of good work in conserving much that is beautiful in our country. It owns many acres of magnificent scenery, mediaeval castles, ancient monuments and historic houses which it preserves for all time for the enjoyment of present and future generations. Much of this has been given to the Trust by public-spirited people and much by money donated to it. The latest project is the preservation of those parts of our 3,000 mile coastline which are of outstanding beauty and this amounts to about 1,000 miles, the remainder either being industrialised or of no scenic beauty. Under the **National Parks and Access to the Countryside Act**, some parts of the country not owned by the Trust have been designated as *National Parks* such as the Lake District, Snowdonia, Exmoor and Dartmoor. Other parts have been designated as *Areas of Outstanding Beauty,* the first of these being the Quantock Hills in West Somerset. In all these areas development must be strictly controlled to ensure that the character and beauty of the landscape remain unimpaired. Thus pollution of our land by ugliness and other forms of spoliation is avoided.

We must now consider the various forms of pollution, their effects on the environment in which we live and the ways in which they are controlled.

Air Pollution

Smoke

Ever since the industrial revolution air pollution has been a major problem in Britain and one of the chief offenders has been smoke but conditions have been very noticeably improved as the result of the application of the terms of the **Clean Air Acts** of 1956 and 1958. Smokeless Zones have been created and smoke control orders made which control or prohibit the emission of smoke from chimneys in particular areas. The emission of dark smoke is forbidden and the height of chimneys must be approved and grit and dust arrestment installed. Control of domestic smoke and simple industrial emissions lies with local authorities and they create smoke control areas in which the emission of smoke from all chimneys, including those on domestic premises, is banned. Over 1,000 tons of grit and dust fall on every square mile in heavy industrial areas every year and 6 to 7 tons elsewhere. In 1952 4,000 people in London died as the result of inhaling smog (fog caused by smoke). There has been no smog since 1962, thanks to smoke control, and Winter sunshine in central London has increased by 50 per cent. and in Glasgow by 60 per cent.

Since the early 1950s the average levels of smoke and sulphur dioxide at ground level in England have fallen by about 60 per cent. and 40 per cent. respectively and in Scotland by 6·5 per cent. and 25 per cent. respectively and visibility in Winter has increased from $1\frac{1}{2}$ to 4 miles. Sulphur dioxide dissolves in rain forming sulphurous acid which is soon oxidised to sulphuric acid and this corrodes both vegetation and buildings. Emission from Cement and electricity works has fallen by 80 per cent. in the last ten years.

Vehicles

Pollution from motor vehicles presents a serious problem. Avoidable smoke is forbidden and all petrol-engined vehicles must be fitted with a crankcase ventilating device. Heavy goods vehicles are tested annually and the Department of the Environment carry out more than 200,000 spot checks a year on roads and at operators' premises. Exhaust smoke from diesel engines manufactured since 1972 must be barely visible under normal operating conditions. New regulations were introduced in the same year controlling the emission of carbon monoxide and hydrocarbons in accordance with those of the U.N. Economic Commission for Europe and still more rigid controls will apply to all such vehicles used in the United Kingdom from 1st April 1977. The maximum lead emission from petrol-engined vehicles was reduced from 0·84 gm. per litre to 0·45 gm. per litre by the end of 1975 as the result of agreement between industry and government.

Noise

This form of air pollution which is now becoming a major problem is not new for Queen Elizabeth I did not allow drunken cries after 9 p.m. and husbands were forbidden to beat their wives after 10 p.m.!

One definition of noise according to a World Health Organisation writer is "undesirable sounds producing an auditory sensation considered to be annoying". It induces a state of stress and emotion. Ability to concentrate is adversely affected by noise and sudden or intermittent noise is more harmful than continuous noise. There is no question that noise can have serious effects on the nervous system. The sympathetic system is affected and an increase in tension in the organs it controls results from this.

Whereas ordinary conversation produces some 60 to 70 decibels, your telephone produces 70, an alarm clock 80, pneumatic drills used in road works about 120, a siren 150 and a rocket launched into space 170 decibels. During sleep there should not be more than 30 decibels in the bedroom. It is obviously essential that there should be some statutory control of noise where this is possible. One instance is that produced by motor vehicles and this is as follows:—

Heavy goods vehicles	92 decibels
Light goods vehicles	88 ,,
Private cars	87 ,,
Motor-cycles up to 50 cc.	80 ,,
Motor-cycles above 50 cc.	90 ,,

New E.E.C. limits come into effect in 1976. Further limits are being discussed in Brussels to call for a more substantial reduction for vehicles manufactured in the early 1980s, particularly for commercial vehicles and 'buses. A further measure of controlling the nuisance is by restricting road access and specifying definite routes for lorries.

When we come to consider aircraft noise, the question of reduction and control is no easy matter. It is the responsibility of the Department of Trade in accordance with international agreements and further means are being studied. Means of noise abatement such as minimum noise routes and power cut-back for departing aircraft are in use at major U.K. airports to reduce disturbance due to aircraft noise. Summer night jet take-offs are banned at Heathrow and other specified airports, for aircraft noise prevents people from sleeping in towns near airports. Sonic booms create an increase in ground air pressure by as much as 4 lb. per sq. ft. which rapidly decreases by 4 lb. per sq. ft. and then returns to normal and the area normally affected by Concorde is estimated at 50 to 80 miles in width and the distance of the supersonic flight in length.

The principal control of all noise emission other than that from vehicles and aircraft was originally covered by the Noise Abatement Act of 1960 which was principally concerned with noise from fixed sources such as industrial and domestic premises but this Act has been replaced by provisions in the Control of Pollution Act 1974 which gives local authorities greater powers and will enable them to designate noise abatement zones in which they will be able to control noise emission levels from premises. Houses, factories and workshops should be so designed that noise is reduced to a minimum and individuals should exercise restraint in creating noise, for example from radio and T.V. sets, for the sake of others to whom it may be distasteful or even harmful. A code of practice for the reduction of the exposure of employed persons to noise has been officially published. It gives advice on how best to keep noise within specified limits and how to protect workers exposed to noise above those limits. It has been proved that when noise level is lowered productivity increases, mistakes are less frequent and accidents fewer.

Water Pollution

Fresh Water

According to surveys conducted by the World Health Organisation 90 per cent. of rural people in developing countries are using unsafe water all the time, faeces and household refuse spreading disease by water and soil contamination and the most urgent need is therefore better sanitation. But it is not only in these countries that there is need for improvement.

A new **Water Act** came into effect in 1974, making 10 **Water Authorities** in England and Wales responsible for the development, management and control of rivers and all uses of the water associated with them. Discharges from industry and sewage works must be authorised by them and 77·4 per cent. of the total length of non-tidal rivers in England and Wales is now free from significant pollution. The Control of Pollution Act provides controls to cover virtually all discharges to inland and coastal waters and introduces severe penalties for the offence of polluting water. Prevention of water pollution in Scotland is covered by the **Rivers (Prevention of Pollution) Scotland Act** and the **Secretary of State**, advised by a special committee has the responsibility for promoting the cleanliness of rivers and other inland and tidal waters in Scotland. There are 9 **River Purification Boards** covering the mainland whose members consist of representatives of local authorities and others and, mostly in N. and W. Scotland which are less populated, there are 12 local authorities while island councils are responsible for their own territories. A survey of Scotland in 1970 showed only 8 per cent. of Scotland's rivers, including tidal stretches, were grossly polluted or of poor quality but in areas of the 4 boards in the centre belt about 20 per cent. of the rivers were in this category.

It is the utmost importance that catchment areas should be free from the risk of contamination and, as already stressed, rigid rules concerning human habitation and animal grazing are enforced. Water in reservoirs must be protected against infection and strict rules of hygiene apply to workers, who are examined periodically to see if they are carriers. Regular chemical and bacteriological tests of water supplies have to be made and wells, springs and water supply pipes must be so constructed as to avoid contamination from their surroundings.

More and more tons of nitrogen fertilisers are put on fields every year. Rain washes the nitrogen compounds from chemical fertilisers into the water of rivers which often contain a high amount of phosphate from sewage. The nitrate and phosphate then react and upset the river's normal action stimulating plant growth and the water surface becomes covered with algae. These, of course, consume oxygen and the water then becomes foul. In 1969 40,000,000 fish were killed in the Rhine by endosulfan having been accidentally tipped into the river thereby causing massive poisoning of the water.

Sea and Coasts

Marine pollution is an international problem and one of the worst forms of pollution in the sea is oil pollution. This is covered by several international conventions. The Torrey Canyon disaster in 1967 brought considerable pollution to the Cornish coast. The beaches were sprayed with detergent but this, it seems, did more damage to marine life than did the oil. Local coastal authorities now have emergency organisations to deal quickly with the problem whenever it arises.

Over-fishing has caused a marked decline in the world's harvest of fish and contamination of the human food chain has occurred due to toxic substances and the discharge of waste, garbage and sewage into the sea. The Mediterranean is one of the world's most seriously polluted seas.

Local authorities are also responsible for dealing with dangerous cargoes washed ashore including chemicals and other wastes. Sewage is discharged into the sea by 149 coastal local authorities through 330 outfalls. Discharges and dumping within the 3-mile limit is regulated by the Sea Fisheries Committee and outside that limit by the Ministry of Agriculture, Fisheries and Food in accordance with the Acts of 1974.

Thousand of sea birds, most of them guillemots, were killed in the Irish Sea in 1969 and washed ashore along the coast. Investigation eventually revealed that the cause was waste products from a plastic industry discharged into the sea.

The Control of Pollution Act and the **Dumping at Sea Act** of 1974 forbid the depositing of any noxious materials into the sea without a licence. As already stated on p. 398 untreated sewage when discharged into the sea must be carried out by pipes to a sufficient distance beyond low tide level to prevent its being washed ashore by the tide. The Regional Water Boards are responsible for the hygienic disposal of all coastal sewage.

Land Pollution

Land

Pollution is partly due to natural processes and partly man-produced. At least 3,200,000,000 hectares* are potentially suitable for farming. 1,500 million hectares are cultivated but only 44 per cent. of the world's potentially arable land is cultivated, the remaining 56 per cent. remaining uncultivated because of problems which are inherent in the soil and created by man. In Europe 90 per cent. of potentially arable land is fully utilised but in Africa it is only 22 per cent. India's food problems could be solved if the 40,000,000 hectares of irrigated land could be used to produce 3 tons of food grains per hectare.

Pesticides

These have, unfortunately, caused a great deal of damage and a voluntary control scheme run by the Ministry of Agriculture, Fisheries and Food has phased out a large number of the more persistent ones. D.D.T., once thought to be harmless to human beings, has disastrous effects on some kinds of wild life and may

*1 hectare = 2·471 acres.

be linked to some human illnesses. It is therefore banned or restricted in use in many countries and it is no longer supplied here for use in homes and gardens. Bird populations which had declined as the result of the use of persistent pesticides are now noticeably recovering.

Toxic Wastes

The waste disposal authorities in England and Wales are the county councils and in Scotland the districts and they are responsible for the proper disposal of all waste in their areas with special attention to toxic and other hazardous forms of waste which require special care. Heavy penalties can be imposed for depositing any poisonous, noxious or other polluting waste of any kind which might cause damage to persons or animals or pollute water.

Soil Erosion

This has been caused in many countries by deforestation, over-grazing and overcultivation. The destruction of hedgerows and woodland between fields which act as windbreaks allows the top soil to be blown away by the wind and deprives much wild life of natural habitats. In Great Britain alone 50,000 acres of top-soil are lost every year as the consequence of this.

By the removing the natural habitats of wild creatures man can set in motion a cycle of events which lead to self destruction. Birds which feed on insects disappear as a result and the insects breed at alarming rates. These attack man's crops. Man then sprays the crops to kill the insects and by so doing destroys bird life and, maybe, himself in time. Another form of crop pollution is lead from petrol-engine exhausts and three-quarters of anti-knock lead settles on surrounding vegetation.

Dutch elm disease is taking a heavy toll of these trees both at home and abroad. The present epidemic here began in 1968 in what is now known as Avon, Somerset, Gloucestershire and Wiltshire and it is spreading to the Midlands, Northumbria and Scotland. In Somerset alone 350,000 elms are already dead and a further 350,000 are infected and dying. In Gloucestershire 170,000 were lost in 1974 and we are losing over 100,000 a year by death from this disease. It is thought that infected timber imported from N. America where the disease is rife is responsible for the present epidemic. It did not originate in Holland as the name suggests. This name was given to it because most of the early research into the disease was done there. The cause is a fungus, the spores of which are carried by the elm bark beetle, *Scolytus*. The beetles fly to young trees to feed, boring into and weakening the bark.

The fungus then grows and causes internal rotting. There are two different strains one, left from a 1930 epidemic, is non-agressive and this does not usually cause death of the trees. The other, which grows twice as quickly, is the agressive strain and it is this which is causing so much havoc and death of the trees. Treatment by injection is very costly and often ineffective. The only method of attack on the disease is by felling the infected trees and burning the wood and this, too, is costly. Trees form part of the ecological balance of the countryside. The English landscape is a most valuable heritage and the death and felling of elms is making an unfortunate and devastating change in its appearance. Speedy replanting of other trees is essential to ensure its ultimate recovery.

Radiation

Ionising radiation as opposed to other forms such as infra-red and ultra-violet radiation can have very definite ill effects. We are always subject to radiation from cosmic rays, thought to originate from interstellar space and entering the earth's atmosphere, and from plants, water and rocks but this is at a very low level and does not affect us adversely. Radiation from *nuclear power* is quite another story, however, and as more and more use is being made of this form of energy—at least 10 per cent. of our electricity in this country is supplied by it—there is considerable danger from the *nuclear waste* unavoidably produced in the process. The unit of measurement of the absorbed dose of radiation is the **rad** and the radiation dose from radioactive fall out is measured in **milli-rads**. Once this enters the food chain and is absorbed by plant and animal tissues it becomes a considerable hazard to human beings. For example Strontium 90 linked with calcium can enter grass eaten by cows; this gets into their milk and the bones of children fed on it contain Strontium 90. People living in Arctic regions have a higher percentage than people living further south because arctic animals feed on lichens which absorb large quantities of radio-active fall-out. The short-term effects depend on the extent of the amount received. Atomic explosions cover a radius of several hundred miles and can affect the intestinal tract causing anaemia, haemorrhage and even death. In the long-term more serious effects result. For example there have been more cases of cancer, particularly leukaemia, amongst the survivors from the atomic bombs on Hiroshima and Nagasaki than was normally the case, many dying years after the event. Other effects have been fibrosis of the lungs and kidneys, cataract and serious disease of the liver. More than a hundred thousand died when the first-ever atomic bomb was dropped on Hiroshima in 1945.

There is, so far, no evidence of any effect on the genes and therefore of future generations but, as already mentioned on p. 216, children of women pregnant at the time of the bombing have suffered, some being on the average shorter and with smaller heads, some being mentally retarded and there have been cases of such children being unable to speak at the age of five. There is, of course, always the possibility of damage to the gametes before fertilisation which could induce mutations and malformations. This has already been mentioned when dealing with mutations on p. 216.

Clearly great care must be exercised in the peaceful use of atomic energy. The strongest controls on the disposal of the nuclear waste from power stations are exercised by the Department of the Environment and the Ministry of Agriculture, Fisheries and Food and the handling of radio-active material within nuclear power stations is controlled by the **Nuclear Installation Inspectorate** which is responsible to the Secretary of State for Energy: The nuclear waste, which is in liquid form, is highly radio-active and it will take an enormous number of years to decompose to safety level. It is therefore stored in double walled stainless steel vessels enclosed in concrete lined with steel. Research continues on the effects of Plutonium which is another radio-active metal in nuclear fission. Taken overall the risks to human beings from nuclear power are small as the chances of accidents might be described as minimal.

Chapter XXXII

PERSONAL HYGIENE

We have effective schemes for protecting our food and water supplies, for the disposal of refuse, for our own personal protection and comfort in our homes and in the places in which we work and for the treatment of the sick but in the long run responsibility rests on the shoulders of the individual. The health of the community as a whole is dependent on the health of its individual members and they will not remain healthy unless they pay proper attention to personal hygiene.

This means that every individual should have a properly balanced diet, have sufficient fresh air, sunlight and exercise, indulge in good and regular habits, have an adequate amount of rest and sleep, be properly clothed and keep himself or herself clean. Nor must recreation be overlooked. Let us consider each of these in turn.

Diet

Meals should consist of properly cooked food providing a correctly balanced diet and they should be taken at regular intervals. The food should be properly masticated and allowed to digest by the avoidance of exercise too soon after a meal.

Fresh Air and Exercise

Exercise, preferably in the open air, for its effects are most beneficial when taken out of doors, should be taken regularly by everyone though the nature and amount will vary according to occupation and age. It should be stressed that while sunlight is beneficial to the body, excessive exposure to the sun can definitely be harmful and can be a predisposing agent of skin cancer brought about by the ultra-violet rays. After the Winter, exposure should be for gradually increasing periods but they should never be too prolonged. Exercise is most important during childhood, adolescence and in early adult life. Heavy manual workers often get sufficient exercise of one kind or another by virtue of their occupation, though this may not necessarily be out of doors. Sedentary workers on the other hand get little so to them the taking of exercise is most important. Exercise has the effect of keeping all the systems of the body functioning as they should and not least the muscular system. Muscles atrophy if they are not used but become well developed with exercise.

Competitive games and other sports are best for the young as they not only provide the necessary exercise but also develop the team spirit. For older people something less strenuous such as walking, golf or even gardening is more suitable while for elderly people a little walking or light gardening should suffice. Under no circumstances must too vigorous exercise such as rowing, athletics or a London to Brighton walk be taken unless the body is trained for it or more harm than good will result. The necessity for training has already been emphasised in Chapter VI (The Muscular System). Nor must any vigorous exercise be undertaken immediately after a meal, particularly if it is a large one, as the blood supply to the alimentary system will suffer and consequently digestion be impaired owing to the heavy demands of the muscles.

Rest and Sleep

Relaxation is essental in order that the tissues may be able to recover from the unavoidable wear and tear which occurs during activity. A manual worker needs more physical rest than one whose occupation is sedentary but the brain worker needs to rest his brain and he does well if he has some entirely different occupation or hobby in his spare time. Recreation is essential to keep mind and body rested from work. Outdoor games, gardening, music, painting and other arts, reading, sewing and so on all provide pleasant relaxation and at the same time give mental stimulation which is a good thing in itself.

Sleep is the most complete form of rest as the activity of all the body functions is lessened and the muscles are relaxed. The amount of sleep necessary for health varies with age and, indeed, with the individual as has already been pointed out in the chapter dealing with the nervous system (Chapter XXII). Bedrooms should be quiet and well ventilated and bed clothes light and warm.

Regular Habits

It has already been stated that meals should be taken at regular times but regularity in other habits such as defaecation are equally important and should be acquired during childhood and continued throughout adult life. Habits develop from constant repetition and bad habits such as excessive smoking, and drinking, overeating and drug-taking, must be avoided and any tendency towards them should be "nipped in the bud". Habits which are conducive to good health on the other hand, should be encouraged and inculcated in the young.

Clothing

The chief purpose of clothing, apart from the necessity to conform to the laws of decency, is either to protect the body from loss of heat or, alternatively, to promote loss of heat from it, depending on environmental conditions. Heat loss from the body is reduced by clothing which allows an insulating belt of air to remain between it and the skin but this kind of clothing is obviously unsuitable in hot climates or hot weather when it is desirable to lose heat from the body. The amount and nature of clothing will therefore vary in different climates and at different times of the year. A great variety of materials is used for clothing but all should conform to the following conditions:—

Clothing should be:—

(i) Light in weight. If suitable material is used there is no need for unnecessary weight.

(ii) Loose, in order to allow for a layer of air next to the skin.

(iii) Absorbent if next to the skin to allow sweat to be absorbed.

(iv) Porous in order to allow sweat to evaporate.

(v) Sufficiently warm, particularly in elderly people, to avoid too great a heat loss from the body.

(vi) Non-inflammable.

Clothing Materials

Wool is the hair of sheep but that of other animals is also used. Wool fibres are tubular and covered with overlapping scales and these are spun into yarn. This is very absorbent and loses its water slowly though it does not feel wet to the touch, the surface being somewhat oily. A great deal of air is enmeshed amongst the fibres and wool is therefore a bad conductor of heat and very suitable for use in cold weather. It is rather rough and, consequently,

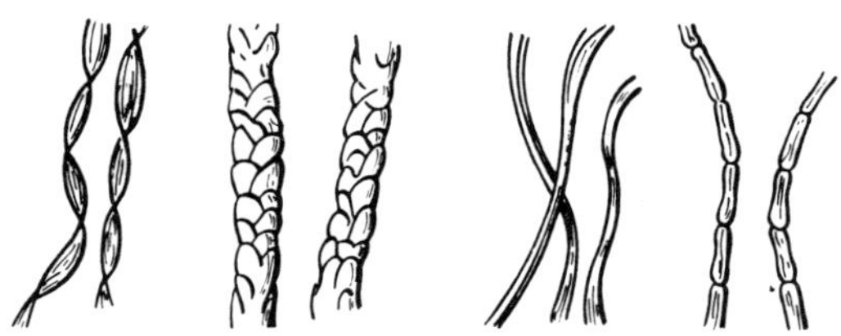

Fig. 32.1. **Clothing fibres.**

some people find it unsuitable for wearing next to the skin. Unfortunately it tends to shrink and "felt" or harden when washed but methods have been devised to overcome this and woollen garments should never be washed in really hot water or rubbed too hard.

Cotton is made from the hairs on the seed coat of the cotton plant. Unlike wool, the fibres are flat and twisted and these are spun into yarn. Little air is entangled amongst its fibres unless it is made into *cellular cotton* and this is the best form for underwear. Cotton is not very absorbent and is a better conductor of heat than wool. Consequently if worn next to the skin it may cause chilling. This is because it gives off such moisture as it is able to absorb quickly. It does not shrink but it has the disadvantage of being inflammable. Less care is necessary in washing cotton fabrics and it is cheaper than wool.

Linen is made from the phloem fibres in the stem of the flax plant and these are spun into yarn. The fibres are rounded in cross-section and are composed of a series of joints. It is stronger than cotton, is a better conductor of heat and, like cotton, is not very absorbent. Furthermore it is more costly.

Silk is an expensive material derived from the larvae of the silk moth. These so-called "silk worms" spin cocoons and it is from these that silk is obtained and woven. The material produced has a very fine mesh, has a smooth surface and is a bad conductor of heat. It is light in weight and relatively warm. It is mainly used for women's stockings, dresses and underwear but owing to its high cost these are now usually made of synthetic materials. Care is needed in laundering silk.

Rayon, formerly called "artificial silk", is a material composed of synthetic fibres mostly obtained from wood pulp by one of three different methods. It has the same attractive appearance as silk and is very much cheaper but it is inflammable. It is not very absorbent, is a fair conductor of heat and is used mainly for stockings and underwear. *Brushed rayon* is warmer and more porous.

Nylon is another synthetic fabric and is much stronger than rayon. It is therefore often mixed with other fibres in clothing manufacture. It is a poor absorbent and dries quickly. When laundered it does not shrink and needs little, if any, ironing Hence it is used for "drip-dry" shirts, blouses, nightwear and underwear.

Terylene, also made of synthetic fibres, produces yarns which are very strong and durable. Heavier types can be used for making suits and skirts and lighter ones for shirts, blouses, etc. It is almost non-absorbent, easy to wash and does not shrink. It is often mixed with wool in the making of suits and socks.

Fur consists of fibres which entangle a great deal of air and is therefore very warm.

Leather is made from skins which have been tanned and dressed. It is the most suitable material for shoes and boots, though plastics, some very good imitations of leather in appearance, are now being used for cheaper footwear.

Rubber, obtained as a latex from the rubber tree in tropical regions, is waterproof and therefore well suited to the making of protective garments and Wellington boots. Its great disadvantage is that it does not allow heat to escape and retains moisture.

Personal Cleanliness

It goes without saying that cleanliness of the body and clothes is not only essential for the good health of the individual but also for the sake of other members of the community. The oily secretion of the sebaceous glands of the skin collects excretory products from the sweat, scales from the horny layer of the skin and dirt. This encourages the growth of germs, blocks the pores of the skin, thus interfering with the natural process of sweating, and creates an unpleasant odour. Hence the necessity for keeping the skin clean by washing. Soap is used because it forms an emulsion with the

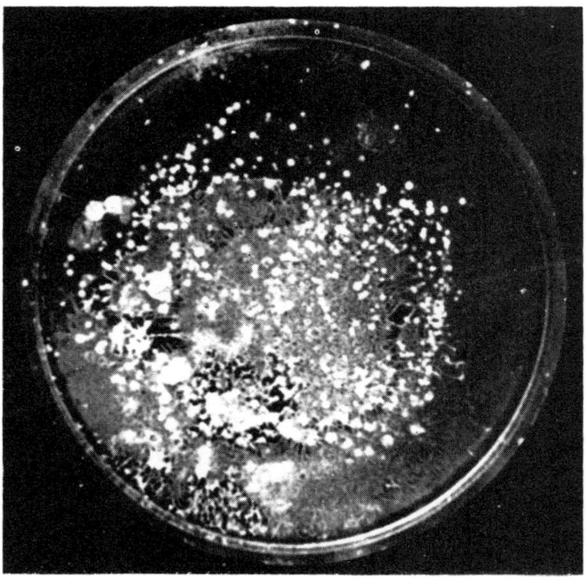

Fig. 32.2. **Bacterial colonies growing on an agar plate over the surface of which fingers have been rubbed.** *In the author's laboratory.*

dirty sebum. Hot baths should be taken once a week at the very least and preferably once a day. The best time for these is at night as they open the pores of the skin and if taken during the day, chilling is liable to occur unless precautions are taken to avoid it. Hot baths should never be taken after a meal. Cold baths taken first thing in the morning with the friction of the towel which follows stimulate the circulation but they are neither agreeable to nor suitable for everyone. If a hot bath is not taken daily, places where the skin is in contact with other places such as in the axilla as well as the external genitals and anus should be washed in hot water.

Hands should always be washed before meals and handling food and after evacuation to prevent the spread of disease. Nails must be kept clean and short. Attention must also be paid to the feet. The hair should be brushed and combed twice daily and washed once a week to prevent the accumulation of scales from the epidermis known as *scurf* and *dandruff* and to keep it free of parasitic infection.

The teeth should be brushed first thing in the morning and last thing at night and brushing should be away from the gums and the backs of the teeth should not be neglected. Many dentists advocate that the gums should also be brushed to stimulate circulation. The cleaning of teeth is to prevent the retention of food particles which will decompose as the result of bacterial action. Acids are produced which attack the enamel and eventually cause the dentine to be exposed to further attack by bacteria. Any sign of *caries* or dental decay indicates the necessity for an immediate visit to the dentist. In fact, the teeth should be examined at regular intervals of a few months. Decayed teeth can interfere with digestion and cause other disorders of the body. *Pyorrhoea* is an inflammation of the gums with a purulent discharge wich causes *halitosis* or bad breath and can lead to disease. Saliva has a cleansing action on the teeth and in between them but tiny food particles may get lodged there and the use of a tooth brush in the manner indicated is therefore necessary. The part played by flourine in preventing dental decay has already been discussed in an earlier chapter.

Chapter XXXIII

HEALTH SERVICES FOR THE COMMUNITY

It was not until the cholera epidemic of 1831, to which reference was made in a recent chapter, that any real responsibility for public health was taken by the State in this country and then only in the case of this disease. But gradually during the 19th century it became realised that some form of control of water supplies, sanitation and so on was necessary in the interests of the health of the community. There had been earlier legislation concerning some diseases such as bubonic plague but it was spasmodic and specific rather than general. Knowledge steadily accrued as time went on, Pasteur discovered the germ origin of disease, Lister antiseptic surgery, and so ultimately in the 19th century various health services came into being which culminated in the introduction of the National Health Service in 1948. The idea had been suggested, carefully studied and a general scheme outlined long before this but the war intervened and so it fell to the first post-war Government to bring it into being. This Parliament introduced the scheme but it did not conceive it. In 1974 it was entirely re-organised. During the previous twenty-six years the N.H.S. had expanded to an enormous degree in personnel, in services and in cost and its various departments and responsibilities had been organised and supervised by different bodies. This led to complications and interfered with the smooth and successful running of the service. So a new arrangement was worked out putting all the various departments under one authority. This enables co-operation between the various services to be easier and more efficient.

Quite apart from any National organisations, there is a **World Health Organisation** which has been in existence since 1948 and which obtains information on health matters from all parts of the world and is able to provide valuable information to all countries. The W.H.O. in conjunction with various departments of the United Nations does very valuable work of this kind and there are many international bodies, some entirely voluntary, such as Oxfam and the International Red Cross. Health is defined by the World Health Organisation as "a state of complete physical, mental and social well-being and not the mere absence of disease and infirmity".

THE NATIONAL HEALTH SERVICE

Let us now examine our re-organised N.H.S. as it stands at present. Doubtless further changes will be found to be necessary as time goes on and will be implemented as far as conditions permit.

The **Secretary of State for Social Services** is responsible to Parliament for the running of the National Health Service **(N.H.S.)** which comes under the **Department of Health and Social Services (D.H.S.S.)**. The N.H.S. costs the country over £4,000,000,000 a year and nearly 70 per cent. of this is incurred in salaries and wages, for it employs over 900,000 people. The organisation in **England** is first described below.

Regional Health Authorities (R.H.A.)

The country is divided into 14 R.H.As. In London there are 4 Metropolitan R.H.As. The members are appointed by and responsible to the Secretary of State for the planning and organisation of the requirements of their Regions and of the services which used to be organised on a larger scale than can be effectively done by the authorities of the smaller units into which the Region is divided. The **Ambulance Services** and the **Blood Transfusion Service** are examples of these. In each Region there is a **Regional Medical Officer** and at least one Medical College associated with a University responsible for medical and dental teaching and research. Medical Consultants are appointed by the R.H.A.

Area Health Authorities (A.H.A.)

The Regions are divided into Area Health Authorities and there are ninety of these corresponding to local government counties and in London to groups of boroughs. Members of each A.H.A. are appointed by the R.H.A. of which they form a part and they are responsible for the administration of all the health services in their Area. Each has an **Area Medical Officer (A.M.O.)** and an **Area Nursing Officer (A.N.O.)**. General Practitioners, dentists, opticians and pharmacists remain as independent self-employed contractors and a **Family Practitioner Committee** set up by each A.H.A. administers these contracts. Though some doctors have not joined the N.H.S. and continue to practise privately the majority have done so, a number taking private patients as well. Each receives a capitation fee for every patient on his list for which a maximum number is laid down. Altogether we have about 58,637 doctors in this country. **Ambulance Services** also come under the A.H.A. unless provided by the R.H.A. Areas with substantial facilities for medical and dental teaching, although individually administered by A.H.As., are finally administered by R.H.As., as already stated, to ensure unification.

Districts

As some of the A.H.As. are rather large they are subdivided into Districts but in the smaller ones this is not necessary. Each District consists of population of between 150,000 and 300,000 and the number of Districts in each Area varies between one and six. A **District Management Team** is responsible for the organisation, running and co-ordination of the health services in its District apart from G.Ps., dentists and others who come under the Family Practitioner Committee appointed by the A.H.A. as already explained. Each District Management Team is composed of a **District Community Physician** (corresponding to the Medical Officer of Health in the original organisation), a Consultant, a G.P., a District Nursing Officer, a District Administrator and a District Finance Officer and the Team is in a position to make all local arrangements. There is at least one **General Hospital**, probably several **District Hospitals**, separate Maternity Hospitals, Mental Hospitals, Geriatric Hospitals for the aged and Hospitals for Mentally Handicapped children as well as a number of Health Centres and Clinics.

In each District, **Community Health Councils** are appointed by the R.H.A. Each consists of 20 to 30 members nominated in part by the R.H.A. and in part by the local authority and local voluntary organisations such as the Red Cross. The function of the Community Health Council is to look after the health interests of the people living in the District. They have powers to secure information, visit hospitals and other institutions and their views, including complaints, are conveyed to the A.H.A. to whom they have direct access. Local authorities, district councils and London boroughs retain their former responsibility for environmental health such as sanitation in general, food hygiene in factories, food shops, cafés and restaurants. District Nurses, Health Visitors and other social workers are provided by Local Authorities.

Medical Research is largely under the wing of the **Medical Research Council** and investigation into all aspects of disease is also carried out in bacteriological and virulogical laboratories which are under the aegis of the **Public Health Laboratory Service**. The medical profession generally comes under the **General Medical Council** and nurses under the **General Nursing Council**. There are still a few private hospitals which are administered by their own governing bodies and to which the only patients admitted are members of some particular organisation or society but the majority of our hospitals come under the National Health Service. There are also private nursing homes outside the scheme.

The **National Blood Transfusion Service** comes under the Regional

Health Authorities and **The Port Health Authority** is responsible for the control of health at seaports and airports, medical assistance being provided by the Area Health Authorities. There are approximately 2,700 hospitals with some 506,000 beds and over 350,000 nurses in Britain and of these there are approximately 2,170 hospitals with 416,000 beds and some 320,000 nurses in England. The number of hospital medical staff in Great Britain is around 33,000 and the number of G.Ps. approaching 26,000. Overall, the N.H.S. employs 850,000 people.

Hospital Hygiene

Definite regulations have been made concerning hygiene in hospitals and the general outline of these is given below.

1. A minimum of 100 sq. ft. (93 sq. m.) of floor space and 1,200 cu. ft. (34,000 l.) of air must be provided for each bed and the ward must be ventilated without draughts so that the air is changed at least three times an hour.

2. Floors should be made of non-absorbable material and all corners should be rounded to enable absolute cleanliness to be maintained.

3. The temperature of wards should be kept at about 15°C (60°F.). Heating is usually by radiators.

4. Lighting of wards should be by reflection from the ceiling and each bed should have its own individual light within the control of the patient. Lighting of the tables in operating theatres should be shadowless.

5. Excreta must be emptied into special sluices in a separate room, the outlet from the sluices leading into a soil-pipe.

6. Dry refuse should be burned in an incinerator.

7. In Isolation Hospitals for infectious diseases, even greater care must be exercised, particularly in the disposal of excreta and in the matter of ventilation.

8. Diet should consist of properly cooked and attractively served food suited to the patient's condition. Unfortunately hospital food often leaves much to be desired. No doubt, staffing difficulties are largely the cause of this but the importance of diet to the sick cannot be overlooked.

So far we have been dealing with the N.H.S. in England. In Scotland, Wales and Northern Ireland the arrangements are somewhat different.

In **Scotland** responsibility rests with the **Secretary of State for Scotland.** There are no R.H.As. but there are 15 **Areas**, each having a **Health Board** assisted by the **Scottish Health Service Planning Council** and a **Common Services Agency.** The contracts of G.Ps.

dentists, pharmacists and opticians are administered by the Health Boards and there are no Family Practitioner Committees as in England.

In **Wales**, the **Secretary of State for Wales** is responsible for health services and the **Welsh Office** for organisation. There is also a **Welsh Health Technical Services Organisation**. As in Scotland there are no R.H.As. and the country is divided up into 8 **Areas.**

In **Northern Ireland** Health and Social Services are organised together and come under one authority, the **Department of Health and Social Services**. There are no R.H.As. and the country is divided into 4 **Health Areas** each under **Health and Social Services Boards** with whom G.Ps. contract. The Areas are divided into **Districts** as in England and these are administered by **District Committees** which correspond to the Community Health Councils in England.

The Health Service is available to everyone in the country without exception and regardless of nationality. Under E.E.C. Security Regulations, nationals of all E.E.C. countries (and this, of course includes Great Britain) who are employees and entitled to benefits in their own country are entitled to medical treatment in any other E.E.C. country on the same basis as the nationals in that country. Self-employed and non-employed persons are not covered by these E.E.C. regulations at present except in Denmark and the German Federal Republic. There are also reciprocal arrangements with Austria, Bulgaria, Malta, New Zealand, Norway, Poland, Romania, Sweden and Yugoslavia. When visiting any country abroad, particularly countries not mentioned above, it is always advisable to take out a medical insurance covering the period of stay before leaving this country as, although some countries have a modified health scheme, medical attention can prove to be very costly in some instances.

Many organisations such as factories have their own medical officers and nurses and employees are trained in first aid. Some have special sick rooms. Boarding schools have their own medical officers, infirmaries and sanatoria. The Public Health Service also provides such facilities as maternity and child welfare clinics and in some places women are able to have cervical tests. Travelling X-ray units also move around to different districts and are freely available to the public for chest X-rays.

The State of Health of the Population

It is often said that statistics can be made to prove anything to suit one's views and opinions but what are known as **vital statistics** are certainly of value as an indication of the health of the population. These include such items as the size of the population, the birth rate, the marriage rate, the morbidity (disease) rate and the death rate.

Population

This is obtained in Great Britain by a census taken every ten years when householders are bound to complete a form giving the age, sex, marital status and occupation of everyone in the household. The information is collated by the Department of the **Registrar-General** and from it the population of the whole country and of individual districts is obtained, together with further details of the numbers of both sexes and the number of married and single persons.

The last census was in 1971 when the total population of Great Britain was found to be 53,978,540 of whom 26,197,605 were males and 27,780,935 females. Estimates are made during the periods intervening between each ten-year census and for 1974 the estimated population was 54,421,500. Of the total population of 54,421,500 the maximum number were in the 5–9 age group and this was 3,986,500. The number of males of pensionable age was 2,736,530 and the number of pensionable aged females was 6,073,580, making a total of 8,810,110.

The estimated Home Population of England and Wales for 1974 was 49,195,100 of which 23,941,000 were males and 25,254,100 females and that for Scotland was 5,226,400 of which 2,512,700 were males and 2,713,700 females. Over the last ten years nearly 5,000,000 people (about 10 per cent. of the population) moved in and out of this country. In spite of the arrival of 26,000 Asians from Uganda in 1972 and 1973, the population of the United Kingdom decreased by more than 500,000.

The population of India is 547,949,800 and this is increasing by about 12,000,000 a year. It has been estimated that 200,000,000 of the population suffer from chronic malnutrition. Another country suffering severely from mulnutrition and disease is Ethiopia which has a population of about 27,000,000.

Live Birth Rate

This is expressed as the number of live births per 1,000 of the population in one year *i.e.*

$$\frac{\text{Number of live births per annum} \times 1{,}000}{\text{Population}}$$

To be of value, of course, the number of married women of child-bearing age (15–45) must be considered and this is known as the **Fertility Rate**. It is expressed as

$$\frac{\text{Number of legitimate births} \times 1{,}000}{\text{Number of married women between 15 and 45 years}}$$

In addition there is a large number of unmarried mothers in the country and the **Illegitimate Birth Rate** is expressed in a similar way

$$\frac{\text{Number of illegitimate births} \times 1{,}000}{\text{Number of unmarried mothers between 15 and 45 years}}$$

The total number of births in England and Wales in 1973 was 675,953 of whom 348,678 were male and 327,275 female and of these 617,856 were legitimate births but 58,097 were illegitimate, a percentage of 8·59 per cent. The Live Birth Rate for the year was 13·7, in 1974 it was 13·0 and in 1975 it fell still further to 12·2, an all-time low rate in peace time. This is probably due to careful family planning and more widely practised abortions.

There is also a **Still Birth Rate** which is the number of still births per 1,000 of the total births (still and live). In 1974 it was 11·3.

Marriage Rate

This is expressed as the number of marriages per 1,000 of the population per annum.

$$\frac{\text{Number of marriages per annum} \times 1{,}000}{\text{Population}}$$

The number of marriages in England and Wales in 1974 was 384,658. This was a marriage rate of 15·6.

Morbidity Rate

This concerns disease and the only figures obtainable are those for notifiable diseases such as smallpox, typhoid fever, diphtheria, measles, tuberculosis, typhus, plague, cerebro-spinal meningitis, poliomyelitis and whooping cough.

Cases of these diseases must be reported to the Area Medical Officers of Health. Thus the morbidity rate is really a notifiable disease rate and is expressed in the usual manner

$$\frac{\text{Number of cases of notifiable disease} \times 1{,}000}{\text{Population}}$$

In 1972, the latest available, it was 80·3.

A few instances of recorded notifiable diseases indicating a noticeable decrease in the last eighteen years are given below, the figures in brackets being the totals in 1974.

Tuberculosis of the respiratory system. . 30,246 (8,110)
Poliomyelitis. . 3,527 (5)
Typhoid Fever. . 571 (132, of which 99 cases were contracted abroad)
Diphtheria. . 59 (3)

Unfortunately venereal diseases are not notifiable in this country but known cases show a very high increase over the past few years. Both syphilis and gonorrhoea show a steep and continuous rise in most countries of the world. Some European countries have recorded increases of 60 per cent. to 400 per cent. in reported new cases of syphilis between 1957 and 1971 and of 90 per cent. to 500 per cent. in such cases of gonorrhoea. These were mostly amongst teenagers and young adults.

Death Rate

This is indicated by the number of deaths per 1,000 of the population in each year

$$\frac{\text{Number of deaths} \times 1{,}000}{\text{Population}}$$

In 1974 the total number of deaths was 584,531 and with an estimated population of 54,421,500 this gives a death rate of 11·9.

This vital statistic is clearly a valuable indication of the health of any particular county or district, a low death rate indicating a healthy population. But other things must be taken into account such as the age of the population and the nature of the district *i.e.* rural or industrial.

Infant Mortality Rate

This is the number of deaths of infants under one year of age per 1,000 live births each year

$$\frac{\text{Number of infant deaths} \times 1{,}000}{\text{Number of live births}}$$

In 1974 the number who died was 584,531 and the Infant Mortality Rate was 16·3.

At the other end of the scale, people are living longer as the result of the better health of the nation with the improvement in hygiene and medical care.

A **Death Certificate** must be issued by a medical practitioner before burial or cremation is permitted. If he feels unable to issue such a certificate a **Coroner's Order** is necessary.

The total number of deaths from *cancer* in the U.K. in 1960 was 122,648, an increase of 15,500 in ten years. Of these 30,483 were due to lung cancer and this was 9,000 more than in 1950. In 1974 the total number in England and Wales alone was 121,711, cancer of the lungs bronchi and trachea accounting for 33,028. Diseases of the *cardiovascular system* have also shown a serious increase, the number of deaths from this cause in 1974 amounting to 195,876 an increase of 53 per cent. in 5 years. On the other hand diseases showing a marked decrease are *respiratory tuberculosis* (57%), *typhoid* with only 2 deaths and *poliomyelitis* and *diptheria* with none.

Deaths from *accidents* in England and Wales in 1974 totalled 11,679 of which 6,320 were road casualties and 5,359 occurred in the home. Deaths from *natural causes*, excluding violence, were 563,863. It is worthy of note that in the U.K. in addition to 6,890 killed in road accidents, 82,100 were seriously injured and 235,900 slightly injured. However, the number of fatal accidents was 50 per cent. lower than in the previous year.

Chapter XXXIV

MAN IN THE WORLD

CONCLUSION

We have now made a fairly comprehensive study of this creature, man, of the ways in which he has learned to develop the natural resources at his command and to control to some extent the environment in which he lives for the benefit of his health and comfort. Every single man and woman is but a tiny atom in the complex body of the human community and it must not be forgotten that not only has the community a duty to the individual but that the individual has a duty to the community. There can be no room in a healthy community for work-shy lay-abouts and law breakers who are parasitic on the community. It is surely a fallacy to contend that the State owes everyone a living. Every individual member must justify his or her existence by playing his part in the general scheme of things, working not only for himself but for the benefit of all, in the family and in the community as a whole, at the same time making a useful contribution to the life of that community.

The State must arrange for the provision of proper water supplies, heating, lighting, sanitation and so on and it should provide for education, sickness, old age and misfortune. But these things cost money (the National Health Service alone cost this country over £4,000 million in 1975) and members of the community must therefore contribute to the upkeep of the services of all kinds which the State provides. However, the opinion is very widely held that if any individual chooses, in addition, to incur extra expenditure, assuming he is in a position to do so, for particular kinds of education for his children (often by considerable self-sacrifice) for private medical treatment and to make financial provision for old age, he surely has a perfect right to do so. There must be freedom of opinion, freedom of choice and individual freedom in all things, provided, of course, that such freedom does not interfere with that of others. With this proviso there must not be any infringement of individual liberty. Control there must be to maintain justice, law and order and to ensure a fair deal for everyone. But what an incredibly dull and uninteresting society we should be if we allowed ourselves to be regimented along identical lines into identical channels so that we became stereotyped units in a homogeneous body! As human beings we have the right to the freedom to develop our own individualities and to guide our children to do likewise.

Every member of a community must have a fair and equal chance of attaining his ambitions with regard to his place in the community but it would be nothing short of ridiculous to imagine that every individual is capable of or adapted to the attainment of the same goals. We all have our separate parts to play in life, whether they be great or small, whether they be humble or otherwise and each and all of us are essential links in the chain of man's life on earth.

Man has achieved some remarkable things for the benefit of his existence here. He has made, and is still making, far-reaching discoveries about his body and of the ways in which he can prevent and cure the diseases with which it may become inflicted. He has learned how to develop and control the natural resources which are available to him. He has created much that is beautiful in music, literature and the arts for the satisfaction of his mind, made discoveries and inventions in abundance for the benefit of mankind and devised means of efficient and rapid communication between men, however distant they may be from one another, by telephone, radio, television, satellites and the aeroplane. He has discovered how to obtain energy from the atom and means of using it for the benefit of humanity. He has walked in space, set foot on the moon and brought samples of it back to earth and even kept a timed appointment in space with fellow men from a country far distant from his own and returned safely to earth. These are wonderful achievements. Unfortunately in many things he has also created ways of destroying mankind.

There is so much good in the world and so many good deeds are performed by man throughout the world. Yet he has still to learn how to live at peace with his fellow men regardless of colour, race or creed. Greed, envy, lust for power, terrorism, murder, religious and political intolerance and persecution, injustice, cruelty, brutality, rape and violence are but a few of the blemishes on human society throughout the world and from which it is always the innocent who suffer most. Venereal disease is rife and on the increase and drug trafficking is a serious problem throughout the world. Overcrowding, poverty and famine are constantly present in some countries and millions suffer from malnutrition while others live in abundance. Wars and aggressors are forever with us and as long as man is attacked he must perforce defend himself. There can be no denial of the fact that many of these things have been brought about by man himself. It would almost seem that he is bent on the destruction of the very civilisation he has created.

Has all this come about because man has become too materialistic? Could it be that he is too intolerant and dogmatic and has allowed his political creeds to play too large a part in his life to the exclusion of religion? Could it be that he concentrates too much on his many grievances and fails to appreciate adequately the benefits which are his and which he is not always ready to share with those who are less fortunate? Does he realise that the lives of future generations will be affected by what happens now and that by many of his actions he may be endangering the future of mankind, for is not the present the fulfilment of the past and the foundation of the future? Or is it simply that man has not yet had sufficient time to master all these things which cause such discord in his life? Perhaps it is just "human nature" but all these matters affecting man's life on earth provide food for serious thought. It is surely not beyond his power to pause and think and to make a conscious and deliberate effort to overcome those evils of his own creating which threaten to overwhelm him and, perhaps, ultimately to destroy him.

How little man is, yet in his own mind how great!

EDMUND BURKE

APPENDIX I

EQUIVALENTS AND CONVERSIONS

1 cm.	= 0·3937 in.
1 m.	= 39·37 in.
	= 3·2808 ft.
1 sq. m.	= 10·76 sq. ft.
1 hectare	= 100 sq. m.
	= 2·471 acres
1 litre	= 1·76 pints
	= 0·22 gal.
	= 0·0353 cu. ft.
1 gram.	= 0·0353 oz.
	= 0·0022 lb.
1 kg.	= 2·205 lb.
1 micron (μ)	= 10^{-3} mm.
1 millimicron ($m\mu$)	= 10^{-6} mm.
1 micromicron ($\mu\mu$)	= 10^{-9} mm.
1 Ångström (Å)	= 10^{-7} mm.

$$°\text{C to }°\text{F} = °\text{C} \times \frac{9}{5} + 32$$

$$°\text{F to }°\text{C} = °\text{F} - 32 \times \frac{5}{9}$$

APPENDIX II

SUPPLIERS OF SPECIALISED MATERIALS FOR PRACTICAL WORK

BIOVIEWERS and BIOSETS

The Banta Company, 279 Church Road, London SE19 2QQ

THE LENHAM CHROMATOGRAPHY KIT

(Devised by Dr. T. J. Weston M.Sc., Ph.D., M.I.Biol.)

* T. Gerrard & Co., Gerrard House, Worthing Road, East Preston, Sussex BN16 1AS

ELDON CARDS FOR BLOOD GROUPING

(Produced by Nordisk Insulinlaboratorium, Gentofte, Denmark)

T. Gerrard & Co. Address as above

* Phillip Harris Biological Ltd., Oldmixon, Weston-super-Mare, Avon BS24 9BJ

MULTISTIX, CLINISTIX, QUANTAN CI and QUANTAN DT FOR URINE TESTS

(Produced by Miles Laboratories Ltd., Stoke Poges, Slough, Bucks.)

T. Gerrard and Co., and Philip Harris Biological. Addresses as above.

* These firms also supply microscopical slides and general biological apparatus and materials.

INDEX

INDEX

B

C

D

E

F

O

Q

R

X

Y

Z